Thomas Harriot

Thomas Harriot

A LIFE IN SCIENCE

Robyn Arianrhod

OXFORD
UNIVERSITY PRESS

OXFORD
UNIVERSITY PRESS

Oxford University Press is a department of the University of Oxford.
It furthers the University's objective of excellence in research, scholarship,
and education by publishing worldwide. Oxford is a registered trade mark of
Oxford University Press in the UK and certain other countries.

Published in the United States of America by Oxford University Press
198 Madison Avenue, New York, NY 10016, United States of America.

Library of Congress Cataloging-in-Publication Data

Names: Arianrhod, Robyn, author.
Title: Thomas Harriot : a life in science / Robyn Arianrhod.
Description: Oxford ; New York, NY : Oxford University Press, [2019]
Identifiers: LCCN 2018024943 | ISBN 9780190271855 (hardback : alk. paper)
Subjects: LCSH: Harriot, Thomas, 1560-1621. | Scientists—Great
Britain—Biography. | Explorers—Virginia—Biography. |
Virginia—Discovery and exploration—English.
Classification: LCC Q143.H36 A75 2019 | DDC 509.2—dc23
LC record available at https://lccn.loc.gov/2018024943

1 3 5 7 9 8 6 4 2

Printed by Sheridan Books, Inc., United States of America

Things are in such a state here that it is not lawful for me to philosophize freely; we are still stuck in the mud.
—Thomas Harriot to Johannes Kepler, 1608

CONTENTS

Thomas Harriot

PROLOGUE

ON APRIL 9, 1585, a three-masted sailing ship named the *Tiger* left the safety of Plymouth on the southwest coast of England and headed for the open ocean. Queen Elizabeth had loaned the 149-ton vessel to Sir Walter Ralegh. It was the flagship in his small fleet bound for North America, and there was a rough crossing ahead: cramped, unhealthy shipboard conditions and the Atlantic's treacherous waves and sudden storms would test the nerves and stomachs of the hardiest sailors. Of the 160 passengers and crew on board the *Tiger,* however, one passenger in particular remained undaunted throughout the voyage. He was Thomas Harriot, a twenty-four-year-old Oxford graduate and a promising mathematician and scientist. As Ralegh's expert navigational theorist, he was up on deck even in the foulest weather, advising the captain and his chief navigators, taking astronomical readings, and gathering firsthand information for improving the design and use of navigational instruments. In his spare time, he observed the sailors and their conditions, recording their patois in his notebook as they shouted instructions to one other. No occupation or task was too humble to attract his interest, and few details escaped him.

His interest in shipboard culture and dialect foreshadowed Harriot's pioneering contributions to phonetics and American ethnology. His role in this expedition reflected his wide-ranging interests and abilities, for he was not only Ralegh's navigational expert; he was effectively the first European diplomat to, and scientist-in-residence in, what is now the United States. He was among the first Europeans to acquire a working knowledge of a North American language—in this case, North Carolina Algonquian—and by means of it to understand and record indigenous culture at the time of first contact with Europeans. Outgoing and amiable, he made friends with the people, hunted and feasted with them, learned their methods of agriculture, canoe building, and fishing, and clearly enjoyed much about their way of life. As a general rule, he recorded what he saw with the detachment of a physicist and the engagement of a linguist and ethnologist, describing rather than judging religious practices and cultural ceremonies that were completely alien to him, and observing in context the details of Algonquian life.[1]

"Physicist," "ethnologist," "linguist," and "scientist"—these are all anachronistic terms that would have seemed as strange to Harriot as an Algonquian ceremony. With no road map in these yet-to-be-formalized disciplines, he had to invent his own methods of recording and interpreting what he saw and heard. Not surprisingly, his assessments of America, like his ideas on physics, were not always complete or entirely accurate. That is the way with pioneering work. Nevertheless, his eye for detail and his intellectual rigor make him an almost unique guide to an age before the world had quite become modern—before the tragedy of a new wave of colonization that would soon unfold, and before modern science had emerged fully from its medieval and mystical past.

What makes Thomas Harriot's story especially compelling is that it helps fill in some of the gaps in our picture of that transitional era, because he is one of those proverbial forgotten geniuses. He has a place in history for his work in America alone, but his scientific achievements offer a much bigger story and suggest that he should have a far bigger place in the scientific pantheon.

His career began a century before his countryman Isaac Newton set the keystone for modern science with his 1687 masterpiece *Principia*, which gave us the universal theory of gravity and pioneered modern theoretical physics and astronomy. Newton is arguably the greatest mathematical scientist of all time, although, as he famously told Robert Hooke, if he had accomplished anything it was because he had stood on the shoulders of giants. The best known of these are Harriot's Continental contemporaries Galileo Galilei and Johannes Kepler. Galileo published the first empirical law of gravity—that of freely falling bodies near the surface of the earth—while Kepler derived the laws of planetary motion, including the fact that the shapes of the planetary orbits are ellipses about the sun.

Galileo and Kepler were working at a time when belief in magic was common, and science was struggling to free itself from the confusion of superstition and the constraints of scriptural literalism. Their discoveries, and their battles with authorities, are deservedly famous, but much less is known about what was happening in England at that time. There was the legendary Dr. Dee, of course—the Elizabethan "conjuror," alchemist, mathematician, and astronomer. And this was also the age of William Shakespeare, Christopher Marlowe, Edmund Spenser, John Donne, Ben Jonson, and Philip Sidney and his sister, Mary, as well as Ralegh and so many other luminaries. These were writers, dramatists, and poets, all of them figures of profound culture. Dee notwithstanding, what seems missing from this stellar line-up is a standout scientist—an English bridge to Newton, or someone who personifies the scientific culture that ultimately produced him.

That's where Harriot comes in, the man modern scholars consider the greatest British mathematical scientist before Newton.[2] There were mathematicians closer to Newton's time who were perhaps more sophisticated in their techniques than Harriot—notably James Gregory and John Wallis—as well as major scientists such as Edmond Halley and Robert Hooke. But Harriot stands alone, because of the quality and versatility of his best work, which, like Newton's, spanned a range of topics, mathematical, theoretical, and experimental. He discovered the law of falling bodies and formulated fledgling laws of motion independently of Galileo. He outpaced Edward Wright—one of the most accomplished and well-known navigational experts in early seventeenth-century Europe—in the sophistication and priority of his practical and theoretical navigational work. He built his own telescope and studied the moon and the motion of sunspots, again independently of Galileo, and made some of the most accurate astronomical calculations of his time.

The list of achievements goes on. Harriot discovered the law of refraction and the secret of the rainbow decades before Willebrord Snell and René Descartes. He discovered the dispersion of light more than half a century before Newton and pioneered the "Newton-Gregory interpolation formula" long before Newton and Gregory. He discovered binary arithmetic nearly a century before Gottfried Leibniz and made a quantitative study of population growth versus food supply two centuries before Thomas Malthus. He produced the first universal phonetic alphabet and the first fully symbolic algebra. He did all this and more, and yet who, outside of academia, knows much about him? It is only relatively recently that historians them-selves have begun to appreciate the wealth of material he left behind.

Yet this long-lost pioneer had lived and worked at the heart of English intellectual and political life during one of his country's most celebrated ages. Born in 1560 (so far as can be discerned), he was an almost exact contemporary of Shakespeare and witnessed the spectrum of Elizabeth's reign, both its glories and its conflicts. He cheered the English defeat of the Spanish Armada in 1588; he explored the New World and learned some of its language and culture, as mentioned; he attended some of the plays of Shakespeare, and of his own close friend George Chapman, when they were first performed, and he read and conversed with other literary and scientific friends and acquaintances. He kept fully abreast of religious debates, too, which flourished and flared during the Elizabethan Age as it struggled to reconcile its religious identity and loyalties. Religious pamphlets were a favorite means of contributing publicly to the debate, but some of them included diatribes against Harriot and Dee—the supposedly "atheist" rational man of science and the "conjuring" mathematician-astrologer—and Harriot became caught up in the potentially deadly politics of patronage and the snares of sectarian strife. He was also affected, directly, by the repercussions of the Gunpowder Plot early in the reign of James I. During all this turmoil, excitement, drama, and danger, he worked, relentlessly, tirelessly, exploring nearly every facet of mathematics and physics. His scientific reputation in his own time was so great that Kepler had sought his advice, while others admired him as one of the fathers of modern algebra.[3]

It is no wonder, then, that for a growing number of scholars Harriot has become a cause, and restoring him to his rightful place in scientific history a powerful desire. It remains an unfinished task, though—a work in progress, like his own research. Indeed, the challenges of giving Harriot his due are daunting. He had no close family who might have passed down records and stories, and few of his letters have been found. Even the identity of the sitter in the portrait and engraving that were once thought to have been of him has been called into question.[4] Most frustrating of all, he left behind no published edition of his scientific work—no polished, public account that would have made clear what his thousands of calculations and pains-taking observational data meant to him, and how he wished them to be read. This was partly a matter of circumstance. Still, it was not unusual at that time to circulate manuscripts rather than published works, and it is understandable that a perfectionist such as Harriot didn't rush into print. He did publish a short work on America, and his executors eventually published an algebraic treatise based on some of his mathematics. Important as they were, these, too, were incomplete, because many of Harriot's American notes were lost at sea, and his executors failed to understand his algebra sufficiently to publish a suitable exemplar.

This relative paucity of primary material explains why there has been no popular scientific biography of Harriot. What remains is his manuscripts, disordered and unfinished; together with his American booklet and a handful of documents they are the only direct evidence on which to base a biography.

To find his life, then, it is necessary to explore his writings—although it is something of a miracle that these exist at all. Within a few decades of his death in 1621, he had all but disappeared from the scientific record, and his manuscripts lay hidden, presumed lost, for nearly a hundred and fifty years. Then, in 1784, some eight thousand pages of his scientific and mathematical research were discovered at the ancestral home of Lord Egremont, a descendant of Harriot's patron Henry Percy, 9th Earl of Northumberland, a major figure in Harriot's story, as we'll see. Thanks to the generosity of Lord Egremont's family, most of the papers ended up in the British Library.

When I visited the library to see these for myself, the Manuscripts Reading Room felt like a place of pilgrimage. After all, it's both thrilling and humbling to hold the handwritten treatises, jottings, and calculations of someone who died four centuries earlier—especially when these are virtually all that we have to reconstruct a life. I was amazed that Harriot's thick, rough-edged foolscap pages were so well preserved. Unfortunately, his small, quill-penned handwriting and laconic style made deciphering his notes difficult, while the stops and starts in his working made following his trains of thought a challenge. Nevertheless, I soon came to admire the breadth and depth of these manuscripts, this lifetime of unpublished effort for the sake of science and mathematics. I understood why Harriot had attracted so much scholarly attention in recent years—and also that he will likely remain one of the most fascinating of history's scientific figures partly because he is so elusive.

Yet these papers convinced me that it would be possible to write this biography. As I read through Harriot's pages, sometimes puzzling over them, sometimes smiling with joy at his tours de force, I was moved by the intimacy they afforded. I began to feel the same enthrallment expressed by Voltaire—author of a popular book on Newton's theories—as he alternately struggled with and delighted in the great Englishman's work: "This strange man turns my head!"

I am, of course, not the first to feel this way about Harriot. One of the first modern scholars to study his papers was the American historian John Shirley, who published an invaluable scholarly biography in 1983. Shirley and his fellow "Harrioteers" inspired new generations of enthusiasts, who are spearheading Harriot's gradual restoration to history after centuries of obscurity. I am indebted to all these researchers, whose work I gratefully reference in my bibliography and notes. My goal here is to join their efforts by bringing Harriot to a general readership.

Because this is a popular account of Harriot's life and work, I have occasionally taken what scholars might rightly see as liberties. I've already mentioned anachronistic terms for professions such as "scientist," and I'll use modern nationalities such as Italian and German, although in Harriot's time Continental political power and identity generally lay in city-states and regions rather than today's nation-states or a unified Europe. (On the other hand, in speaking of ancient and medieval "Greek" and "Arabic" science, these descriptors refer to the scholars' culture and language, not to modern notions of ethnicity and nationality.) I'll also follow the common practice of modernizing spelling in my quotes from the period—and occasionally I'll use the terms "Renaissance" and "Reformation," although they belie the complex, humanist origins and ongoing nature of the intellectual, cultural, and theological developments of the fifteenth and sixteenth centuries. Finally, since one of my aims here is to give nonspecialist readers insight into the ideas and concepts that led to modern science and mathematics, I will sometimes move ahead in time to highlight differences and similarities between Harriot's work and the corresponding modern view.

Still, this is a book that endeavors to put Harriot's achievements into their context. This means offering a more general story about his life and times, and about the evolution of science through both his work and that of his contemporaries and forerunners. So sometimes I'll also move back in time, especially in the early chapters, in order to highlight the legacy on which Harriot and his contemporaries built. I'll also illustrate the vexed connection between the rise of science and the economic and imperial development of the Western world. I hope to give a feeling not only for Harriot's scientific and mathematical achievements but also for the contingencies, people, and events that helped shape them—and that shaped Harriot himself.

Above all, though, Harriot's story is one of insatiable human curiosity. Imperialism and the economic imperative were not the only impetuses for the gradual but transformative process now known as the Scientific Revolution. Curiosity, and where and how far it can take us in understanding our world, was a key driving force. So was the courage to question received wisdom, to challenge orthodoxy, even when doing so meant running personal risks. As Galileo and Harriot discovered to their cost, ongoing and often toxic fallout from the Reformation and Counter-Reformation meant that authorities were ever alert for "heretical" opinions. At the same time, the impact of the Renaissance also played its part. Rediscovered works of ancient Greek science and mathematics—together with medieval treatises by Middle Eastern and Indian scholars—were emboldening European physicists and mathematicians, liberating the bravest of them from the need to fit scientific discovery within a traditional theological and philosophical framework. Harriot read Latin and Greek fluently, and for all his willingness to put his life at risk in New World expeditions, ultimately he preferred intellectual to geographical adventuring. Not that he was always able to lead the life of quiet contemplation he craved, given the excitements, uncertainties, and ever-shifting alliances of the day. He moved in the most glittering of Elizabethan circles, and the most dangerous of Jacobean ones.

As for his scientific legacy, many of those thousands of manuscript pages contain false starts and dead ends, the detritus of trial and error that is the way of science. Had he published the gems of algebra, optics, astronomy, navigation, and mechanics that are embedded in the rough matrix, however, he would have earned lasting fame and influence. As it is, some of the reasons he didn't publish add drama, intrigue, and tension to scientific history. They provide a human face to intellectual discovery. And in place of polished publications, with Harriot's papers we can watch thought unfold. His works in progress are a window into the workings of an extraordinary mind, and they offer up not a metaphorical monument in stone but a life as it was lived—a life in science and mathematics during one of the most fascinating eras in history.

CHAPTER 1

Harriot's London

IN THE WINTER OF 1583–84, when twenty-three-year-old Thomas Harriot moved into Durham House on London's fashionable Strand, he entered a world of unaccustomed splendor. The house itself was fit for a king: according to contemporary accounts its great hall was "stately and high, supported by lofty marble pillars," and there were sumptuous furnishings of soft green velvet and silver lace.[1] Thirty uniformed attendants were on hand, serving meals on silver plates and generally catering to the occupants' every whim.

Harriot was not used to such a lavish lifestyle. He had been born in Oxfordshire of "plebeian" parents, according to the description of his background in the record of his 1577 admission to Oxford. This document, together with official references to his 1580 graduation, is the only extant evidence about his life before he moved into Durham House, and it tells us that his parents were commoners who worked for a living. His father was likely a craftsman or a yeoman (a small farmer), because it was not unusual for Elizabethans of this class to save sufficient money to send their sons to university—fees were reduced for plebeian scholars.[2] Either way, we can surmise that Harriot had arrived in London with an Oxford bachelor of arts degree and little else. But his wide-ranging intellect—especially his knowledge of the principles of navigation by the stars, the hottest scientific topic of the day—had soon attracted a wealthy patron: the tall and dashing Walter Ralegh.

No one knows when or how Harriot and Ralegh first met. It may have been through the Reverend Richard Hakluyt, an Oxford lecturer interested in the New World. Temperamentally, though, the two men could not have been more different. Harriot was an intellectual with relatively little interest in fame or fortune, and in keeping with his vocation, he dressed in the understated black suits favored by many scholars at the time. The ambitious Ralegh, by contrast, was a notoriously flashy dresser whose very shoes were jewel encrusted. Harriot was steadfastly loyal to his friends, kept his opinions to himself, acted cautiously, and won the admiration of an eclectic group of friends and colleagues. The rash and brash Ralegh, on the other hand, had a penchant for upsetting people in power.

Harriot may have been studious, but he was also sociable, confident, and an astute judge of character. As for Ralegh, despite his ostentatious ways he was interested in literature, science, and mathematics, and he and Harriot were bonded by these pursuits. Harriot, of course, was doubly bonded: with no means of his own, he was dependent on his patron for his livelihood. Patronage was a key source of employment for innovative scholars and writers at the time. Nevertheless, accepting Ralegh as his patron meant that Harriot's destiny was tied to his employer's—and, as it turned out, Ralegh's fate was encoded in Durham House's stone tower, backing as it did onto the River Thames just a couple of miles upstream from the Tower of London, where Ralegh would spend his last days.

In early 1584, however, thirty-one-year-old Ralegh's star was rising, thanks to his ability both to flatter Queen Elizabeth I, then nearly fifty, and to inspire her with his vision for England.[3] Raised among the seafaring folk of Devon, he was a younger half brother of the pioneering North American explorer Sir Humphrey Gilbert, who had "claimed" Newfoundland for the British Crown just a few months earlier. Gilbert's overreaching colonial ambitions had ended in disaster: his main supply ship was wrecked and more than eighty men drowned; Gilbert was forced to abandon his expedition and to order the remainder of his fleet to head back toward England. But the weather had turned terrifyingly wild, and Gilbert and his ship simply disappeared, swallowed up by the furious Atlantic.

Ralegh had contributed financially to the ill-fated venture, and now he was determined to build on his brother's legacy. Like Gilbert, he dreamed of making Protestant England mistress of the oceans, displacing Catholic Spain from its position of naval supremacy, through which it dominated the New World. It ruled more than its share of the Old World, too—notably Portugal (since 1580), Belgium, and much of the Netherlands, although Dutch Protestants continued to rebel, with covert English aid. So Ralegh wanted to make England not only rich, through a trading base in the New World, but also safe, able to defend itself in home waters if the growing threat of a Spanish invasion materialized. It was for these ends that he had employed young Harriot: to teach him astronomy, so that he and his captains and pilots would know how to navigate more surely.

During their study together, the respect between the two men had grown into friendship. Sometimes they conversed in Ralegh's study in Durham House's impressive lantern tower, which looked south across the Thames to the peaceful countryside beyond. On the river, boats busily plied their trade, ferrying goods and passengers—although sometimes they might offer less savory spectacles, such as when some petty offender, or some unfortunate woman deemed a "harlot" or a "shrew," was condemned to punishment by "dragging" across the river.[4] Generally, though, the view across the Thames from Durham House was, as Ralegh's seventeenth-century countryman John Aubrey put it, "as pleasant, perhaps, as any in the world."[5]

While Ralegh's New World dreams would prove to be just the starting point for Harriot's diverse career, they would eventually destroy Ralegh himself. But in those early, golden years of the 1580s, his energy and vision inspired those around him—including the queen, who had shown her admiration for her new favorite by giving him the use of Durham House. It seems a prescient as well as a generous gesture, for the mansion's history offered a warning: not to take lightly the privilege of living there. Two of its most famous former occupants, Lady Jane Grey and Anne Boleyn, had been executed, victims of religious fanaticism and royal caprice.

In fact, the house's provenance symbolized the religious and political tensions not only in England but also throughout post-Reformation Europe. Built in the thirteenth and fourteenth centuries for the bishops of Durham, it was occupied by generations of Catholic bishops until 1536, when Bishop Tunstall turned it over to Elizabeth's father, Henry VIII, in exchange for a less prestigious property.[6] In 1531, Henry had proclaimed himself head of the Church of England. By 1540, he had suppressed all Catholic monasteries, diverting most of their revenue to the Crown. On his death in 1547, Durham House passed to Princess Elizabeth (his daughter with Anne Boleyn), in accordance with his will. After Elizabeth's Catholic half sister, Mary, became queen in 1553, however, the house was soon returned to Bishop Tunstall. Mary Tudor was the daughter of Henry's first wife (Catherine of Aragon), but she died just five years into her reign. In those five years at least three hundred

English Protestants were burned at the stake. Many commoners had fanned the flames, believing that by inflicting physical suffering on these unbelievers they were doing God's will. On her accession to the throne in 1558, Elizabeth took Durham House back, so that once again it was in Protestant hands. And although the new queen resolved to create an atmosphere of religious tolerance, by the 1580s, Catholic uprisings and plots to depose her had eroded her will.

For instance, in December 1583—around the time Harriot and Ralegh moved into Durham House—Francis Throckmorton was caught with coded messages to Mary Queen of Scots, a Catholic who had been under house arrest in England for the past sixteen years. (Having lost Scottish support after her supposed complicity in the murder of her former husband Lord Darnley, she had abdicated her throne and escaped to England, where she was kept safe, albeit as a "guest-prisoner." Rebellious English Catholics believed that, as a great-granddaughter of Henry VII, Mary had a better claim to the English throne than the "bastard" Elizabeth—Catholics did not recognize Henry VIII's divorce from Catherine, after which he had married Anne Boleyn—and so Mary was under house arrest both for her own safety and for Elizabeth's.) Throckmorton was the son of a distinguished English Catholic family and had been acting as an intermediary between Mary and the Spanish ambassador, Bernardino de Mendoza. His papers revealed tentative plans for a Spanish invasion of England that would place the Queen of Scots on Elizabeth's throne.

Under repeated torture on the rack Throckmorton confessed to the plot. Mendoza was expelled, Anglo-Spanish tensions mounted, and Throckmorton was taken to Tyburn, located just outside the city's gates, for execution. The very name of this notorious place inspired terror. Bells would ring and crowds would line the streets for miles to jeer—or cheer—the prisoners as they were carted on their long and excruciating journey from London to agony and oblivion. A traitor's death was particularly horrific, with the victim being partly strangled and then drawn and quartered while still alive.

Just a few months after Throckmorton's execution in mid-1584, his Protestant cousin Bess Throckmorton would become one of Elizabeth's ladies of the privy chamber, a trusted inner group of the queen's personal attendants. It was a fateful appointment, for she was destined to play a dramatic role in the life at Durham House. Yet for all the queen's willingness to entrust another Throckmorton, Elizabeth was being forced into a corner. A key trigger for Catholic unrest and Protestant paranoia had been the 1570 papal bull excommunicating Elizabeth and instructing her Catholic subjects to pledge loyalty not to their queen but to Rome. Her government's ultimate response included making the act of converting anyone to Catholicism treasonous, and punishable by the ghastly form of execution administered to traitors. When Cardinal William Allen complained it was wrong to turn a matter of conscience into treason, he was only half right: until England was free of Roman interference, and of threats to Elizabeth's legitimacy by supporters of foreign Catholic rulers, religion and politics would remain intimately entwined.[7]

Freedom of speech was another matter—it effectively did not exist.[8] Like others in his position, Allen was writing from the safety of exile. He was right to be cautious; during Elizabeth's long reign, 189 Catholics would be executed for their beliefs.[9] The Jesuit missionary Edmund Campion was one of these. He had complained to a colleague that Protestant "heretics" had spies everywhere in England, so that he had to change his disguise frequently as he went about the country ministering and proselytizing.[10] Elizabeth's ambivalence about the deteriorating course of religious politics meant she sometimes showed a degree of mercy, as she did to Campion, allowing him to be hanged until dead before being drawn and quartered. But a number of onlookers would have been disappointed: some Elizabethans enjoyed a gruesome execution, just as some enjoyed seeing animals fighting each other in gladiatorial combat. The queen herself was partial to a good bearbaiting.

It seems hard to believe anything ennobling could come from such a barbarous and fearful time, yet this was also one of the most dazzling periods in English history. In all times of momentous change, and perhaps in all exceptional individuals, contradictions coexist and moral, sometimes Faustian, bargains are made. The boldly original playwright Christopher Marlowe would soon become the first to dramatize the legendary pact with the devil, in his *Tragical History of Doctor Faustus* (1589), but first he worked his way through Cambridge as a courier for Elizabeth's spymaster, Sir Francis Walsingham.[11] Many well-placed scholars sometimes led double lives, gathering useful intelligence. John Dee—the most colorful and well-known English mathematician of the day, who would later become a friend of Harriot—is famous today for using the code name 007, although this is unproven. It's also been conjectured that the Italian mystic and quasi-Copernican Giordano Bruno was the spy who helped Walsingham uncover the Throckmorton Plot.[12]

Poets also made their bargains. Edmund Spenser was weaving beautiful Homeric and Arthurian allegories from a stolen Irish castle, having first worked for Lord Grey, the commander who, in 1580, had brutally crushed an Irish Catholic uprising that was part of the so-called Second Desmond Rebellion. Ralegh, too, profited from and participated in the ruthless English conquest of Ireland. His family background was fiercely anti-Catholic, and he had fought for his religion early in life. Inspired by a Huguenot (French Protestant) kinsman, he had joined a band of volunteer Devonshire soldiers who went to France to fight on the Huguenot side in the religious war that was tearing that country apart. He was about fifteen years old. Twelve years later, he had found himself fighting Catholics in Ireland.

The English had been trying for decades to "civilize" the "wild Irish"—to use an anti-Irish term that was common at the time—and to appropriate their land. Finally, desperate Irish leaders had begun to seek help from Catholic France and Spain, and the pope himself was only too willing to supply troops, weapons, and moral support in a "holy" war against the heretic Elizabeth. Ralegh's involvement in such a war had begun in the summer of 1580. As a hotheaded young newcomer at court, he'd earned himself a reputation for brawling, but he was already someone to be noticed, and sympathetic statesmen such as Walsingham had hoped that a tour of duty in Ireland would teach him some discipline. Later that year, in preparation for what turned out to be one of the bloodiest battles in the Desmond Rebellions, Ralegh was given a post as captain of a hundred soldiers. His commander was none other than Lord Grey. Grey was a fanatical Protestant who showed no mercy even when papal troops, trapped in a fort, raised a white flag of surrender after a four-day siege. Along with the other officer on duty that day, Ralegh and his men carried out Grey's orders, butchering with their swords hundreds of terrified Continental soldiers who had surrendered in the belief that their lives would be spared.[13] Edmund Spenser, Grey's secretary, approved: the Irish had been "much emboldened by those foreign succors," he said, so the defeat of the pope's army had to be complete.[14]

Spenser did show some compassion for the common people of Ireland, thirty thousand of whom had died in six months—slaughtered, starved, or diseased. Hunger left survivors too weak to walk, so that they emerged from the woods "creeping forth upon their hands [and knees]. They looked like little anatomies of death: they spoke like ghosts crying out of their graves." They were reduced to eating rotting corpses and weeds, and Spenser wondered if anything was to be done. "Should the Irish have been quite rooted out? That were too bloody a course: and yet their continual rebellious deeds deserve little better."[15] Amid such unspeakable suffering, Ralegh and Spenser had become friends. Later, at a more sorrowful and reflective time in his life, Ralegh would write of the bitter pains inflicted when "one nation...labors to root out the established possessors of another land."[16] Back in 1580, however, he'd been prepared to do whatever it took to be noticed. His reckless courage, together with his newly gained knowledge of Irish affairs, and his fearless criticism (in his reports to court) of incompetent or brutal superiors such as Grey, did indeed launch him on his career as a courtier.[17] And soon the charismatic Ralegh had won over the queen herself.

YET THOMAS HARRIOT's and Walter Ralegh's destinies were forged not only by their era's religious and political turmoil, its ethical compromises, and its Faustian bargains but also by its creative energy and sense of possibility. This was, after all, the Renaissance, and outside the battle lines of sectarian conflict there thrived a "pagan" love of life. Ordinary folk still celebrated ancient festivals— even the queen enjoyed the rites of spring—while poets such as Marlowe and Spenser were devouring Greek and Roman classics and transforming them into lyrical poetry, vehicles for celebrations of nature and life, and for meditations variously erotic, satirical, and philosophical.

New ideas—from home and abroad—were finding outlets through a growing (if closely monitored) publishing industry.[18] And for the first time, local translations of literary and scientific classics began to appear. The first English translation of Euclid's legendary mathematics textbook, the *Elements*, was published in 1570; Euclid had written it in Alexandria around 300 BCE, and John Dee, a promoter of education in English, wrote a preface for the new translation. Another friend of Harriot, George Chapman, would soon begin work on his translation of Homer's *Iliad* and *Odyssey*.

English versions of contemporary treatises were appearing, too. In the field of navigation, Richard Eden's *The Art of Navigation*, a translation of Martin Cortes's Spanish treatise on the subject, was published in 1561. By the 1570s, English experts were beginning to publish their own such manuals, notably William Bourne's 1574 *A Regiment for the Sea*—which he wrote for everyday navigators, not for academic astronomers. In 1584, Harriot was writing a new navigation manual especially for Ralegh. He called it *Arcticon*, and his goal was to update and correct earlier works, including Bourne's and Cortes's. Instructional books on all topics were popular in increasingly literate Elizabethan England, whether the required instruction was professional, practical, or spiritual.[19]

At the same time, the first commercial theaters were being built—much to the horror of Puritans, radical Protestants who regarded idle entertainments as sinful—and these would give paid work to such great playwrights as Marlowe and Shakespeare and to actors such as Edward Alleyn and Richard Burbage. There were economic problems, of course, notably rising inflation, but the new theaters were signs of commercial growth and hope. This was due partly to reforms initiated in the 1530s and 1540s and developed by early Elizabethan advisors such as Nicholas Bacon (father of Harriot's contemporary Francis Bacon) and especially William Cecil, Elizabeth's skillful chief secretary of state. These men had continued to support the ideal of a commonwealth that promoted the well-being of all citizens, through economic prosperity and a fair legal system.

Growing affluence meant that London's busy marketplace—"crowded with citizens and foreigners, abounding in riches and goods," according to a contemporary sightseer—was a must-see for visitors from abroad.[20] London was rapidly becoming one of the most exciting places in Europe, and Durham House was at the heart of it. It was also close to the queen's palace at Whitehall and to the House of Commons at Westminster (where Ralegh would soon sit as the elected Member of Parliament for Devon[21]). The fine buildings that stretched along the Thames, together with its colorful river traffic, made this part of London, some said, the rival of Venice. In fact, with its population set to reach two hundred thousand by the end of the century, London would soon overtake Venice to become one of the world's most prosperous and powerful cities, the third largest in Europe, after Paris and Naples. As a visiting Continental official put it, "London is an excellent and mighty city...Londoners are magnificently appareled and extremely proud and overbearing."[22] More charitable observers might have used the word "confident" rather than "overbearing," but either way, for those who could afford it—and who were able to avoid or ignore sectarian strife—this was a time of outward-looking optimism.

Indeed, the visiting official continued, London's citizens were busily employed "buying and selling merchandise, and trading in almost every corner of the world." From the riverbank near Durham House you could see ships moored in the Thames while wharf workers unloaded cargoes of spices and silks from the East, furs from Russia, Persian and Turkish carpets, Spanish silver and gold looted from South America, Venetian glass, Mediterranean wine, and other luxuries.[23]

England's premium export was high-quality woolen cloth. Cecil had advised value-adding at home rather than exporting the raw cloth, but he also favored in general the Continental idea of granting monopolies as a way of stimulating business, and Elizabeth had recently given the monopoly for issuing licenses for the export of woolen cloth to Ralegh. In May 1584, she would also give him the monopoly on licenses to sell imported wine. In both cases, these benefits were given in exchange for Ralegh's overseeing and managing the trade. To some, though, it seemed as if he were simply skimming off profits for himself. He certainly wasn't popular with trading companies, or with their customers who paid higher prices for goods— or, for that matter, with Jesuits who opposed his lavish lifestyle as well as his Protestantism. But that was the way the queen rewarded useful courtiers and business investors, and it kept them both loyal and dependent on her.[24]

By 1584, English companies were trading legally in Russia, Venice, Turkey, Spain, and North Africa. Trading illegally meant being a pirate or privateer, the latter being a more respectable version of the former. In England, piracy was illegal and conviction meant death by being "hanged on the shore at low water mark," where the bodies were left "till three tides have over-washed them," as one contemporary put it.[25] But privateers were private traders whose governments licensed them to raid enemy ships in times of warfare. It was a lucrative if dangerous business, and Ralegh would soon do well from it. It would help him finance his New World dreams. In 1584, however, English privateering against Spanish ships in particular was doubly controversial, because Spain and England were not yet technically at war. Nevertheless, the Throckmorton Plot had shown just how tense the relationship between the two countries was. Until Henry VIII had broken with the Catholic Church, Spain had allowed the English to trade legally in Spanish territories in South America and the Caribbean, but for years now English merchants had chafed under the Spaniards' increasingly anti-English trade policy in the New World. To assert their monopoly there—a monopoly the pope himself had granted, believing he had God-given power over the fate of "heathen" lands—Spanish seamen often harassed non-Spanish merchant ships, sometimes stealing cargo and seizing Protestant sailors in the name of God. In turn, English adventurers sometimes raided Spanish treasure ships en route from South America. In 1574, Elizabeth and her former brother-in-law, Spain's King Philip II, had tried to sort things out with a nominal peace treaty. (Philip had been married to Mary Tudor.) The treaty allowed English traders to

continue legally visiting ports in Spain to exchange English cloth
and tin, and Newfoundland fish, for Spanish-controlled goods, but
the Spanish made it clear they regarded Englishmen attempting to
trade in the New World itself as "corsairs." This did not stop English
forays into the forbidden territory, where some Englishmen did indeed
continue to behave as pirates—and in 1585 Philip II would ban all
English trade with Spain.[26]

If Catholic Spain was unwilling to establish an ongoing trade
agreement with Protestant England, the Ottomans had had no such
qualms. In 1580, negotiations between Elizabeth and Sultan Murad
III had resulted in an agreement allowing England to trade in North
Africa and the eastern Mediterranean. This provided a ready market
for England's cloth, tin, and lead, because the Ottomans were still
expanding their empire and needed the woolen fabric for uniforms
and the metals for weapons.[27] The pope was horrified at such com-
merce with the infidels: the Ottomans had already conquered much of
Eastern Europe, as well as the ancient eastern capital of Christendom,
Constantinople. Economic considerations trumped religious differences, however, and the trade continued. It also initiated Anglo-
Ottoman diplomatic ties. The legacy of religious tolerance under
Murad III's predecessor, Suleiman the Magnificent, meant religion was
not really a barrier; moreover, Protestants and Muslims, unlike
Catholics, shared important theological principles such as the rejection
of idolatry.

Ralegh's license for the trade in woolen cloth no doubt increased
in value because of the Ottoman markets, but England was still a
small player on the world stage, and he was keen to open up new territory.[28] The Spanish and Portuguese had already laid claim to Central
and South America, and the Spanish and French had a tentative
hold in "Florida," which extended further north than the state of
Florida does today; in fact, the Spanish nominally claimed the entire
eastern coast of North America, although they had not succeeded in
settling much of it. The Portuguese and Spanish also dominated
European trade with the African Gold Coast (which is now Ghana).
The Ottomans controlled North Africa and much of Eastern Europe
and the Middle East, as mentioned, while Russia's Ivan the Terrible
had recently secured the Volga River trade route. The Muslim
Mughals had taken much of northern India and were still expanding
their influence there, and the English were vying with the Spanish,
Portuguese, Dutch, Arabs, and other long-established traders in the
East Indies. China had thrown out its Mongolian conquerors two
centuries earlier, although the Manchu conquest was still to come;

meanwhile, Portuguese and Spanish merchants dominated—often illegally—Western trade with China.[29] North America seemed like the last available opportunity, and Ralegh was ready to take it.

He had begun planning a reconnaissance voyage in search of an American location for a permanent English trading base. As he gathered around him men who could help bring his dream to fruition— notably Harriot, and more recently his chosen sea captains, Philip Amadas and Arthur Barlowe, and his chief pilot, Simon Fernandes— Durham House, during the winter and spring of 1584, was abuzz with excitement.

CHAPTER 2

Sea Fever

AMID THE ANTICIPATION AND ACTIVITY of Ralegh's preparations for the expedition to North America—the financing and provisioning of ships, the hiring of crew, and the applying for exploration patents—Harriot was hard at work, gathering astronomical data for *Arcticon,* his book on navigation, and instructing those among Ralegh's captains, pilots, and shipmasters who needed specialist training. No one knows just how he went about his teaching, but the topics he taught are clear from his surviving notes. And although no one knows how he met his patron, it is worth asking what was the broader context for Ralegh's bold move in employing a young unknown with an obscure background—a man who had never even been to sea—to teach himself and his men.

The short answer is that Harriot was brilliant, and had studied mathematical astronomy in unusual depth, even for a graduate. A university degree was something the restless Ralegh had not quite pulled off; he'd been a student at some time during the early 1570s, but the regimented life of an Oxford college was not for him, and he had left after a year or two. After another year studying law in London, he'd set about maneuvering his way to a position at court.

Harriot had chosen a more modest route to success. Obtaining a university degree was a way of rising above the class one was born into—if one was a man and could afford the fees and living expenses. Scholarships existed—Christopher Marlowe had one—but still it was expensive to live at the university colleges; Harriot had resided at St. Mary Hall, which was later merged with Oriel College. During the half century or so that spanned Harriot's lifetime, around 6,700 "sons of plebeians" registered at Oxford.[1] Not all of them would have graduated, but those who did would rank on a par with a gentleman (a man of some means, who did not have to work for a living), a priest, or a merchant. This put the former plebeian a step above a landowning yeoman or a craftsman and just below a knight. Knights occupied the lowest rung of the aristocratic ladder; when Ralegh became "Sir Walter" in 1585 he would join several hundred others at this rank, which was a rung below the barons, of whom there were around forty during Elizabeth's reign, and who, in turn, were below the twenty or so earls. There had been one Elizabethan duke—dukes were second in rank only to the queen—but he (the Duke of Norfolk) had been lured into an intrigue with Mary Queen of Scots at the time of yet another plot to depose Elizabeth (the so-called Ridolfi Plot), and he had been executed in 1572.[2]

One of the more influential scholars at Oxford during Harriot's time there was the Reverend Richard Hakluyt, a geography lecturer.[3] Geography at that time showcased history-in-the-making as the geographical boundaries of the so-called known world expanded with each new voyage across the ocean. This kind of geography also fostered the peddling of dreams and the making of heroes. Hakluyt had become a tireless promoter of the idea that New World trade and influence were vital to England's security and economy, while students and the public alike were enthralled by tales of fabulous South American wealth and of courageous men risking their lives on treacherous seas. The North Atlantic was especially dangerous, with its icy waters and waves as high as fifty feet—the kind of waves that had consumed Humphrey Gilbert and his crew.

Hakluyt fueled this public passion for exploration and the New World by publishing English "travel books," beginning in 1580, when he arranged for the English translation and publication of the Frenchman Jacques Cartier's reminiscences of his voyages of 1534–36. Cartier's account of his first contact with the exotic North Americans, whom he described as friendly and eager to trade for European goods, captivated English readers. Not that present-day Quebec had been his intended destination: like Christopher Columbus and John Cabot before him, Cartier had been heading for Asia. Columbus had been hoping to reach Japan, and when he came upon the West Indies instead, he initially thought he'd arrived in India. Similarly, Cabot and his sons had come upon Nova Scotia and Newfoundland instead of Asia, while Cartier had ended up in the St. Lawrence River seeking passage further west.

By the 1570s, English explorers, too, had begun searching for a "North-West Passage," a sea route from Europe through North America to Asia. Overland caravans were all very well for ancient traders, and even for Marco Polo at the dawn of the fourteenth century, but now European merchants wanted a quicker route to the silks and spices of Asia. Sea fever was in the air in England, and especially so after Francis Drake's triumphant return from circumnavigating the globe, in the very same year Hakluyt's first travel book appeared.

Drake had led the first successful round trip since the voyage initiated by the Portuguese navigator Ferdinand Magellan sixty years earlier.[4] Not surprisingly, he soon became legendary in England, and his ship, the *Golden Hind*, would eventually be destroyed by adoring tourists, each chipping off a bit of the boat for a souvenir.[5] At first, though, Elizabeth was not pleased with Drake: her enthusiasm for his achievement was muted by her concern—and the Spanish ambassador's fury—over the Spanish treasure ships he had plundered en route. After some equivocation, she decided to go ahead and knight him.[6] After all, the courage required for such a venture was encapsulated in the statistics: Magellan had set out with five ships and 270 men, but only one ship and fewer than 20 men returned; Magellan himself had been killed in the Philippines. As for Drake's expedition, three ships had set out on the voyage; one sank with all on board, another turned back to England, and only Drake's *Golden Hind* completed the trip.

Ralegh's goal was to establish a base in America itself, not just to bypass it en route to Asia. Between northeast Canada and Florida lay a vast territory that was virtually unknown to Europeans, and Ralegh's new captains, Barlowe and Amadas, had been employed to lead a reconnaissance voyage to explore the east coast of what is now the United States. Harriot was to oversee their training, but as for the broader context of his employment, in the wake of Drake's success and Hakluyt's publications, private tutors and new private schools had already begun teaching practical mathematics to prospective navigators. John Dee had pointed out that the Spanish and Portuguese had long been providing such training, and in order to catch up and surpass their expertise Ralegh had chosen the most mathematically gifted tutor in England.[7]

Harriot was not the first mathematician to provide such advice in Ralegh's circle, though: he had been preceded by Dee himself. Notorious as a seeker of spirits and master of invocations, Dee was also a respected mathematician and theorist on navigation. In the 1570s, in addition to his polemical writings on the importance of "the mathematical arts," he had taught these arts to Martin Frobisher before his attempt at finding a North-West Passage. He had also been a navigational advisor to the Muscovy Trading Company, and he had advised Humphrey Gilbert. But Dee had eventually grown weary of giving practical advice to "ungrateful" businessmen, as he put it. Instead, he wanted to be the official court philosopher and astronomer.[8] For many years he had acted as Elizabeth's informal advisor on matters astronomical and philosophical—and on matters mystical, too, beginning in 1558 when one of Elizabeth's chief advisors asked him to choose an astrologically propitious date for her coronation. By the early 1580s, he felt it was time she offered him a permanent position.

One of the more interesting matters on which Elizabeth had asked for Dee's advice concerned the sudden appearance, in November 1572, of a "new" star in the constellation Cassiopeia. This was the same year as the Duke of Norfolk's execution for his apparent role in the Ridolfi Plot against the queen, who was understandably nervous at any new ill omen. It was also the year of the gruesome St. Bartholomew's Day Massacre of Protestants in France, so officials on both sides of the Channel were worried about the astrological significance of the new star. But Dee's opinion was reassuring: he believed it indicated the finding of the fabled "philosopher's stone"—the elusive substance that would enable alchemists to turn ordinary metals into silver or gold. Five years later, when called to advise on a "blazing star" that terrified people throughout Europe, Dee reassured Elizabeth that it signified her ascension, not her downfall.[9]

Today we know that the 1577 "blazing star" was an unusually bright comet, while the "new" star of 1572 was a supernova—the brilliant last hurrah of a dying star that previously had been too dim to see with the naked eye. No one knew then about supernovae, so the new star was an astronomical as well as an astrological mystery, whose solution was sought by the best astronomers of the day—including Dee and his former ward and student, Thomas Digges. Like astronomers on the Continent, notably Tycho Brahe of Denmark, Dee and Digges made careful measurements of the star, which they also hoped might help to settle a larger issue—namely, which of two theories of the known universe was correct.

The two theories in contention were the one proposed by the Polish astronomer Nicolaus Copernicus, who, in 1543, had published a model in which the earth and other planets moved around the sun—and that of the Greek philosopher Aristotle, who, around 350 BCE, had argued that the sun, stars, and planets rotated about the earth on crystalline spheres.[10] Christian teaching still held that Aristotle's view—and its second-century geometrical adaptation by Claudius Ptolemy—was consistent with biblical references to the "moving" sun and to the idea of heavenly perfection.[11] But Digges, unlike Tycho, favored Copernicus. For computations, at least, so did Dee (to whom Digges referred as his "second mathematical father"; his father, Leonard Digges, had been a mathematician).

Harriot, too, was a Copernican.[12] English scholars were particularly receptive to the idea: Anglicans were perhaps not so concerned as Catholics, Calvinists, and Lutherans about squaring science with biblical references. In 1556, barely ten years after the publication of Copernicus's opus *De revolutionibus orbium coelestium (On the Revolutions of the Celestial Sphere)*, a Welsh-born Anglican mathematician, Robert Recorde, had published a book on astronomy that included a brief dialogue between a Copernican and a supporter of the earth-centered model of the heavens. This was three-quarters of a century before Galileo's famously "heretical" *Dialogue concerning the Two Chief World Systems—Ptolemaic and Copernican.* Then, in 1576, Digges had published the first English translation of Copernicus's main argument, and for the next two centuries it would remain the only known non-Latin version of a significant part of *De revolutionibus.*[13]

Copernicus's model had certain conceptual advantages, but Dee and Digges knew that astronomers needed more tangible evidence. In their quest to determine the correct theory of the heavens, they needed to estimate how far away the new star was from earth. To do this they measured its parallax—the change in its apparent position when measured from two different locations. It is the same kind of change you see if you hold a finger in front of your eyes and notice how it appears to move relative to the background as you close first one eye and then the other. The distance between your eyes means that each observation is made from a slightly different position in space, so that for each eye, the line of vision between your eye and your finger points to a different object in the background. Analogously, stellar parallax arises when observing a particular star's position with respect to the background of more distant stars, but the difference in space arises when astronomers make observations at two different *times.* This is because during the interval between observations, the earth has actually moved through space as it orbits the sun—according to the Copernicans. For Aristotelians, it was the stars that had moved. Either way, the alignment between the earth, the particular star, and more distant stars in the background changes over time, so that two readings at two different times give an indication of the star's parallax.

Digges, Dee, and Tycho were interested in the fact that parallax is greatest for nearby objects: it is very pronounced for your finger, but less so for a tree or a post outside your window. They therefore concluded that the new star of 1572 really was a faraway star, because they could detect little or no parallax with respect to the background stars. (This is why Harriot would teach his students to account for parallax only in relatively nearby celestial objects, such as the moon and the sun.[14]) Tycho also found that the unusually bright comet of 1577 was further away from earth than the moon.

Results such as these began to chip away at the mainstream Aristotelian view, in which suddenly appearing stars and comets were considered inconstant, and so were deemed imperfect—which supposedly meant they must inhabit the corruptible, perishable earthly region between the earth and the moon. The moon and the stars, on the other hand, were supposed to be the embodiment of heavenly perfection as they rotated serenely, far above the earth, on their perfect crystalline spheres. Tycho Brahe's observations so impressed his king—Frederik II of Denmark—that he was given the island of Hven in the Danish archipelago, along with royal support to build the finest observatory in the world. By contrast, Digges's patron, William Cecil (recently ennobled as Lord Burghley), seems to have been more interested in astrological prognostications than in astronomical theory. Not that Digges was any more bashful than Dee when it came to giving astrological advice. In *Alae seu scalae mathematicae,* his 1573 book on testing Copernicanism via parallax, Digges described the new star as "God's true messenger." Tycho feared it as a harbinger of war and chaos, but Digges, who was a radical Puritan, welcomed the star's message: he believed it signaled the end of the world, when God's chosen ones would ascend to heaven.[15]

Harriot, for his part, kept his religious and astrological beliefs separate from the experimental and mathematical work that fills virtually all of his surviving manuscripts, so that it is difficult to know exactly what beliefs he held.[16] But this also illustrates the anticipation of science as a discipline in its own right, methodologically separate from philosophy and religion. Such a separation was rare in the profoundly religious sixteenth and early seventeenth centuries. It would be a hallmark of Isaac Newton's 1687 *Principia,* and thence of modern physics, in which theories are based on experimental evidence and quantitative, experimentally testable hypotheses. Although Newton was intensely religious, he knew that religion and mysticism treat different kinds of knowing than science. So, it seems, did Harriot. In choosing his scientific advisor, perhaps Ralegh was drawn to Harriot's down-to-earth experimental and mathematical approach.

Dee, on the other hand, had seen no contradiction in giving precise mathematical advice to navigators while reading their destinies in the stars and conjuring spirits to aid their journey. Almost everyone believed in the existence of spirits, especially the Holy Spirit, angels, and the devil, which were central to mainstream Christian belief at that time. For those religious leaders who disapproved of astrology and the conjuring up of spirits, the problem was not the supernatural forces themselves but the idea that people might try to control such powers for their own personal ends. In response to such criticism, Dee had insisted that "natural magic" was an experimental subject, whose goal was an understanding and mastery of the mysterious forces of nature—just like physics. While most people believed, simply as an article of faith, in the existence of astrological and spiritual "rays," Dee adduced the science of *optics*—the study of light—as a model for these unseen influences. He reasoned that if curved mirrors and lenses could focus light, so crystal balls could focus spiritual energy.[17] It was an inventive analogy, but in hindsight, the trouble with his argument is that he assumed the existence of these spirits at the outset and offered no means of testing this assumption. Instead he relied on the self-proclaimed medium Edward Kelley to "hear" messages from the conjured spirits, so his experiments with crystal balls in the early 1580s were far from scientific.

Harriot, by contrast, would become a leading expert on optics, but he was not interested in conjuring spirits. He appears only to have dabbled in astrology—possibly at the request of his patron, just as Kepler sometimes cast horoscopes for a living—but for a while, at least, he was interested in alchemy.[18] This was not the turning-lead-into-gold kind that Ben Jonson would satirize in his 1610 play *The Alchemist* but the alchemy of physical processes of decay, combination, and transmutation. Two centuries before the birth of modern chemistry and biology, this kind of alchemy had its roots in an earlier, more unified way of seeing the world, in which all changes were linked, metaphorically if not causally—be they everyday rust and decay or the grand mysteries of birth and death. Although most of Harriot's recorded "alchemical" experiments involved what we would now see as the scientific use of substances traditionally used by alchemists—experiments designed to find such things as specific weights and angles of refraction—he read widely if critically on magical, alchemical, and astrological topics.[19] But these interests did not overlap with his experimentally and mathematically based advice on navigation or astronomy—advice that we therefore now call scientific.

HAVING THE CONTROVERSIAL RALEGH as patron meant that Harriot's position would not always be secure. But Digges's scholarly career had ended almost as soon as it began. With no luck in finding a patron who would support his scientific research, by 1579 he had become an administrator in the army, working on military rather than astronomical matters.

As for Dee, by 1583 Elizabeth still hadn't appointed an official court philosopher or astronomer, unlike Frederik of Denmark and other great Continental princes (and unlike her father, too: Henry VIII had had a court astronomer[20]). Ralegh had tried to help Dee's case by remembering him to Elizabeth and reminding her of his marvelous library of over four thousand books and manuscripts, which Dee encouraged scholars to use freely. Dee recorded in his diary that thanks to Ralegh, the queen did call in when she was out riding. Three months later, he also noted a letter from Ralegh, conveying the queen's good wishes.[21] But she never offered him a permanent job. Perhaps she'd become wary of his recent obsession with spirits.

If crystal-gazing was too much for the queen of England, however—or if she simply wasn't interested in giving a permanent position to a philosopher of any kind—some Continental rulers, including the Prague-based Holy Roman Emperor, Rudolf II, were receptive to Dee's mystical "natural philosophy." (Rudolf was no more interested than Elizabeth in crystal-summoned angels and their outlandish prophecies, but he *was* keen to see Dee's famous "magic mirror" with its bizarre optical illusions.[22]) And so, around the same time Harriot and Ralegh had moved into Durham House, the fifty-six-year-old Dee had left England to take up positions at the courts of Cracow and then Prague.

Now it was the twenty-three-year-old Harriot's turn to shine, and he would soon eclipse Dee and Digges to become the best and most famous English mathematician and astronomer of his age. But first he threw himself with his customary vigor into developing navigational and related astronomical theory under Ralegh's patronage—and teaching Ralegh's reconnaissance team to find their way on the open ocean with only the sky to guide them.

CHAPTER 3

The Science of Sea and Sky

T HERE WAS STILL MUCH TRIAL and error involved in sailing, as Ralegh had discovered during his first ocean voyage in 1578. He had been a captain in the first of his half brother Humphrey Gilbert's attempts to found a colony in North America, but he and his chief navigator, Simon Fernandes, had found themselves at the mercy of unfavorable winds and a ship that was later deemed unseaworthy, so they did not manage to sail far enough to reach America. Luck, good and bad, would always play a part in a sailing voyage. But science could help to shorten the odds, which of course is where Harriot came into it.

No one knows precisely who attended his classes, but we do know, from a statement by Richard Hakluyt, that by 1587 Harriot had taught "many" of Ralegh's captains. It is almost certain that these included Philip Amadas and Arthur Barlowe: their sailing experience was limited, and Amadas, if not Barlowe, too, was living at Durham House prior to his American voyage.[1] It was important to train the captains, as well as their pilots and shipmasters, because ultimately it was the captains who directed the course of a voyage. Ralegh knew that when the weather turned rough, provisions became moldy and rat infested, and land seemed a perpetual mirage, pilots, officers, or crew were easily tempted to turn back home. Under such circumstances, captains had to be able to make informed decisions, in order to avoid chaos.

For millennia, people had been sailing the seas and using the sun and stars to steer by, so Harriot and his contemporaries had inherited an astonishing body of knowledge. In fact, the study of astronomy was the greatest success story of ancient science, and in Harriot's time this ancient knowledge remained very much alive, thanks to the Renaissance, and to the legacy of medieval Arabic-speaking scholars and their Persian and Syriac predecessors. Most of the authors of "pagan" Greece and Rome had fallen out of favor nearly a thousand years earlier, with Christians and Muslims alike. Many ancient manuscripts, including those in the legendary Alexandrian library, had been destroyed over the centuries by war, religious fanatics, accidents, and especially mold and pests; few religious scholars had been interested in copying those that did remain. By the ninth century, however, a very different intellectual climate prevailed.[2] In the Islamic world, under Caliph al-Ma'mun Baghdad had become a new Alexandria, a place where scholars gathered and sought to translate, study, and improve upon as many ancient Greek works as they could find. It was a high point in the great translation era initiated after 754 by Caliph al-Mansur.[3]

Europe had its own center of learning in the eighth and ninth centuries, at the court of the Frankish warrior-king Charlemagne and his successors. Following in the footsteps of his grandfather Charles Martel, Charlemagne had conquered and Christianized much of pagan Western Europe, introducing various political and cultural reforms. He had a special interest in astronomy—initially for dating holy days by the moon, then for its own sake, too—and encouraged its study in schools and monasteries. Ancient Greek astronomical treatises were still unknown in Europe, though, so science in the "Carolingian Renaissance" was driven by the study and preservation of surviving Roman works, by writers such as Pliny the Elder and Martianus Capella. (The term "Carolingian" derives from Carolus or Charles.) Charlemagne attracted scholarly priests from Ireland and England as well as Continental Europe, while Caliph al-Ma'mun sent translators as far as Constantinople, Persia, and India. Roman-based Carolingian astronomy did not reach the sophistication of its Greco-Arabic counterpart, however. Nevertheless, innovative Carolingian planetary diagrams and astronomical commentaries were used until the thirteenth century, and presumably they helped lay intellectual foundations for the reception of Greek astronomy when it was introduced to Europe, thanks to the Middle Eastern scholars who had salvaged and developed it.[4] Indeed, many ancient Greek treatises survived only in Arabic translations, commentaries, and reworkings, which eventually found their way to Europe via Moorish Spain—together with surviving or transcribed

Greek works, Arabic translations of Indian mathematical manuscripts, and Arabic-speaking scholars' own discoveries.

Like the Arabic translation movement itself, this stage in the process of transmitting ancient knowledge is a tale of multicultural cooperation. By the thirteenth century, Spanish Christians had taken back most of the territory they had lost to Muslim invaders nearly five centuries earlier. (In the 730s, Charles Martel's military ambitions in France had helped halt the Moorish conquest of Western Europe.) The final Spanish Muslim province, Granada, had fallen to the Christians in 1492, the year Columbus set sail for the New World. His patrons were Isabella and Ferdinand, the sovereigns of Castile, Aragon, and Navarre—he had not managed to find sponsorship in his native Italy. In the years and decades after 1492, the rise of Spain as a unified Catholic state was accompanied by forced conversions or expulsions, first of Jews and then of Muslims. But during those previous transitional centuries, Spain had been a hub of scholarly activity as Europeans traveled to re-Christianized regions to study the precious Arabic manuscripts for themselves. In particular, sometime around 1140 Gerard of Cremona had traveled to Toledo, where he stayed for the rest of his life. Among his dozens of translations from Arabic to Latin, the most important was *Almagest*, by Claudius Ptolemy of Alexandria (c. 90–c. 168), the greatest ancient exponent of scientific astronomy. Gerard's was not the only translation or paraphrase of Ptolemy's treatise, but it was the most influential; in particular, it was the basis of the first printed version of *Almagest*, which was made in Venice in 1515. A Greek version—Greek was the language Ptolemy wrote in—was first printed in 1538, at Basel. (In his 1984 English translation, G. Toomer worked primarily from a Greek text, but also from medieval Arabic versions.)[5]

Harriot's foundation for his teaching and his own navigational research was Ptolemy, in the sense that *Almagest* had set the keystone for mathematical astronomy. In its modern-seeming focus on observational evidence, logic, and mathematical proofs, Ptolemy's treatise is a precursor to Newton's *Principia*. (In fact, Newton deliberately chose to present his revolutionary book in the time-honored axiomatic, geometric manner of ancient Greek works such as Euclid's and Ptolemy's.) Of course, logic is only as good as its premises, and not all of Ptolemy's assumptions were correct: in particular, he believed that the earth was at the center of the universe. Consequently, the Ptolemaic model of the heavens was less elegant than Copernicus's, and Harriot was not interested in this aspect of Ptolemy's work. Rather, he was concerned with trigonometric methods of tracking the sun and stars—as viewed from earth, the way sailors did—which had been adapted over the centuries from Ptolemy.

Ptolemy in turn had drawn on an older body of knowledge; after all, it had taken thousands of years for early stargazers to unravel the secrets of the heavenly patterns. Discovering even the most obvious celestial cycles must have required years of patient observation. For instance, as the setting sun sinks below the horizon—entering the land of the dead according to many ancient peoples (or a realm akin to hell according to the North Carolina Algonquians, as Harriot would discover)—it does so at a point on the western horizon that changes throughout the year. Yet there is a pattern to these changing directions: in summer, the sun sets toward the north of due west, and in winter, it sets toward the south of due west. (In the southern hemisphere, the sunset tends more northerly in winter and more southerly in summer.)

Anyone watching diligently every evening in June will notice that a time comes when the setting sun has traveled as far as possible in its northerly journey. This is the summer solstice in the northern hemisphere, the day, too, on which the noonday sun reaches its highest point in the sky for the whole year. Then the sunset arc begins to swing slowly back toward the west, and the noonday sun sinks lower in the sky. The setting sun reaches due west on the day of the autumn equinox, when day and night are of equal length, and then moves on to reach its most southerly point on the western skyline at the winter solstice—the shortest day, when the noonday sun is at its lowest height for the year.[6] Then it returns, via the spring equinox, to the summer solstice, and the pattern begins again.

There is a wealth of astronomical information hidden in this everyday natural phenomenon—information that would prove indispensible to navigators, and that Harriot was now teaching the ship captains and pilots in his charge. To gain a feeling for these astronomical facts, and how they helped navigators at sea, it is worth exploring how the ancients teased out such complex information from such seemingly simple observations, and how they turned sky watching into such a sophisticated science.

NO ONE KNOWS JUST WHEN ancient humans began seriously taking notice of the changing directions of sunrise and sunset. For thousands of years, though, solar phenomena such as the solstices symbolized not so much an intriguing astronomical fact as the power of the sun god, or of those rulers who identified with him. Egyptian pharaohs, Chinese emperors, rulers of the Incas and Aztecs, and many hunter-gatherer elites—all drew on the power of the sun to enhance their status.[7] Even in Harriot's time, many people connected God with the sun. Kepler used this metaphor as a theological argument in support of his sun-centered model of the heavens.[8]

Many ancient people also believed that the planets—or the deities ruling them—had a direct influence on the world, which could be divined through the art of astrology. As Dee's story shows, this ancient art was still alive and well for many people in Harriot's time. Astrology was also sometimes used as a diagnostic tool in conventional Elizabethan medicine, but it played a uniquely central role for the self-taught astrologer, physician, and occultist Simon Forman; as his meticulous consultation records attest, he continued to do a roaring trade despite the College of Physicians having him frequently fined or jailed for practicing medicine without a license.[9] Naïve mysticism aside, planet-based astrology was quite an achievement in its own way. Although a rudimentary form of sun-and-moon-based divination goes back at least five thousand years, it had taken much longer for ancient stargazers to single out the planets from the stars—to recognize that while the stars' rising and setting and wheeling through the sky was like clockwork, five of the brightest celestial bodies seemed to wander throughout the year. Our word "planet" comes from the Greek "wanderer." Indeed, when viewed from earth, the planetary paths are so erratic that sometimes they appear to change direction and move backward. (This is largely why Ptolemy's earth-centered geometrical model was more complicated than Copernicus's heliocentric one.) Such wayward motions suggested to many that the planets had a godlike will of their own—and of course we still refer to the five planets that are visible to the naked eye by the names of ancient Rome's planetary deities.

To propitiate these gods, however, it was necessary to study their ways carefully. Long before the Roman Empire—and even before the rise of Greece—the sky-worshipping Mesopotamians had made the momentous discovery that the planets, along with the sun and moon, seem to restrict their capricious celestial journeying to a specific band in the sky.[10] They had identified this "zodiacal" region—which they called Anu's Way, in honor of their chief god, Anu—by 700 BCE.[11] (In figure 1 in the appendix, you can see how the ancients would have discerned this zodiacal belt.[12]) The discovery of the zodiac eventually led to a most important concept in astronomy: the ecliptic, the apparent yearly path of the sun, whose navigational importance Harriot was about to teach his pupils.

The Mesopotamians had developed not only astrology and observational astronomy but also the art of astronomical prediction. From their detailed record of centuries of observations, they knew the number of days between recorded sightings of such phenomena as eclipses and solstices, or between successive risings and conjunctions of various celestial bodies, whose relative positions in the sky they measured by holding up a finger or a forearm and counting off the distances.[13] From this data, they deduced arithmetical relationships between various celestial events and the time of year—and these relationships enabled astronomer-priests to estimate expected times not only for solstices and seasons but also for eclipses, those eerie warnings from the gods.[14] (The tenth-century Mayans may have had a similar arithmetical model for predicting the motion of Venus, a planet that played a key role in their rituals. To take just one more example of New World astronomy, some traditional Australian Aboriginal stories suggest an understanding of the fact that solar eclipses are caused by the moon god covering the sun goddess.[15])

In their bid to understand the will of the gods, the Mesopotamians had obtained such detailed astronomical data that later Greeks used and built on it in their development of what we now call scientific astronomy. For many Greek-speaking astronomers, it was not enough to observe the skies and to divine the decrees of the gods. They also wanted to understand the world—to know *what* it was made of and *why* things appeared the way they did. The astronomer-mathematician Eudoxus of Cnidus (408–353 BCE), for example, was said to have declared, "Willingly would I burn to death like Phaethon, were this the price for reaching the sun and learning its shape, its size, and its substance." Fifty years earlier, the philosopher Democritus of Abdera had expressed the desire to know *why* when he said, "I would rather discover one cause than gain the kingdom of Persia."[16] He was born around 460 BCE, half a century after the Persians had conquered Mesopotamia and three decades after they had tried to invade mainland Greece. (Democritus's approach to finding causes will play a significant role in the story of Harriot's later work.) Of course, the Greeks were not the only ones to ask why. Mythology itself is often an attempt to understand why the world exists or why nature behaves as it does—and it can also encode astonishingly detailed observational knowledge of the natural environment. But science as we know it, and also as Harriot was teaching it, offers a more detached way of knowing that owes much to ancient Greece.

Eudoxus is sometimes called the father of scientific astronomy, and he was not satisfied with the Mesopotamians' *numerical* predictions and portents. Reputedly in response to a challenge from his teacher, Plato, he envisaged a *geometric* picture of the motions of the planets, sun, and moon—a series of linked concentric spheres centered on the earth, each rotating at its own rate. It was an inspired first attempt at imagining how the motions of the heavenly bodies were related physically. But the exciting thing from a conceptual point of view is what happened to Greek astronomy when later astronomers developed more accurate Eudoxus-style geometric models.[17]

This new phase in astronomy began around 150 BCE, three hundred years before Ptolemy, with the work of Hipparchus of Nicaea. In particular, he systematized the application of geometry to the study of the "celestial sphere." The ancients had long referred to the sky as a sphere centered on the earth: the sun, moon, and stars traced celestial circles as they rose and set each day, so it was natural to look up and imagine the sky as a huge hemispherical dome, with the celestial bodies sprinkled across its rotating ceiling. Nearly two centuries before Hipparchus, Aristotle—inspired by Eudoxus's model—had proposed his crystalline spheres.[18] But Hipparchus was different, because he studied this sphere mathematically. To do this, he first had to decide how to divide the sphere into mathematical units—and he seems to have been the first to systematically use a 360-degree circle to make precise geometrical representations and calculations.[19] Unfortunately, almost none of his work survives. We only know about it from an extraordinary book that did survive, thanks to the work of all those medieval translators and copyists: Ptolemy's *Almagest.*[20]

By the middle of the second century CE, Ptolemy had gathered together all the astronomical and mathematical knowledge of his forerunners that he could. He was heir to the work and traditions of ancient Mesopotamian and Mediterranean scholars; he knew nothing of astronomical observatories in Asia, Australasia, and the Americas. Modern astronomers have recently formally recognized contributions from these latter cultures.[21] But it is the Mesopotamians and Greeks whose discoveries led to the astronomy that Harriot was researching and teaching, because Ptolemy's collection would influence Western, Middle Eastern, and Central Asian astronomical study for nearly fifteen hundred years.

Intriguing evidence of this influence can be seen in the three star catalogs known in Europe in 1584. (Observations for the fourth such catalog were currently being taken afresh by Tycho Brahe, who would publish his results in 1600.) First there was Ptolemy's, which was an extension and updating of an earlier catalog by Hipparchus. Ptolemy recorded the relative positions and the relative brightness of over a thousand stars, grouped in constellations. He began with the constellation of Ursa Minor (the Little Bear, also known as the Little Dipper)—a constellation of vital importance to navigators, as we'll see. Ptolemy listed seven stars in the Little Bear, such as "the star on the end of the tail," "the one next to it on the tail," "the one next to that, before the place where the tail joins [the body]," and so on. Next came the constellation of Ursa Major, and so on through 48 constellations and 1,022 stars.[22] The effort involved in observing and recording such details had been enormous. Not surprisingly, the next two surviving star catalogs—that of the Persian astronomer 'Abd al-Rahman al-Sufi, in 964 CE, and that of the Tartar Ulugh Beg, around 1437—appear to be indebted to *Almagest*, just as Ptolemy was indebted to Hipparchus (who was no doubt indebted to *his* forerunners...). Ulugh Beg was a grandson of the warrior king Timur (or Tamerlane), who had conquered Persia, Syria, northern India, and much of central Asia.[23] Christopher Marlowe was about to turn the tyrannous Tamerlane into the protagonist of his first play, *Tamburlaine the Great* (1587–58), and, in the process, to set the benchmark for the great Elizabethan dramas to come. Back in 1437, Ulugh Beg had ruled his grandfather's kingdom from Samarkand (in Uzbekistan), and his astronomers had laboriously checked and updated Ptolemy's measurements, just as al-Sufi had done.[24]

Almagest contains far more than a star catalog, though. Its name derives from an Arabic word meaning "the greatest," which was the term widely used since antiquity to distinguish Ptolemy's collection from others. Ptolemy himself had called it a "mathematical synthesis" (*Mathematike syntaxis* was its original title). With all the knowledge of his forerunners, not to mention his own considerable skill (and that of subsequent translator-correctors), Ptolemy was able to formulate mathematical astronomy in a form that, in essence, still serves astronomers and navigators. The genius at the heart of it lies in its rigorous development of Hipparchus's program of representing the sky as a sphere or a circle and proving mathematical theorems about the angular positions and motions of the sun, moon, and stars. Angles were much easier to measure than distances, because with angles, the radius of the "celestial sphere" did not matter; it was assumed to be so large as to be almost infinite.

By representing the locations of celestial objects geometrically, Ptolemy tied the stars in the sky to the geographical location of earthbound observers—including those on a ship at sea—because in another influential book, *Geography*, he used an analogous system to visualize the earth. This is a more familiar example today: most people have seen a model globe of the earth that is ringed around the middle by the equator, with its lines of latitude running parallel to the equator, while lines (or "meridians") of longitude encircle the globe in arcs issuing from the North Pole to the South. Any place on earth can be referenced in terms of these two coordinates: latitude and longitude.

Model globes of the earth were not common in Harriot's time, and in the early 1590s he would act as a technical advisor to the designer of the very first such globe to be constructed in England, Emery Molyneux.[25] Nevertheless, despite the lingering idea of a "flat earth" in some quarters, educated Europeans had accepted that the earth was spherical long before Magellan's crew had circumnavigated the globe.[26] Many ancient Greeks had believed it since the time of the Pythagoreans two and a half thousand years ago, although half a millennium later Ptolemy had thought it necessary to give logical arguments for why this is so. For instance, "the sun, moon and other stars do not rise and set simultaneously for everyone on earth, but do so earlier for those more towards the east, later for those towards the west...If the earth's shape were anything other [than a sphere], this would not happen." He clinched the argument with what every sailor knew, too: that the top of a ship's mast would appear first over the horizon, rather than the whole ship.[27]

Nearly four centuries before Ptolemy, Eratosthenes of Cyrene—who had been head of the great Alexandrian library, and who had developed a prototype version of latitude and longitude for map-making—had worked out what turns out to be a very good estimate of the size of the earth's circumference.[28] So the spherical earth had a definite size and shape—it was concrete, tangible—and it was relatively straightforward to visualize it (as in figure 2 in the appendix), with the equator as the zero line: all points on the equator have zero latitude, while all other latitudes are measured in degrees north or south of the equator. Longitude also needs a zero reference line from which to begin measuring; since there is no obvious geometrical feature like the equator that could serve as such a reference, early mapmakers often chose the line of longitude passing through their own town. Today, of course, by international agreement longitude is the number of degrees east or west of the meridian through Greenwich, England.

But where were the center and "equator" of the imaginary, intangible "celestial sphere"? The answer was not straightforward: navigators needed to know *three* sets of celestial coordinates. The most obvious set begins with the observer. For each of us, no matter where we are on earth, *we* seem to be at the center of this imaginary sphere. We look into the distance and observe the roughly circular horizon, and we look up at the hemispherical dome of the sky above, and it seems as if the horizon, extended far out into space, defines the middle of our own celestial sphere. Using this framework, the celestial analog of latitude for a star or other celestial body is its "altitude"—the angle it makes with the horizon, as in figure 3.

If a sailor uses the horizon as the zero reference line for celestial latitude, then the coordinates assigned to the stars depend on the location at which the observations are made. These coordinates are different for observers at different locations, and so they also change as a sailor's ship sails from one place to another. It would be much more useful to record the positions of the stars in a way that can be tabulated and used by any observer at any time. One natural way to do this is to mimic the terrestrial grid of latitude and longitude. If the center of the earth (rather than the position of the individual observer on the surface of the earth) is taken to be the center of the celestial sphere, then the circle around the middle of this imaginary sphere is the "celestial equator," and it is parallel to the earth's equator. The celestial analogs of latitude and longitude in this coordinate system are shown in figure 4. (They are called the "declination" and "right ascension.") Figure 5 shows the relationship between these two systems. It contains the basic geometrical information needed for Ralegh's navigators to convert their own observations, made with respect to their own particular horizons, to those uniquely tabulated with respect to the equator.

Harriot also taught a third way of recording the positions of the sun and stars—and, like the "celestial equator" framework, this third method was independent of the terrestrial location of the observer. In this system—which Ptolemy had used in his star catalog—instead of the reference line being the celestial equator it is the ecliptic, the line representing the sun's apparent yearly path through the zodiac. The intriguing question is how the ancients knew where to draw this line on the celestial sphere.

It all goes back to the behavior of the rising and setting sun, which moves along the horizon during the course of a year in arcs that pivot about due east and due west, respectively. To illustrate the way we understand this phenomenon today, and how it explains the seasons and ultimately leads to pinpointing the ecliptic, imagine a horizontal tabletop with a bowling ball "sun" sitting in the middle and a golf-ball "earth" moving around the bowling ball, tracing out its "orbit" on the table. This is essentially the way Harriot and his Copernican colleagues would have understood it, too. The tabletop represents the plane of the ecliptic, that is, the plane of the earth's orbital path about the sun (or equivalently, the sun's apparent path around the earth). The earth spins on its north-south axis each day, in the direction from west to east. If the sun were to set due west all year—the configuration in which there would also be no seasons— the golf-ball earth would have to move around the bowling ball with its North Pole pointing straight upward, above the table. In other words, the equator would have to be in the same plane as the ecliptic. The picture would then be symmetrical (neglecting the slight asymmetry because the earth's orbit is slightly elliptical rather than circular, as Kepler was about to discover): no matter what time of year, any particular location would always remain in the same position with respect to the sun, so there would be no seasons. And the bowling-ball sun would always rise due east and set due west (as in figure 6).

The fact that there are seasons and the sun doesn't always set due west can be explained very elegantly if the earth's axis of rotation is tilted with respect to the plane of its orbit (as in figure 7).[29] The plane perpendicular to the tilted axis is the plane of the earth's equator, while the tabletop still represents the plane of the earth's orbit around the sun, that is, the plane of the ecliptic. (I am using the term "orbit," although the ancients had spoken of literal celestial spheres or "orbs": the term "orbit" would not be used until 1607, when Kepler introduced it in connection with his groundbreaking revision of Copernicus's model.[30]) Determining how to place the ecliptic on the celestial sphere is equivalent to determining the angle of the earth's tilted axis.

Today we speak readily and easily of the earth's tilt, but in the late sixteenth and early seventeenth centuries, most people did not accept the Copernican idea of a tilted, moving earth. Yet around 260 BCE, Aristarchus of Samos had also proposed a heliocentric system, and a few decades later, Eratosthenes had actually estimated the tilt of the earth's axis needed to produce the seasons. Most, however—from antiquity to Harriot's time and beyond—believed that the earth was motionless at the center of the universe while the sun really did move along the ecliptic. Ptolemy had given a detailed, rather passionate argument "explaining" why this must be the case. He said that the earth was so small compared with the rest of the universe that it was like a point at the center, being "pressed in equally from all directions to a position of equilibrium." If this were not so, he argued, then the earth would fall "out of the heavens" under its own weight, leaving its less heavy living creatures behind, "floating in the air."[31] Harriot, like Galileo, would later challenge this intuitive belief that heavy objects fall more quickly than lighter ones. This work had nothing to do with Ptolemy's argument, though, because Harriot and Galileo already accepted Copernicus's model.

On the other hand, their older contemporary Father José de Acosta agreed with Ptolemy. Acosta was a Spanish Jesuit missionary in Peru, with an interest in science and a questioning mind. Writing in the late 1580s—just a few years after Harriot began his work for Ralegh at Durham House—he used his own experience of sailing and living in the southern hemisphere to challenge various physical and cultural misconceptions, including the view of St. Augustine, who, in around 400 CE, had argued that the Antipodes could not be inhabited. But Father Acosta was a man of the Church, and for all his references to observational evidence—and his willingness to challenge even St. Augustine—he remained in the tradition of Augustine and Thomas Aquinas, trying to fit reason, experience, and scripture into a seamless whole. In this vein, he reconciled astronomical evidence with biblical statements by arguing that the sun and stars were fixed to the heavenly celestial sphere, which circled around the earth of its own divinely assisted accord.[32] (Not that scripture hampered all early scientific enquiry: many early modern scholars were ordained. Copernicus himself was a canon, and Kepler initially trained to be a Lutheran minister; he only turned to mathematics when he was assigned a teaching post in a Lutheran school.)

Acosta was writing nearly fifty years after Copernicus had revived Aristarchus's sun-centered theory in his *De revolutionibus*, published as he lay dying in 1543. (He had been holding off publication for fear of "babblers" who, "although completely ignorant of mathematics," would distort scripture for their own purpose and ridicule his theory.[33]) When it came to understanding the seasons and the ecliptic, though, Acosta and Ptolemy would still have understood the basic configuration that arises in the golf-ball analogy. What matters is that the plane of the apparent yearly orbit of the sun about the earth (the plane of the table in the analogy) is at an angle to the plane of the equator, which is the plane of the sun's apparent daily rotation about the earth. In other words, for navigational and early observational astronomy, it didn't matter whether the earth's axis is really tilted with respect to its orbital plane, and whether it really rotates around this tilted axis to produce the illusion that the sun moves through the sky each day from sunrise to sunset. Nor did it matter whether the earth really orbits the sun each year. All that mattered was that the daily and yearly "motions" of the sun were in different planes, whose angle of intersection could be found by observation. So the ancients established an equivalent of the tilted-earth/golf-ball model simply by drawing the ecliptic at the correct angle to the equator on the celestial sphere. And they determined this angle—known as "the obliquity of the ecliptic"—simply by observing the behavior of the sun.

In *Almagest*, Ptolemy described how he went about this task.[34] The key idea was to measure the difference in the angle between the zenith (the point directly overhead) and the midday sun at the winter and summer solstices. Nearly a millennium and a half later Harriot would make his own check of the obliquity of the ecliptic, using the same method as Ptolemy but with more modern astronomical instruments.[35] What led to the choice of the solstice days for measuring the midday angles was the symmetry of the direction of the setting sun, which reaches its most northerly and southerly points at the solstices. Ptolemy used this symmetry to conclude that the angle between the ecliptic and the celestial equator was half the difference between the summer and winter noon solar measurements. (His reasoning is explained in the endnote and in figure 8.[36]) He found—confirming the estimates made several centuries earlier by Eratosthenes and Hipparchus—that this angle is about 23.5 degrees. (Actually, neither the ancients nor Harriot used decimal point notation, which was a later development. Harriot recorded angles in degrees, minutes, and seconds.) Once this angle had been measured, the ecliptic could be drawn on the celestial sphere, as in figure 9.

Then, putting everything together, figure 10 summarizes all the key astronomical facts that Harriot's pupils needed to know before he could get down to the practical side of his navigation classes. Not that he had access to the kind of diagrams used today (such as I have drawn in the appendix). In the sixteenth century, it was difficult to render diagrams into a printed form, and the modern reader may well wonder how inexperienced students were expected to visualize abstract three-dimensional concepts with just a few words of clarification and definition, such as those given in the opening pages of William Bourne's 1574 *Regiment for the Sea* and in Edward Wright's more advanced *Certain Errors in Navigation*, which would be published in 1599 and make Wright famous throughout Europe.

Unfortunately, the manuscript of Harriot's *Arcticon* has not survived. And it was never published: it would have been a trade secret, with handwritten copies made available only to Ralegh's employees.

The whole point of Ralegh's patronage was not to foster Harriot's fame but to give his own ventures an edge against his competitors— and Harriot's students were indeed the best trained in the world.[37] But we know that *Arcticon* existed because Harriot's surviving navigational instruction manuscripts from the 1590s refer to it and give an idea as to what it contained—although they do not seem to have included diagrams. Instead, he may have shown his students an "armillary sphere" to help make celestial geometry concrete. This was an instrument known since antiquity; Ptolemy had described how to make one, and Tycho had used one to observe the 1577 comet. Some armillary spheres (including their pedestals) were taller than a person. They looked something like the bones of a large globe. Rather than being a solid sphere, they were made up of movable concentric rings representing the ecliptic, celestial equator, observer's meridian, and horizon. An additional ring had a 360-degree scale marked on it, and another had sighting holes so that the rings could be moved to align with various celestial objects, whose angles could be read off the scale.[38]

No one now knows the level of Barlowe's and Amadas's prior education before they joined Harriot's classes. Amadas was from a well-to-do Plymouth family, but he was only eighteen or nineteen when he moved into Durham House prior to leading the reconnaissance voyage to America. (Many of his sailors would have been even younger, some not even in their teens.) Barlowe had served under Ralegh in Ireland, during the bloody English responses to the Desmond Rebellions in 1580, and he had gained a little sailing experience during a voyage to the eastern Mediterranean.[39] This time, though, he would have to contend with the open ocean and an uncharted route. Once he, Amadas, and their navigators had the basic astronomical and geometrical concepts in hand, Harriot could get down to the business of teaching them the hands-on practice of reading the stars and navigating their way to the New World, then returning safely home.

CHAPTER 4

Practical Navigation (and Why the Winds Blow)

H ARRIOT MAY HAVE BEGUN HIS classes by asking Barlowe and Amadas to imagine being blown off course in the
middle of an uncharted ocean. All they would see was end-
less sea and sky. To find out literally where on earth they were, they
would need to know their latitude and longitude. In short, they
needed to learn how to tie the celestial sphere firmly to the terrestrial globe.

If the night was clear and they could see the star Polaris, they
would be in luck, because it points almost exactly toward the
North Celestial Pole. Consequently, it is known as the Pole Star or
the North Star. According to his surviving notes, Harriot told his
pupils that "the elevation of the pole is always of one in quantity with
the distance of your zenith from the equator, [which is equal to] the
latitude of your place."[1] By "elevation of the pole" he meant the altitude—
the angle above the horizon—of the North Celestial Pole, so you
could determine your latitude by finding either the altitude of the
pole or the angle between your zenith and the equator. (These
angles are shown in figure 10.)

Finding the altitude of the pole meant first finding the altitude
of Polaris—but the catch lay in knowing how to correct for the fact
that Polaris isn't *exactly* in line with the North Celestial Pole. Rather,
it is at the top of the handle of the Little Dipper (aka the Little Bear,
or the constellation Ursa Minor). From his own observations, Harriot

knew that published values of Polaris's distance from the pole were inaccurate. In fact, he told his charges that even "the King of Spain's cosmographers" were still using older, inaccurate values, and that these inaccuracies led to errors in the estimate of latitude by at least one degree. One degree of latitude, measured along a meridian, translated to an error of about 69 miles or 110 kilometers in a ship's estimated position.[2] Such errors made it clear to many navigators that the North Star, despite its seemingly ideal location near the North Celestial Pole, was a far from ideal way to find latitude in practice. With his new measurement of Polaris, however, Harriot made his own tables for Ralegh's captains and pilots, complete with clear instructions on how to convert their observations of Polaris into the "altitude of the pole," which would be equal to their latitude.[3] Edward Wright would provide similar tables in the 1610 edition of his *Certain Errors in Navigation*, so Harriot's work on the Pole Star almost twenty-five years earlier than Wright was ahead of its time.

If Polaris was not visible, then sailors would need the alternative method that Harriot mentioned: finding the angle between their zenith and the celestial equator, which depended on the declination of a visible star (that is, on the angle between the star and the celestial equator, as in figures 4 and 10). Unlike the planets, most stars are so far away that if you observe them over many nights from the same spot and at the same time of night, they appear in the same place in the sky. They do appear to move over the course of a night as the earth makes its daily rotation, and the closer stars appear to drift slightly over the year, because the earth's position in space changes with respect to the background stars as it moves in its yearly orbit around the sun. But the stars' positions with respect to each other, and with respect to the celestial equator, do not change appreciably (unless observations are compared over many hundreds or thousands of years). In particular, each star's declination remains the same for long periods of time and can be tabulated.[4]

Without such tables, finding latitude in this way is more difficult and less accurate, although Polynesians had traveled for many hundreds of miles on the open ocean by using the same principle. For instance, to sail due east or west, latitude has to be kept constant, and this can be done by steering a course such that each night, a particular star keeps reaching its highest point directly overhead—that is, at the observer's zenith. Since the boat is steered so that the star always culminates overhead, the angle between the zenith and the equator remains constant, and hence what Harriot called "the latitude of your place" remains constant. Arab traders sailing along the Red Sea, Torres Strait Islanders navigating the waters between Australia and New Guinea, and many other ancient peoples had used a similar approach.[5] But Ralegh wanted his navigators to have the best training available, so Harriot taught them about the apparent motions of the stars, the meaning of their declination, and how to apply the tables that made their job easier and their calculations more accurate.

If there was no handy star at the zenith, navigators could measure the altitude of any known star as it crossed their meridian (that is, the imaginary arc from north to south that passes directly overhead). This was not straightforward: instead of simply looking up a table of declinations and thus automatically learning the ship's latitude, navigators first needed to know how to combine their altitude measurement with the star's declination—a procedure that Harriot spelled out in detail when he taught this same method using the sun instead of a night star.

Using the sun was necessary when clouds hid the stars, but it was an even trickier matter because the sun's declination, unlike that of the stars, changes every day. This is because there are 360 degrees in a circle and 365 days in a year, so the sun "moves" about 1 degree along the ecliptic each day. This apparent motion, and the tilt of the earth's axis, is why the sun rises and sets in different directions throughout the year (as you can see in figure 10)—but it also means that the sun's declination oscillates, too, from zero degrees at the equinoxes, when the sun crosses the celestial equator, to 23.5 degrees north or south at the solstices (as in figure 9). In order to find their latitude, navigators needed to measure the sun's altitude above the horizon and then consult tables of solar declination.[6] However, it was not just a simple matter of subtracting one reading from the other; rather, there were thirteen possible combinations or orderings of the declination and altitude numbers, depending on such factors as the time of day and whether the navigator was north or south of the equator. To make life easier for Ralegh's navigators, Harriot listed these alternatives in an easy-to-follow chart—rather like a modern computer algorithm flowchart of the "if yes, do this; if no, do that" kind.[7] It is an early illustration of his distinctively concise, logical way of thinking.

Harriot's contributions to navigational practice did not end with charts, tables, and calculations. He also made suggestions for reducing the difficulties and errors in using the practical instruments needed, along with the tables, to find the way at sea. One of his favorite astronomical instruments was the cross-staff, which John

Dee had introduced to England after his Continental travels forty years earlier.[8] It consisted of a rigid piece of wood or brass, with a scale marked on it so that it looked like a long ruler. The ruler passed through a slot in the middle of another, smaller rectangular piece of wood or metal that acted as a movable transom, which could be slid along the ruler or staff. The astronomer or navigator would hold one end of the staff at his eye and point the other end roughly in the direction of the relevant star. To measure the angle of the star above the horizon, the transom was moved along the staff until the bottom edge of the transom was aligned with the horizon and the top edge with the star. (You can see a schematic diagram of the setup in figure 11.[9]) The position of the transom on the staff corresponded to a reading on the scale marked on the staff, and this reading gave the star's angle above the horizon. Figure 11 shows that the scale was calibrated using simple trigonometric ratios, which relate the distance along the ruler to the altitude angles.

When using the cross-staff to observe the angle of the sun rather than a star, Harriot cautioned his pupils, they needed to align the top of the transom with the top of the sun, so that the rest of the transom blocked out most of the sun and thus avoided "offending the eye." Then they had to subtract the angular radius of the sun from their measurement, so that the final angle was taken from the center of the sun.[10] The Greek mathematician Thales of Miletus in the sixth century BCE had measured the sun's angular radius to be 15 minutes, where a minute is 1/60th of a degree. (Actually, the 360-degree circle was not then in use, so instead of minutes or degrees Thales's measurement for the radius, or half the diameter, was half of 1/720th of a circle. He had apparently calculated the diameter of the sun by timing how long it took for the sun's disc to rise fully above the horizon, then dividing this time by the 1,440 minutes needed for the sun to make the whole of its daily 24-hour circuit.[11]) In his 1574 navigational manual, Bourne had used Thales's time-honored measurement of 15 minutes (where minutes are denoted by a prime symbol, as in 15′). Harriot, however, measured the sun's radius to be 16′. A difference of 1 minute may hardly seem worth worrying about, but Harriot was already quite a perfectionist. He was also very accurate: NASA has the sun's angular radius at 15.9′.

Next Harriot told his pupils that the cross-staff was calibrated with the supposition "that the end thereof [that is used in] observing should stand at the center of the sight"—that is, at the center of the observing eye. Yet, he noted, experienced seamen believed "there was little error in putting it close under the eye" no matter where they put it, whether "under the utmost corner of the eye, on the cheekbone, or on one side of the bridge of the nose." It was again a matter of parallax, so that slightly different readings would be obtained each time the navigator placed the staff in a slightly different position relative to his eye. This meant that when using the staff to measure solar and stellar altitudes, the error could be as much as 1.5 degrees, which was "so palpable as that it cannot be borne withal, without artificial correction—that is, an allowance or abatement, which hitherto have not been made."

Harriot was not arrogant, but he knew his own worth. Earlier in his notes, he had not held back from critiquing the Spanish theorist Martin Cortes, whose 1551 treatise contained inaccurate measurements of Polaris. Now, in noting that corrections for ocular parallax had "hitherto not been made," he was reassuring Ralegh and his men that he was providing cutting-edge advice. To make such corrections, he recommended that each of his prospective captains, pilots, and other key navigators should always rest the staff next to the corner of his eye. By testing various men's eyes, he had found an average measure of how far a staff in this position was from the center of the eye. He called this distance the "excentricity" of the staff. To compensate for it, he instructed his students that after they had taken an altitude reading, they were to move the transom further away by the distance of this "excentricity." Then the altitude would be "as true as if the end of your staff did stand at the center of your sight."[12] It is a remarkably simple adjustment to make, but Harriot was the first to pay sufficient attention to the errors caused by ocular parallax and to provide such a clear manner of correction.

Harriot noted that another error in using the cross-staff occurred when one tried to align it with the horizon. The problem lay in the fact that a navigator on board a ship was actually standing above sea level, so that his instrument was aligned with his *apparent* horizon only. The problem of finding the so-called dip of the true horizon seems trivial in hindsight, although correcting for it required quite a sophisticated knowledge of geometry and trigonometry and, more importantly, the ability to apply such knowledge to practical issues.[13] Harriot saved Ralegh's men from this mathematical work by presenting the results of his calculations in a table, which listed various corrections for dip according to the height of the navigator's eye above the water. Edward Wright was the first to publish such tables, in his 1599 *Certain Errors in Navigation*, but Harriot had produced his own fifteen years before—yet another of many "firsts" that he was destined not to be credited with because *Arcticon* and his other scientific works were never published. He was to be repeatedly "robbed of glory," as one of his loyal friends would later put it.[14]

FINDING LATITUDE AT SEA WAS far from the end of it. Ralegh's men needed to know how to find their longitude, too. To prepare for this task, Amadas and Barlowe—or their chief pilot, Fernandes—would have to select a location from which they would begin measuring. They would likely have chosen the port from which their voyage was to begin, which was probably Plymouth. Longitude en route to America would then be measured in degrees west of the chosen port.

With 360 degrees in a circle and twenty-four hours in a day, the earth rotates from west to east through 15 degrees each hour. In other words, the time difference between two meridians of longitude that are 15 degrees apart is one hour. So the easiest way to estimate longitude at any point during a westerly voyage would have been to compare the times at each location—just as travelers today have to reset their watches during long travels in the east-west direction. This would be easy to do if two accurate seaworthy clocks could be carried on the journey, one keeping the time of the home port and the other adjusted to the time of the current location, where noon could be determined by observing when the sun reached its highest altitude for the day. In 1584, such clocks were not available: John Harrison wouldn't invent his famous timepiece for another two centuries.

Less accurate clocks were available, though: in 1570, Dee had suggested that all ships should be equipped with hourglasses and "spring clocks."[15] With their help, instead of calculating longitude exactly, captains could chart their voyages using "dead reckoning"—that is, by estimating both the time and the distance traveled during any leg of the journey and marking their estimated positions in a logbook as they sailed. Distance was estimated by multiplying the ship's speed by the time of travel, and speed was estimated with the help of a literal log, a piece of wood with a weight tied to one end and a rope tied to the other. Knots were tied in the rope, with a specific distance between them. The log was lowered overboard behind the moving ship, and a sailor would slowly let out the rope so that the floating log kept pace with the ship. By counting the number of knots that ran through his fingers in a given time, the sailor could calculate the ship's speed—and this method is the origin of the term "knots": 1 knot equals 1 nautical mile per hour, where a nautical mile is 6,076.1 feet (an ordinary mile is 5,280 feet), or 1.852 kilometers. Sailing ships could travel at about 2 knots in a light wind, up to about 6 knots in a good strong wind.[16]

At this rate, traveling several thousand miles west across the Atlantic Ocean would take about three weeks sailing day and night with the best wind all the way. But if a ship leaving from Plymouth simply sailed west, it would end up in Newfoundland—where Ralegh's half brothers John and Adrian Gilbert had interests, having inherited their brother Humphrey's patents for the area—so Armadas and Barlowe would first have to sail south until they reached the estimated latitude of their own American destination. It was determined that this should also be well south of the mouth of the St. Lawrence River, where Walsingham's stepson Christopher Carleill hoped to found a trading colony for the Muscovy Company. (In the end, Carleill's colony did not eventuate.) It should be south of "Norumbega," too, a part of today's New England that Fernandes had briefly reconnoitered for Humphrey Gilbert in 1580 and that Carleill also had his eye on. It seems that there were gentlemen's agreements in place so that intending English explorer-entrepreneurs did not compete for New World locations.[17] On the other hand, for political reasons Ralegh's chosen site would have to be well to the north of the Spaniards' territory in Florida. All this diplomacy meant that Amadas and Barlowe were aiming to land on the east coast somewhere between the latitudes 35 and 42 degrees north.[18]

Nevertheless, Fernandes knew that they should initially set sail for Florida or the West Indies, so that they could make use of the southeasterly "trade winds" that fifteenth-century Portuguese sailors had noted in their pioneering trading journeys down the coast of West Africa. On the way home, their route would be more direct, because they could use the westerly winds that blow most reliably at latitudes between about 30 and 60 degrees—a fact that had only been known to Europeans since 1565, when the Spanish Basque friar Andrés de Urdaneta used these winds to sail home from the Philippines, where he had established a mission.

These prevailing wind patterns were so important to sixteenth-century sailors that it's worth noting briefly what people believed about their cause. After all, it was an era of scientific awakening that laid foundations for the Newtonian revolution still to come, so we would expect to find some people asking *why* about such things. José de Acosta, the curious Jesuit missionary in Peru, was one such person. In his *Natural and Moral History of the East and West Indies* (1590), he asked why it was easier to sail from east to west in the tropics than from west to east. Acosta tried to base his conclusions on experience, but when it came to the trade winds this intuitive everyday experience was misleading. He said that the earth didn't move, because it was too heavy, and that the seas didn't move the way winds do because the seas were joined to the earth. But air and fire were lighter, more subtle elements than earth and water (he was using Aristotle's two-thousand-year-old assumption that there were four fundamental terrestrial "elements": earth, air, fire and water), so that fiery comets and airy winds were carried up to the heavens. Once they joined the celestial sphere—which, Acosta believed, literally moved from east to west, as shown by the "evidence" of the "motion" of the sun and stars—the celestial sphere then carried the wind with it, and "hence" we have easterly breezes in the tropics.

Acosta's background physics may have been Aristotelian—Aristotle was still widely considered the ultimate authority on the natural world—but in puzzling further over his experience of the easterly trade winds, he came intriguingly close to anticipating the correct analysis. Acosta claimed that the air at the equator was "swifter and lighter" than air toward the poles, which was "heavier and slower." He illustrated this assertion with the analogy of a rotating wheel, which moves fastest at the outer rim, the "part of greatest circumference."[19] This is analogous to the spinning equator, which fits with Acosta's statement about the "swift" equatorial air. The equator has the largest, and therefore the swiftest, circumference of all the parallels of latitude (swifter because as the earth rotates from west to east each day, equatorial points must travel further during one twenty-four-hour rotation than any other point on the earth's surface). The wheel analogy is, in fact, part of the modern explanation of the Coriolis effect, named after the Frenchman Gustave-Gaspard Coriolis, who in the late eighteenth century analyzed the effect of the earth's rotation on such things as wind and water currents. (A brief explanation is given in the endnote.[20])

In his 1632 *Dialogue concerning the Two Chief World Systems,* Galileo, too, would grapple with the cause of the trade winds. Like Acosta, he would refer to the "lightness" of the air and to the faster motion of the equatorial region, although of course he believed it was the earth that was rotating, not the celestial sphere.[21] Galileo's interest in the trade winds had nothing to do with navigation, however, or even with physics for its own sake; he was casting about for physical evidence to prove to the Catholic Church that the earth really did rotate on its axis.[22] The explanation of the trade winds via the Coriolis effect provides such evidence, albeit indirectly, but although Galileo had an inkling of the importance of the fact that a sphere rotates fastest at its equator, he could not manage to tie it into a convincing explanation. (What he did discover, though, was the reason we don't feel the earth move as it rotates on its axis and revolves about the sun: the principle of Galilean relativity. As explained in the endnote, this principle demolished the seemingly self-evident anti-Copernican argument that a moving earth would produce noticeable physical effects.[23])

In the 1580s, while he was focusing on practical matters such as making astronomical tables and solving the problems of taking measurements at sea, Harriot was not concerned with why the winds blew. Later, though, he would show an interest in the issue by reading Giambattista della Porta's 1610 pamphlet *De aeris transmutationibus* (On the transmutations of the air).[24] Della Porta's essay was not widely known (which highlights Harriot's unusually wide reading), but his use of the alchemical language of "transmutations" illustrates the connection between the old mysticism and the emerging new science, because his pamphlet contains a key element of the modern explanation of the trade winds and westerlies that even Galileo missed: temperature variation.[25]

Aristotle had taught that winds were "dry exhalations" emitted by the earth. Like Galileo and Acosta, della Porta challenged this view, connecting winds to the moving air itself; he also claimed that it was heat from the sun that caused the air to expand and move, and he described an "inverted glass" experiment to back up his claim. An uncorked flask full of air was inverted and placed on top of a flask full of water; when the air-filled flask was heated, the air on its hotter sides expanded, pushing out the colder air, which bubbled through the water in the lower flask. In this he was inspired by the ideas of his contemporary Giovanni Benedetti, although the first to suggest that winds were caused by heated expanding air seems to have been the ninth-century Arab scholar al-Kindi (whose ideas on energy Harriot would later study).[26] The modern, Coriolis-effect explanation of winds does indeed rely on temperature differences, as well as on the rotation of the earth.

Harriot's goal in 1584, however, was to help Amadas and Barlowe's reconnaissance team to navigate safely, and to this task he brought an attention to detail that resulted in improved data and techniques. Ralegh had worked hard to avoid the mistakes of his half brother Sir Humphrey Gilbert. He had secured better financial backing—and therefore better-built and better-provisioned ships—than Gilbert had managed, and in employing Harriot, he had also taken steps to ensure the best training for his captains and crew.

Soon his efforts paid off. On March 25, 1584, Elizabeth approved Parliament's decision to grant her "trusty and well-beloved" Ralegh a "patent" authorizing his activities in the New World. Like the pope, she assumed a God-given right to her authority, which permitted Ralegh to explore and exploit any "remote heathen and barbarous lands" not already under the jurisdiction of any "Christian Prince" or "inhabited by Christian people." The patent granted Ralegh "and his heirs and assigns" authority over such lands and their people—that is, over both the original inhabitants and any future colonists in the new land. It also granted him authority to license traders in his new territory and to "take and surprise by all manner of means whatsoever" any person or ship found "trafficking into any harbor or creek" in the new land—unless they were friendly fishing or trading fleets en route to Newfoundland, or ships of any nation "driven by force of tempest or shipwreck."[27] In light of the fabulous riches colonization had brought to Spain, the queen followed the Spanish Crown's practice of laying claim to one-fifth of any gold or silver that Ralegh's future settlers might extract.[28]

The legal jargon in the patent highlights the commercial nature of such enterprises, and Ralegh's ambition was to establish a colony in the New World that would expand English trade and influence. This was also his shot at solidifying his favor with the queen, for Amadas and Barlowe would sail in search of a site he would later name Virginia, in honor of the Virgin Queen, who needed no husband to secure her realm.

As for Harriot, his primary concern was to support his friend and patron, and during these hectic months of preparation he had done his job uncommonly well. We know that he would sail to "Virginia" the following year, but there is no record of whether he was to be part of the initial reconnaissance voyage. (Some scholars think it likely that he was, although others dispute this.[29]) Either way, his thoroughgoing commitment to the theory and practice of navigation suggests he was as excited as anyone about the idea of sailing across an unknown ocean to a new and exotic land.

CHAPTER 5

America at Last

ON APRIL 27, 1584, RALEGH'S reconnaissance team of
thirty or forty men set sail in two small ships, one of around
fifty or sixty tons, the other of about thirty tons. The ships'
names and dimensions are not recorded, but the larger vessel could
not have been much more than thirty feet long. It was captained by
Amadas and piloted by Fernandes, a master navigator, a pirate, and
a Portuguese Catholic turned Protestant who had married and settled in England. Barlowe was captain of the smaller ship. His subsequent report of the voyage—including his account of the first contact
between Englishmen and North Carolina Algonquians—suggests he
was a keen observer and a good diplomat. He was also deeply religious, recording the date of their departure as "the 27. Day of April,
in the year of our redemption, 1584."[1] Harriot did not use such a
devout form of dating, although it was not uncommon. Both Bourne's
Regiment for the Sea and Wright's *Certain Errors in Navigation* sometimes prefaced dates with "the year of our Lord" (the literal translation of *anno domini*, or AD).

The little reconnaissance fleet followed the route used by
Columbus nearly a century earlier: south to the Canary Islands, then
southwest under the tropical trade winds whose cause would so
intrigue Acosta and Galileo. Columbus had been a brilliant navigator for his time—some said he had supernatural powers, so uncanny
was his ability to read the winds and currents—but he had not deliberately sought the best route to America.[2] After all, he had "discovered" its existence by accident, during his westward bid to reach

Japan. He had ended up in the Caribbean and in Central and South America. Amadas and Barlowe knew where they were heading—somewhere on the east coast of America within a few degrees of the 40th parallel. But they chose the southwest route not only because of its trade winds but also because it had been mapped; Columbus had pledged he would "neglect sleep" so as to make new navigational charts and to carefully record his route via daily estimates of latitude and longitude.[3] They also chose Columbus's route because they did not know how long they would be at sea before finding a suitable place for Ralegh's colony, and they knew they could reprovision in the West Indies. Despite Spain's attempts to keep the New World for itself, many Spanish colonists were willing to sell water and other supplies to the captains of passing ships. And if they were not willing to trade, privateers—notably the explorer and sometime slave trader John Hawkins and the redoubtable privateer Francis Drake—were likely to take what they needed by force.[4]

By the time Barlowe and Amadas reached the Indies, they had been at sea for six weeks. Deteriorating food and sleepless nights would have sapped energy and morale: officers and gentlemen slept in cramped makeshift cabins, while the sailors "slept" wherever they could find a space amid the rigging, on deck, or, as a last resort, in the fetid holds. Everyone was glad of a break on dry land, and they rested for twelve days. Then, having "refreshed ourselves with sweet water and fresh victuals," as Barlowe put it in his report, they weighed anchor once again.

Now they were heading northwest into uncharted waters. It must have been a thrilling and terrifying experience both—the exhilaration of adventure tempered by fear of the unknown and of the vagaries of sea travel. During stormy weather, the deafening roar of tempestuous winds and huge waves could overpower all reason as tiny ships were tossed about like a cat's playthings. On a more clement day, though, and with a competent, united crew, sailing into the unknown would surely have afforded a transcendent thrill. Unfortunately, Barlowe's report is silent on personal matters: its primary purpose was to promote Ralegh's colonial enterprises to potential investors and colonists.

The report is also silent on whether the captains and other key navigators felt more confident because of Harriot's lessons. But others who benefited from them were more openly grateful. For instance, in the dedication to his 1595 publication *The Seaman's Secrets*, the Arctic explorer John Davis would describe both Harriot and Dee as navigational experts who "are hardly to be matched."[5]

And Harriot particularly impressed the geographer and colonial advocate Richard Hakluyt, who in 1587 would publicly praise Ralegh for having the foresight to "have nourished in your household, with a most liberal salary, a young man well trained in those studies, Thomas Harriot." Under Harriot's guidance, Hakluyt continued, Ralegh's "collaborating sea captains [had] very profitably united theory with practice," thereby achieving "almost incredible results."[6] It certainly seems that Barlowe's and Amadas's ships were sailed "with exemplary skill" (in the words of a leading historian of the Roanoke voyages[7]).

After ten days' further sailing, Barlowe recorded that the scent of the air changed: the fresh tang of sea salt gave way to sweet perfume, "as if we had been in the midst of some delicate garden." They had been sailing for nearly two months, so Barlowe's poetic description suggests his relief: this fragrance must have been carried by an offshore breeze, "by which we were assured that land could not be far distant." Two days later they arrived on the coast of America. But land alone was not sufficient; they needed a harbor so that they could safely anchor, and they had to tack up the coast for another 120 miles before finding an inlet that would suffice. The first thing they did after anchoring was give "thanks to God for our safe arrival." Then they went ashore and "took possession" of the land in the queen's name, performing "the ceremonies used in such enterprises." Barlowe's summary of his era's religious and imperial priorities could not have been more succinct.

Next they set about exploring this new land. Barlowe felt they had found the Garden of Eden. The terrain was "sandy and low" near the beach, "but so full of grapes that the very beating and surge of the Sea overflowed them." In fact, there were so many grapes, both here and further afield, that the well-traveled Barlowe enthused, "I think in all the world the like abundance is not to be found." When they climbed the sandy bank, they "beheld the valleys replenished with goodly cedar trees," and not just any cedar trees but "the highest and reddest in the world." They discharged a gun to attract the attention of anyone living nearby, but their only respondent was a huge flock of startled cranes, whose cries "redoubled by many echoes, as if an army of men had shouted all together." Apart from these eerie echoes, all was preternaturally silent. They were alone in a pristine paradise.

AFTER EXPLORING THE AREA, THE Englishmen were surprised to find that they had landed not on the North American continent but on an island about sixty miles long and six miles wide.[8] They did not yet know it, but they had disembarked on one of many islands stretching two hundred miles along the coast—an area known today as the Outer Banks of North Carolina. (Ralegh and Harriot's "Virginia" is in today's North Carolina, not the state of Virginia; in the rest of this book, the name Virginia is meant in its original context.)

For two days the ships were anchored near their island, the men sleeping on board. On the third day, they spotted three Americans in a small boat rowing toward the ships. The canoe stopped in a little cove some distance away, but one of the men waded ashore and began walking up and down as if to signal the foreigners to a parley. Fernandes, Amadas, Barlowe, and a few others rowed to meet him, and Barlowe was struck by the fact that the man showed no fear as they approached him. He spoke to them in a strange language, and after listening politely without understanding a thing, the English party invited him aboard their ship, an invitation he accepted "with his own good liking."

Once on board, he was given gifts—a shirt, a hat, and some trinkets: the same sort of presents Columbus had dispensed in the Caribbean, as had Cartier in Quebec. Then he was offered European food and wine, "which he liked very well." But he did not linger: when he had eaten and had checked out each of the ships, he went back to his own boat, leaving the bemused Englishmen to watch him fishing. In less than half an hour he had caught enough fish to fill his boat so full it could barely float. Then he rowed back to the visitors' ships and calmly proceeded to divide his catch in two, indicating a pile for each of the ships. After that he got into his canoe and vanished.

The next day, however, it became clear that the newcomers had passed their first test: a number of boats came out to meet them, one of which carried "the King's brother." He was, wrote Barlowe, "accompanied by forty or fifty men, very handsome and goodly people, and in their behavior as mannerly and civil as any of Europe. His name was Granganimeo, and the King is called Wingina." Wingina was chief of a number of local groups, known as the Secotan tribes, and he lived on an island some twenty miles away—later called Roanoke Island. He was not fit enough to travel to welcome the visitors, having been wounded in a fight with a neighboring chief. Granganimeo invited the Englishmen to a meeting on shore, where his rank became clear in the way his servants spread out a long mat for him. Four men of slightly lesser rank sat at the other end of the mat, while the rest of the entourage kept a respectful distance. When the Englishmen disembarked, Barlowe was amazed at Granganimeo's serenity:

When we came to the shore to him with our weapons, he never moved from his place, nor did any of the other four, nor never mistrusted any harm to be offered from us, but sitting still, he beckoned us to come and sit by him, which we performed. And being sat, he made all signs of joy and welcome, striking on his head and his breast, and afterwards on ours, to show we were all one, smiling, and making show the best he could, of all love and familiarity.

After this ceremonial welcome, the visitors presented Granganimeo with gifts, which he received "joyfully and thankfully." Then they offered presents to those sitting with him on the mat. This was their first breach of protocol: Granganimeo "arose, and took all [the gifts] from them, and put them into his own basket, making signs that all things ought to be delivered unto him, and the rest were but his servants, and followers." Nevertheless, friendly relations continued, and a few days later, trading began. The English exchanged trinkets such as Venetian glass beads, as well as hatchets, axes, knives, and metal utensils, for dressed deerskins, coral, and dyes. They negotiated "a very good rate" for their goods. For example, Barlowe recorded that a copper kettle was sold for "fifty skins worth fifty crowns"; one 1580s crown, or five shillings, was roughly equivalent to a hundred of today's dollars. A crown was a week's earnings for an Elizabethan craftsman and about two weeks' pay for a soldier.[9]

The exchange rate was even better for English tools, since the Algonquians had no iron-edged implements. But Barlowe was impressed with the indigenous way of making large canoes without the help of such tools. Harriot would later describe this method in detail, noting that "the manner of making their boats in Virginia is very wonderful": despite the lack of iron tools, these boats were made "as handsomely as ours," and for the purposes of fishing and for navigating rivers, they were as effective as English vessels. The first step was choosing a tree of suitable height for the particular boat. The Algonquians felled the tree by building a fire at its base, making sure it was not so hot as to damage the trunk. Once the tree had fallen, they burned off its branches, again taking care not to burn the trunk. Placing the trunk on a frame made of branches, they scraped off the bark with shells and then lit a small fire on top of the trunk to begin the process of hollowing it out. They achieved the desired depth and shape by a careful process of repeatedly burning, quenching, and scraping with shells.[10]

Amid all this companionable trading, the Englishmen remained somewhat wary. Barlowe noted that the people "would have given anything for swords, but we would not part with any." He also commented that Granganimeo himself was a most trustworthy trader, always keeping his promise when the English gave him merchandise, returning "within the day" with payment. Such payment included not only deerskins and dyes but also such diverse and delicious foods that Barlowe was amazed at the apparent fertility of the land. For him, this truly was a land of milk and honey.

After several days of convivial bartering, and entertaining Algonquian visitors on board, Barlowe set off with seven men for Roanoke Island. On their arrival, they found a village containing nine cedar houses, "fortified round about with sharp trees [stakes], to keep out their enemies." Granganimeo's wife had met the newcomers earlier, and now she came running out to welcome them. She informed them that her husband was not then in the village, but from Barlowe's detailed description of her warm and generous hospitality, she seems to have had all the authority, charm, and wisdom of a leader. (He never referred to her by name, though, but only by her apparently high office as the chief's brother's wife.)

All in all, Barlowe liked and admired his American hosts; "a more kind and loving people there cannot be found in the world," he wrote. If he seems like a promulgator of the "noble savage" ideal, he was not alone: Columbus and Cartier had left similarly glowing descriptions of their first contact with indigenous Americans— although Columbus had come upon an island whose people had been terrorized by a supposedly cannibalistic group known as the Caribs.[11] Barlowe did not record any such hostile encounters; perhaps there were none. On the other hand, he was writing to advance his employer's interests, and it suited his purpose—and possibly his temperament—to believe that Englishmen would be welcome in America.

It also suited his purpose, as a devout Christian, to present what he saw as a darker side of Algonquian life. He remarked that Granganimeo's house contained a special place for idols, adding that these were "nothing else but a mere illusion of the Devil." He also noted that their wars were "very cruel and bloody," although he did not say how these could possibly have been more cruel and bloody than the English wars in Ireland, in which he himself had participated. Warming to his theme, Barlowe, who had earlier described his hosts as "void of all guile and treason," now reported, just as credulously, that their neighbors were capable of great treachery—such as luring some of Wingina's people to a feast and, after getting them to relax, suddenly attacking them and killing all the men. The ensuing enmity between Wingina and his perfidious neighbors was such that the Roanoke people tried to persuade Barlowe's men to go with them to the enemy's town, assuring them they would find there a "great store of commodities." But Barlowe preferred not to test whether such information about "great stores" was offered out of friendship or simply to inveigle the Englishmen into avenging their enemies for them. He may have been charmed by the friendly Roanokes, but he made it clear he was nobody's fool.[12]

THE RECONNAISSANCE PARTY STAYED ON the Outer Banks for about a month, after which Barlowe took his ship straight home, returning safely in September 1584. Amadas—or his pilot, Fernandes— apparently chose to detour first on a scouting voyage further up the coast and then on an unsuccessful search for Spanish ships to plunder; they did not arrive home until November.[13] Meanwhile, the excitement when Barlowe turned up at Durham House with his tales of an American Garden of Eden was underscored by the presentation of two exotic visitors who had come back with him: Wanchese, from Roanoke Island, and Manteo, from nearby Croatoan Island (known as Hatteras Island today). In his report, Barlowe did not mention how these two men had been enticed—or coerced—to go with the Englishmen. Most likely they went voluntarily, because Ralegh's policy in the New World was one of peace. Perhaps they went with a sense of adventurous excitement, or as chosen delegates in the hope they would cement trading ties with this distant land of marvelous tools and other goods.

Manteo and Wanchese were not the first exotic strangers to amaze the English. When the Irish tribal chief Shane O'Neill and his bodyguards visited Elizabeth's court in 1562, their long, unkempt hair, long-sleeved saffron gowns, and ancient battleaxes, and the "howling" sound of their native language, had given English courtiers the impression, as a contemporary historian had it, that the Irish were as alien and "primitive" as the peoples on the other side of the world.[14] The first North Americans in England were Inuits, who had been brought back, apparently by force, by Bristol merchants in 1502. This was soon after Bristol-based John Cabot's pioneering voyages had opened up the fishing grounds of Newfoundland. Then, in 1577, Martin Frobisher had been seeking an ocean route to China when he encountered Baffin Island instead; apparently he hoped his failure to find a North-West Passage to the Orient would be mitigated by the curiosity value of an unfortunate Inuit couple he brought back with him. He reported that these people fed only on raw flesh and could kill a duck with a dart from an astonishing distance. Tragically, but not surprisingly, the couple died within a month.[15]

There were other such stories, and not all of them had unhappy endings. In 1534, Jacques Cartier had taken two princes—Taignoagny and Domagaia—from Quebec to France, having managed somehow to convince the chief he would soon come back not only with the young men but also with a new supply of the iron tools and other goods that the people had coveted. The following year he did indeed return, bringing Taignoagny and Domagaia with him.

Manteo and Wanchese, too, would eventually return home. But first they did the rounds of London. A German visitor reported, in October 1584, that although the men's usual habit was a mantle of animal skin, Ralegh had dressed them in brown taffeta. "No-one was able to understand them and they made a most childish and silly figure."[16] But this impression would not remain for long: Wanchese and (especially) Manteo settled into the luxurious life at Durham House, taking up their new roles as students of English and teachers of their languages, which were Carolina Algonquian dialects.[17] The principal teacher-student in this linguistic exchange was none other than Harriot.[18] His *Arcticon* had provided cutting-edge navigational advice for Ralegh's captains in their journey to Roanoke and the Outer Banks, and now he took on a very different role—that of a sort of cultural attaché, liaising between the Old World and the New.

Ralegh's creation of this post for Harriot, and his emphasis on teaching English language and customs to Americans in England, were unusual. Cartier had returned from France with his Canadian princes, but most explorers simply captured "natives" and forcibly "trained" them as interpreters on the spot. Ralegh aimed at a more ambitious and more humane program. Whether his motives were cannily imperial or not, he opened his own home first to Manteo and Wanchese and later to a dozen or so others, almost all of whom would remain steadfastly loyal to him. Even the Spanish—whose spies were itching to learn what Ralegh was up to in the New World—noted that the English treated Manteo and Wanchese well and that the two young men were soon speaking English.[19]

As for Harriot, their teacher and student, his skill in languages was notable even in an age when every grammar-school boy learned Latin. (Girls generally had only a year or two in a "petty," or petit, school, until they were seven.) In addition, Oxford and Cambridge arts students also learned Greek, and some studied a contemporary European language, too. But Latin was the international language of scholars, and Harriot's Latin was fluent, elegant, and sophisticated.

(This is in contrast to much of his written English, and that of most of his nonliterary peers, which sounds clumsy to modern ears, for English was still evolving in style, spelling, and grammar.) Harriot's Greek was excellent, too, and a little later he would act as an advisor to his friend George Chapman, when Chapman was translating Homer's *Iliad*.[20]

What really sets Harriot apart from his contemporaries, though, is that in whatever sphere he entered, he brought a way of applying his knowledge in new ways. In particular, he had an ability to find *general* patterns or solutions. In applying his linguistic talents to Carolina Algonquian, an oral language, he was not content to learn its basics himself: he developed a written system that he hoped would facilitate not only the learning of Algonquian but also any linguistic exchange. He devised his own phonetic system to represent the sounds of an exotic language such as Algonquian—but it was a system applicable to any language, because it was designed to record all the possible sounds of human speech.

Harriot is the first known scholar to develop such a general system. Sixteen years earlier, in 1569, John Hart had begun work on a phonetic way of representing English regional dialects, along with Welsh and Irish. His primary goal was spelling reform, given the haphazard usage in written Elizabethan English: without standardized spelling, he argued, a language was too difficult to learn, and without a literate population, prosperity would suffer and "confusion and disorder" would reign. French scholars, too, had been working on similar reform.[21] No one knows whether Harriot knew of this work. In any case he had a different aim: international communication. To make his phonetic alphabet truly universal, he did not use European letters to represent sounds; rather, he invented his own symbols—strange squiggles that at first sight look like a spell cooked up by John Dee or some other magus.

Harriot created a set of ten symbols for vowel sounds: the five Latin symbols, *a, e, i, o, u,* were clearly insufficient to express the variety of sounds represented by these letters, even in English. So Harriot devised three separate symbols for sounds denoted by *a* in English: one for the sound of *a* "as in all, fall, tall, call"; one for the sound of *a* "as in narrow, man, pan"; and one for the sound of *a* "as in ape, ale, any." (These are Harriot's word selections; his choice of "ale" is a reminder that in Elizabethan England water quality was so poor that ale was often drunk instead.) Similarly, he had three symbols for the different sounds represented by *o,* as well as sounds for *u, i, e,* and *ee.* He then created the consonant sounds simply by adding loops to the

Plate 1: The mysterious squiggle on the title page of this manuscript represents "Thomas Harriot" in his phonetic alphabet.

top and/or the bottom of the vowel symbols.[22] He did this in a carefully thought-out way, grouping and arranging all his symbols on the page in a precise manner according to the way sounds are formed in the mouth.

To illustrate, one of his three groups contains his symbols for nasal sounds—those denoted by our *m*, *n*, and *ng* ("as in king, thing")—together with the fricatives, the sounds made by friction of the breath when breathing out with the mouth almost closed, as when saying the sounds *f* (as in "fish"), *th* (hard as in "the" or soft as in "thing"), and *v* ("as in vine"). Harriot also included here two additional symbols for sounds not made in English, which he described as "gh" and "ch," and the whole group is not only listed separately in his manuscript but also distinguished by the fact that these are the symbols to which he added only an upper loop.

Harriot's other symbols are also grouped—and looped—according to the way their sounds are produced, yet this kind of analysis of sounds (including the word "fricative") really developed only in the nineteenth century. The International Phonetic Alphabet (IPA) used in modern English and other dictionaries dates from 1889. It has twenty-two symbols for vowels, arranged in three groups: short vowels (as in "cat"), long vowels (as in "arm"), and diphthongs (a sound made by two vowels in one syllable, as in "coin" and "loud"). Harriot's arrangements of both vowels and consonants are different, apparently motivated by a desire for conceptual and symbolic economy.

This elegant economy, achieved through an extraordinary attention to pattern and detail, would characterize Harriot's later mathematical work, too. Meantime, when he painstakingly wrote out his systematic phonetic alphabet in 1585, he headed his page "An universal alphabet containing six and thirty letters, whereby may be expressed the lively image of man's voice in what language soever; first devised upon occasion to seek for fit letters to express the Virginian speech."[23] It would have been groundbreaking had it found its way into print; like his *Arcticon*, however, it was destined for private use in Durham House. For Ralegh, at least, that was benefit enough. If he was to succeed in seizing a slice of the New World by establishing a trading base and putting settlers there, he needed every advantage he could muster.

CHAPTER 6

Preparing for Virginia

H ARRIOT WAS NOT EXPECTED TO be a stay-at-home cultural attaché but to travel with Ralegh's first settlers to America. Arthur Barlowe's report had been so encouraging that Ralegh had immediately set about advertising for colonists to maintain a permanent English settlement in the New World.

Such a venture would take many months of planning and fundraising, of course. Providing and provisioning the ships required merchant-investors who were wealthy and bold enough to take part in a venture whose only hope of immediate profit lay in any gold or other valuable commodities found in America, or from proceeds from privateering en route. Ralegh had a difficult job convincing investors of the soundness of his plan, but it was important to choose his partners wisely: Elizabethan businessmen were a litigious lot, and it was not uncommon to find investors in privateering or exploratory expeditions suing each other for a more equitable division of profits. Amyas Preston would soon mount such a case against his business partner in one of the ships that Ralegh hoped would sail to Virginia. In the end, the ship went to Newfoundland on official business—and then sailed south on a piratical expedition that returned considerable profits—and Preston wanted a fairer return on his investment in provisioning it. In another example close to Ralegh, Humphrey Gilbert had sued the merchant from whom he'd chartered a ship for his 1578 attempt at reaching America—the same ship, captained by Ralegh, that had had to turn back midvoyage partly, Gilbert claimed in court, because it proved unseaworthy. (On the witness stand in 1580, Ralegh had been equivocal about the ship's defects.)[1]

While Ralegh's hands were full with such tasks, Harriot was occupied with learning Algonquian from Manteo and Wanchese and teaching them English. The three of them also no doubt helped Barlowe with background details as he was writing up his report: given that while he was actually in America his communication with the local people would have been largely by making signs, the level of detail in the final write-up he supplied to Ralegh and his investors suggests he had additional help once Harriot had helped break down the language barrier.

Harriot was also engaged in teaching navigation to Ralegh's new captains and navigators—this new expedition was to include five more ships than the reconnaissance fleet of 1584—and he was also continuing his navigational research. A pressing topic in the 1580s was magnetism. Acosta may have stuck firmly to Aristotelian cosmology in his explanation of the trade winds, but he noted that Aristotle did not know everything. In particular, neither he nor Pliny nor even St. Augustine had mentioned the magnetic compass![2] The ancients had known that magnetism itself existed because they had noticed the attracting ability of a particular iron-rich black rock that was later known as lodestone. But it seems to have been the Chinese who created a portable north-finding instrument by placing a small piece of magnetic rock on a piece of wood and floating it in a bowl of water. The earliest evidence of the use of compasses at sea dates from eleventh-century China, and over the next century, the idea found its way to the Middle East and Europe. By the sixteenth century, compasses were mounted on a self-aligning mechanism to keep them level against the swell of the sea. By the 1580s, however, a puzzling magnetic anomaly was causing consternation among sailors, and it was occupying some of the best navigational theorists, including Harriot. It was called magnetic variation.

The magnetic compass points to "magnetic north," which is not quite the same as "true north" (the location of the earth's geographical North Pole). This would not be a problem were the difference between magnetic north and geographic north constant, but annoyingly, the compass showed different variations from true north at different places. Such seemingly erratic fluctuations could cost time, if not lives. One of Harriot's examples for his students implies that if a pilot set a course north by his compass but the actual direction of his travel was 9.5 degrees to the west of north, then after sailing a one-hundred-mile leg, the ship would be more than sixteen miles too far west.[3] Left uncorrected over a long ocean voyage, such errors could mount up, compounding the inherent inaccuracy of steering a sailing ship against winds and currents and leading to a ship becoming dangerously lost.

In 1581, William Borough had published a way of accurately determining magnetic variation by comparing observations of the sun's altitude and azimuth, made at any suitable time, with theoretical calculations. But Borough's method required navigators to apply a complex mathematical formula—a task beyond the competence of most seamen—so Harriot came up with a much simpler method.[4] All it required of navigators was a compass bearing of the direction of sunrise or sunset, a measurement of current latitude (which Borough had also required), and solar declination tables, together with a table of magnetic corrections that Harriot had especially prepared. To make this table, he first worked out a formula giving the true directions of sunrise and sunset at a given time and place—an elegant formulation derived from the geometry of the celestial sphere, as figure 12 shows.[5] Then he codified the results in a table whose construction had required hundreds of tedious calculations on his part, in order to make the process easy for Ralegh's navigators. If they were taking a compass reading at sunrise, all they had to do was to look up the table and find the reading corresponding to their latitude and the time of year (given by the sun's declination at sunrise[6]). The table entry then gave the true direction of sunrise at that time and place. The difference between this value and the compass reading gave the magnetic variation.[7]

Despite the hard work involved, it must have been extremely satisfying for someone of Harriot's mathematical bent to deduce, from first principles, a new formula that predicted the behavior of such natural phenomena as sunrise and sunset—and to use it to find a new and simple way of quantifying the effect of another natural phenomenon: magnetic variation. Harriot did not attempt to explain the cause of this variation, though. Ralegh was paying him to come up with practical results, not to solve nature's riddles.

Twenty-five years later, however, Harriot would be pleased to note in one of his manuscripts that his careful quantification had not gone unacknowledged. The man who eventually found a key to the *why* of magnetic variation listed Harriot first among a group of experimenters whose quantitative analyses had helped to clarify this elusive phenomenon. It's worth noting that Harriot's friend and disciple Robert Hues was named next, followed by Edward Wright and Abraham Kendall. Hues had studied at St. Mary Hall, Oxford, so it is quite possible he and Harriot met there. (He was also an all-rounder like Harriot: during the 1590s, Hues would act as Chapman's Greek-language co-consultant, along with Harriot, during Chapman's influential translation of Homer; he would sail with Thomas Cavendish when he circumnavigated the globe in 1586–88; and in 1594, he would publish his *Tractatus de globis,* a book on how to use Molyneux globes in navigation.[8])

This acknowledgment appeared in 1600, in William Gilbert's *De magnete* (On magnetism).[9] This particular Gilbert appears to have been no relation to Ralegh's half brothers and would be appointed to the position of queen's physician in 1601, but his hobby had long been physics. In particular, he experimented with magnetism, and also with "electricity"—a name he coined from the Greek word for "amber," the fossilized resin that acquires static electricity on being rubbed with a silk cloth. His greatest discovery was that the earth itself is a giant magnet. This explained the difference between geographic and magnetic north: evidently the earth's magnetic poles do not lie at the North and South Poles. Gilbert made his discovery when he realized that a compass needle, which tends to dip downward as it orients itself horizontally, is dipping because it is being drawn into the center of the earth—as if the earth itself were a magnet. It was a very "modern" scientific discovery, because Gilbert also tested his hypothesis experimentally, using a spherical magnetic lodestone.[10]

His interpretation of his discovery was not so modern. He believed that without its topographical peaks and troughs—its mountains and seas—the earth would be a perfect, spherical magnet whose poles *would* coincide with the geographic poles. He was evoking the ancient Greek idea that circles and spheres were the perfect shapes, so that only the earth's "imperfections" prevented the magnetic and geographical poles from coinciding. In fact, Gilbert had no idea why the earth was magnetic—no idea of its molten iron-rich core. Instead, he interpreted the earth's magnetism animistically, as if it were an animating force like a "soul"—and he was so struck by the fact that the earth has two natural magnetic poles and two geographic poles that he suggested that this magnetic "soul" was behind the earth's daily rotation about the axis through its geographic poles. Some modern philosophers are not averse to ascribing "agency" to inanimate moving objects, but Gilbert also thought the entire cosmos was alive and "ensouled," and he took to task Aristotle and his followers for treating the earth separately from the heavens, as if it were "savage and soulless."[11]

Today, physicists believe that gravity and the conservation of momentum keep the planets spinning and circling the sun—and have kept them doing so since the planets' constituent matter was flung from the sun during its tempestuous formation. Nevertheless, Gilbert's bold hypothesis about the earth's magnetism causing its daily motion proved inspirational to Kepler and Galileo, in their search for physical evidence that the earth really did rotate and revolve as in Copernicus's model. They did not succeed: Galileo knew, as his attempt at using the earth's rotation to explain the trade winds shows, that since we cannot see or feel the earth moving we need to infer it, but it would take Newtonian dynamics to provide a theoretical framework that enabled the first credible inferences of the earth's spinning motion.[12] Still, Gilbert's speculations did bear some fruit: Galileo may not have proved that the earth moves, but as noted earlier, he did discover the reason we cannot detect this motion, namely, his principle of relativity. And Kepler would use the idea of a magnetic impetus or force in conceiving the notion of planetary orbits.

Harriot seems not to have worried about proving that the earth moves, accepting the rotating heliocentric model as the most compelling—especially with the addition of elliptical orbits, which as we'll see he anticipated before Kepler published his discovery, and which had the advantage of both simplicity and accuracy over the Copernican circular model. As for magnetic variation, the first clue to its cause was Gilbert's discovery that the earth is magnetic. The modern explanation is that the variation in this terrestrial magnetism is due to flows in the earth's molten core and to deposits of magnetic metals in the earth's crust—factors that change over time, causing magnetic variation not only from one place to another but also over time at the same place. Henry Gellibrand had first pointed out this temporal variation, in 1634, although like Gilbert he had no idea why it was happening.

In the 1580s, then, Harriot's accurate practical method for measuring this variation was far more useful to navigators than any speculative theory about its cause.

AS THE MONTHS PASSED AND the time for an early spring departure to Virginia was drawing near, one wonders what Harriot was thinking. Did he see his new assignment—spending an indefinite period in America—as a job or an exotic adventure? These are not the sorts of reflections Harriot offers in his papers. But the creativity and energy he brought to the task of learning and writing phonetic

Algonquian suggests he was exhilarated at the thought of spending time with Manteo's and Wanchese's people. As for the voyage itself, the inordinate patience required to produce cutting edge tables that made sailing more accurate suggest that he couldn't wait to put his teaching and research to the test.

Aside from his excitement and intellectual dedication, however, modern readers must wonder what sort of cultural attitudes or prejudices the twenty-four-year-old Harriot would bring to the New World. There are only small clues. For instance, in his phonetic alphabet Harriot had described his symbols for the non-English sounds he denoted by "gh" and "ch." He did not give examples of how these two special symbols were to be pronounced, noting simply "as in some barbarous words"; but in the case of "ch," he'd added "as in the Greek χ." Was Harriot implying that, like English and French (which he had cited with respect to another symbol that sounded "as in the French j in je, jeter"), Greek was a civilized language, but that other languages with these guttural sounds "gh" and "ch" were "barbarous"?

To most Europeans in that Latin-speaking era, English itself would have seemed "barbarous"—certainly "foreign" (the original meaning of "barbarous") and probably "coarse," too. In fact, it was such a marginal language that English schools themselves favored Latin. To help foster the better use of students' native tongue, in 1582, just three years before Harriot devised his alphabet, the teacher and educational polemicist Richard Mulcaster had published a treatise on elementary education in which he urged that instruction in school be in English rather than Latin.[13] English, he said, was a language both "deep" and "frank"—a "pithy" language in which you could say what you meant with "great plainness." Within the next decade, Shakespeare would begin to immortalize these qualities of the English tongue. A century later, Voltaire expressed his admiration for Shakespeare's "force and fecundity," his genius "natural and sublime, without the least spark of good taste and without the least understanding of [grammatical and theatrical] rules"![14]

It is a short step from regarding foreign languages as "babble" to regarding foreigners themselves as uncultured or uncivilized. While many would have seen the English as uncultured and their language coarse, it went both ways. In 1592, a visiting German duke would note that few Londoners traveled abroad, especially tradespeople, so they "care little for foreigners, but scoff and laugh at them. One dare not oppose them, else the street boys and apprentices gather in immense crowds and strike unmercifully left and right. And because they are the strongest, one is obliged to put up with the insult as well as the injury."[15] Of course, wealthier Elizabethans did travel abroad. It was common for young aristocratic gentlemen to make a leisurely tour of the Continent in order to "finish" their education; the newly knighted Ralegh would later send his own son on such a tour. And scholars such as Dee traveled to meet like-minded colleagues. There is no evidence that Harriot visited the Continent, but given his multicultural reading matter—including the navigational treatises of Pedro Nunes and Martin Cortes, and José de Acosta's *History of the Indies* (which he would read in the early 1590s, before it had been translated from Spanish into English)—it is clear that he was a worldly man, even if in a bookish way.[16] It therefore seems likely that he simply meant "foreign" when he spoke of "barbarous" languages and that he singled out French and Greek examples simply because they were well known.

On the other hand, "barbarous" did take on a negative meaning for many Europeans when it came to discussing some aspects of the New World, especially religion. The generally tolerant Acosta, for example, spoke of the "blind Indians," who, he claimed, had been led astray by the devil, so that in their "barbarous blindness" they worshipped idols instead of the one "true" God.[17]

Most Elizabethans also used "savage"—another problematic word for modern readers—to denote the First Americans (or else, following the Spaniards, "Indians," since Columbus had initially thought he'd landed in India). But it is not so easy to know what individual writers meant by their use of "savage." It was derived from the French word *sauvage*, meaning "wild," and the French had adapted it from the Latin *silvaticus*, meaning "of the woods." By Harriot's time, there was a widespread presumption that New World peoples were "savage" not in a violently uncontrollable way but in the sense of living closer to nature.[18] This view did not necessarily conflate less sophisticated technological achievement with lesser moral or intellectual qualities, although for many Europeans that leap was all too easy. So it is notable that Harriot rarely used the word "savage" or "primitive" at all, preferring "inhabitants" or "people" to "savages" and stressing the Algonquians' technical ingenuity, such as their method of making canoes.

If diplomatic relations between Old and New World delegates at Durham House seemed cozy in early 1585, they were delicate. It was an easy descent from "barbarous," "primitive," and "savage"—no matter how "objective" the intent—into what we now call racism.[19] Religion played a key role in this, as the pope and Queen Elizabeth made clear when they authorized their subjects to exploit any

"remote heathen and barbarous lands" not already under the jurisdiction of any "Christian Prince" or "inhabited by Christian people,"
to use the wording from Ralegh's patent. Devout sixteenth-century
European colonialists sincerely believed they would be saving souls
and changing lives for the better if they could bring Christianity to
the New World.[20]

This missionary zeal had begun with Columbus, whose example
offered lessons for would-be Elizabethan colonizers. Columbus had
begun with benevolent intentions, hoping to convert the friendly,
welcoming "Indians" to Christianity through "love and friendship."[21]
It proved a naïve or, perhaps, disingenuous hope. This was not a
period of religious toleration in Europe. In any case, greed soon
trumped religion. Raised on Marco Polo's tales of fabulous riches in
the East, Columbus had his eye firmly on South American gold—as
did many of the investors and colonists in his subsequent voyages.
Gold mines required labor, and Columbus realized that the indigenous people, lacking serious weapons, could be subjugated with a
few dozen soldiers. Soon he was allocating his men personal indigenous slaves, and by 1512, slavery in the Spanish New World was so
rife that the *encomienda* system was put in place.

Originally a feudal system of land tenure grants, *encomienda* in
the New World included grants of "Indian" labor, akin to the practice of making conquered people pay "tributes," as required of Moors
and Jews during the Catholic Reconquest of Spain. (Exacting tribute
was an ancient custom practiced by many peoples, including the
Moors in Spain, and the Mayans and Aztecs.[22]) In the Spanish colonies, *encomienda* was supposed to regulate the use of indigenous labor
by allowing "ownership" of workers for only a limited time and by
requiring them to be paid something. But Spain was half a world
away, and the colonists tended to make their own rules; in practice,
this system often meant working indigenous laborers to death.[23]

Back in Spain, *encomienda* might have seemed fine in theory, but
it did not stand up to scrutiny when passionate debates about the
treatment of indigenous people eventually erupted. Thanks to the
growing influence of humanism, and to the consciences of colonial
priests such as Bartolomé de Las Casas, educated Spaniards began
debating such things as whether Aristotle was right in suggesting
some people were slaves by nature.[24] And the debate extended beyond
Spain. Renaissance infatuation with the classical legacy had led some
early sixteenth-century commentators—including the Italian political
philosopher Niccolò Machiavelli—to invoke the Roman Empire's
practice of rewarding soldiers with foreign land, using this precedent

as moral support for the idea of planting colonies in conquered territories in Europe and the New World. On the other hand, Thomas More, in his *Utopia* of 1516, imagined South America as the location of a fictional ideal society dedicated to the public good, not to plunder or conquest. However, More's theory was predicated on the idea that it was wrong for people not "using" their land to forbid others to do so, and so taking unused land by force was just in his eyes.[25]

Armchair justifications were not enough for men such as Las Casas. He had sailed on Columbus's third voyage in 1498 and had settled in 1502 as a colonist on the island Columbus had named La Isla Española, which became known as Hispaniola (and which is now Haiti and the Dominican Republic). Initially, Las Casas had participated in *encomienda*, but by 1514 he had become so appalled at what was happening that he gave up his slaves and became an advocate for indigenous rights, both at the Spanish court and through his pen. In 1542, for instance, he wrote a scorching account of the unimaginably brutal slaughter—by Spanish so-called Christians—of "these gentle sheep," who were "without malice or duplicity."[26]

Ralegh and Harriot's friend Richard Hakluyt, the Oxford minister and geographer, had read Las Casas. While Barlowe and Amadas were away scouting for a location, Ralegh had asked Hakluyt to prepare a document on all that was known about colonies in the New World. In it, Hakluyt paraphrased Las Casas to make the point that cruelty was not acceptable; rather, the "Indians" should be treated with "all humanity, courtesy and freedom."[27] Ralegh agreed, and he also apparently accepted another (anonymous) document written to aid him in his planning. It contained a code of laws stipulating that "no Indian should be forced to labor unwillingly," "no-one should strike or misuse any Indian," and "none shall enter an Indian's house without his leave." The stipulated penalties for breaking these rules were, respectively, "three months' imprisonment," "twenty blows with a cudgel in the presence of the Indian struck," and "six months' imprisonment or slavery."[28]

This code was also intended for the soldiers who would be manning the fort—and the first law was that "no soldier shall violate any woman," on pain of death. A fort was needed, according to the document, not so much in case of native hostility (since Barlowe had reported the locals had few weapons and no metal armor) but to defend the settlement from a Spanish invasion if and when the Spaniards learned of its whereabouts. Twenty years earlier they had shown how they could respond to foreigners, especially Protestants, who encroached on their New World patch: when Pedro Menéndez de Avilés founded the town of St. Augustine and claimed Florida for Spain in 1565, he wiped out a nearby French Huguenot settlement. So Ralegh needed to be prepared both philosophically and practically.

In fact, despite his efforts at secrecy, the Spanish were almost onto him. After Ambassador Mendoza was expelled from England for his role in the so-called Throckmorton Plot against Elizabeth, he had landed himself a post in Paris, from where he fed Philip II information gathered by his spies in England. In February 1585, he knew something big was afoot. Elizabeth, he told Philip, had knighted "her favorite" and given him the *Tiger*, one of her own ships, which he believed displaced 180 tons and had five guns on each side. "Ralegh has also bought two [more ships] to carry stores," Mendoza added.[29]

The English had their own spies, of course, and in April, Hakluyt told Walsingham that the "rumor" that Ralegh and Sir Francis Drake were preparing voyages (Drake was heading to the West Indies) "does so much vex the Spaniard [Mendoza] as nothing can do more."[30] Keeping the location of Virginia secret was vital if England was to reap the economic benefits of a New World settlement— benefits in trade and resources that were considered necessary for the country's survival, because England had become uncomfortably dependent on Spain for many basic commodities, including fruits, flax, wines, oils, salt, copper, tallow, and hides. When Philip II banned all English trade with Spain in May, the situation became critical.[31]

The stakes were high for Spain, too. The Spanish not only wanted to keep their monopoly on the New World; they were also afraid the English would establish a base from which to launch more privateering raids. This was indeed one of Ralegh's intentions. He intended to fund his American ventures with the help of privateering profits, now that Elizabeth had relaxed her opposition to the activity (which she had begun to do in the wake of Spain's involvement in the Throckmorton Plot). Fabulous cargoes of spices, sugar, silk, timber, precious stones, gold, and silver could be seized and sold, the profits being distributed among the various investors. It was certainly a profitable business, albeit a risky and morally dubious one.[32]

IN ADDITION TO FINDING FUNDING for the colony by means fair or foul—and one of the fairer means was investment by spymaster Walsingham, no less—Ralegh had to select his personnel.[33] Harriot was key to the technical, scientific side of the venture. In addition to his diplomatic role, he would also be required to map the area in detail. In this he would be working closely with John White, an artist and a veteran of Frobisher's voyage to Baffin Island. White had made dramatic drawings of the local Inuit, including their first encounter with the Englishmen (a hostile one, as Frobisher's unfortunate Inuit captives also attest).

Ralegh also had to choose his captains and crew. Although the queen had given him the use of one of her ships, she would not allow her favorite to undertake such a dangerous voyage himself. Instead, his cousin Sir Richard Grenville was to lead the expedition.[34] Second-in-command in governing the colony was to be Ralph Lane, a soldier, while the vice admiral was Philip Amadas, with Simon Fernandes as chief pilot. Ralegh must have been happy with the performance of the Amadas-Fernandes team on the reconnaissance voyage the previous year, although Amadas's hot temper and Fernandes's independent streak would prove less than ideal qualities for the far more delicate task of establishing a settlement in America.

Amadas would be halfway to America again when his temperament was revealed in a court case involving an incident that had occurred in January 1585. The case seems to have been related indirectly to the fitting out of Ralegh's ships while docked on the Thames. The plaintiff, one John Stile, was one of four oarsmen on a small boat or "pinnace" owned by Ralegh and captained by Amadas. According to a witness for Stile, when Stile's boat overtook another boat containing a certain Hugh Tucker, Tucker "uttered vile and irreverent speeches" against Amadas. Enraged, Amadas steered toward Tucker; failing to get close enough to ram his verbal assailant's boat, Amadas vented his spleen by throwing a heavy staff at Tucker. Tucker retaliated by flinging something back at Amadas, but instead it struck Stile on the head and knocked him to the floor of the boat. His head bled profusely, and he "was so wounded that he was not able to row." Stile claimed that his wound cost him a job on Ralegh's proposed voyage to Roanoke Island, and with it much-needed wages. He was therefore seeking compensation from Tucker. Witnesses were divided over whether Tucker or Amadas had begun the fray, but all agreed Tucker had managed to severely wound Stile.[35] It says something about the Elizabethan justice system that a humble waterman—possibly with support from Ralegh—was able to mount a case in court. And it says something about Amadas's temper.

No one knows whether the level-headed Arthur Barlowe was on the team. He is not mentioned in any of the eyewitness reports of the expedition, although his name appears in a summary of the voyage published two years later in Raphael Holinshed's *Chronicles*; not all the ships' captains were named in the contemporary reports, so perhaps

Barlowe was one of them. Ralegh had to choose seven captains this time, along with ships' pilots, and as mentioned earlier many of these newly employed men had been taking lessons in navigation from Harriot in the weeks leading up to the voyage.[36] Ralegh also had to find nearly three hundred crewmen and a couple of hundred soldiers. In addition, about a hundred men had signed up to stay on as colonists.[37]

By spring 1585, Ralegh had a clear idea of how he wanted his colony to evolve: adhering to the code of conduct, the colonists and soldiers would carry out such jobs as building houses, finding water, and generally establishing the settlement, while Harriot, White, and their assistants would explore and map the territory and make an inventory of its "merchantable" resources. But Ralegh was aware that contingencies would no doubt arise, so he appointed Lane, Amadas, Fernandes, and the remaining pilots, captains, and "gentlemen"— Harriot and White were in this last category—to a twenty-five-member council advising Grenville, should he need to adjust Ralegh's plan. With so much forethought given to the treatment of the "Indians" and to the colony's self-governance, and with provisions more than sufficient to sustain the settlers until they could harvest their own crops and meat, Ralegh's venture was off to a promising start.

CHAPTER 7

Roanoke Island

GRENVILLE'S FLEET OF SEVEN SHIPS left Plymouth on April 9, 1585. Harriot sailed with Grenville on the flagship, the queen's own *Tiger*. Various contemporary estimates placed its tonnage at between 140 and 200 tons, while the two smallest ships were 50 tons. For an idea of how vulnerable to the elements such vessels could be, today's passenger ships displace tens of thousands of tons.[1]

Harriot measured the dimensions of the ship that was carrying him across the open ocean to America, and he knew the shipwright's rule for calculating tonnage: "a ship whose depth [is] 10 feet, width 20, length 50 by the keel is of burden 100 tons." (He later deduced a more precise formula.) He recorded that the *Tiger* was 50 feet long "by the keel" and 23 feet wide at the beam, with a depth of 13 feet, which, according to the rule, made it 149.5 tons.[2] These dimensions mean that, with 160 passengers and crew, the *Tiger* was more crowded than the *Mayflower* on its notoriously hellish voyage thirty years later.[3] Captains and gentlemen had a better time of it than the ordinary passengers and crew, but sleeping rough and cramped in a heaving boat that often reeked of vomit and sewage cannot have been pleasant for anyone. At least the *Tiger* had a large hold, which meant there was plenty of room for provisions.

Harriot's curiosity and intellectual energy seemed undimmed by the rigors of sailing. He was often up on deck acting as navigational consultant, taking astronomical readings, and experimenting with ways to improve his instruments for navigating.[4] When he was not required for these activities, he used the time to observe the ship itself, recording the technical terms for its various parts, noting details of ropes and rigging, mulling over the ideal ratio between a ship's dimensions and the height of its masts, and generally taking note of the day-to-day work of the crew.[5]

Such an interest in shipboard culture was uncommon among men of science. In fact, there was often considerable animosity between practical seamen and theoreticians: practical manual-writers were sometimes sensitive to criticism or patronizing commentary from experts such as the former Cambridge don Edward Wright and deflected it by pointing out that their practical writings were more accessible for ordinary sailors. Not that Wright was an ivory-tower theorist; he was about to head off to the West Indies as a captain in Drake's fleet.[6] But captaining was different from quietly and dispassionately observing the sailors at work, and the plebeian-born Harriot had a sharp eye for class distinctions. "This is not well paid," he wrote of one of the mariners' jobs—and to emphasize his position on the matter, he added, "Pay on more." He also noted the hierarchy of shares in any profits from navigational expeditions, with eight shares for the captain, seven for the master, six for the master's mate and the lieutenants, five for the midshipmen, master gunner, and purser, four for the cook, and so on down to one and a half for the swabber's assistant and one for the soldiers, kitchen hands, trumpeters, and drummers. The surgeon got only two shares, which perhaps reflects the poor state of medical attention at the time.[7]

Harriot's interest in language was evident in his fascination with nautical terminology—he recorded some terms that would not appear in print until several decades later—and with the sailors' maritime patois. For instance, when huge waves dumping seawater on board threatened to sink or overturn the ship, it was all hands to the pump: "To the water!" As the waves continued to wash over them, the tone could turn to panic: "We drink more than we pump!" When the sails refused to catch the wind sufficiently and the ship tilted dangerously to the side, the sailors would shout, "She lies down like a crab!" When it was necessary to lower the sails, they announced, "To bouse is hale. Come shoo bouse here." Through Harriot's record, we can still hear the cries of the sailors as they undertook the skilled and strenuous work of steering a three-masted ship in all kinds of weather.[8]

We can also see something of Harriot himself: his ability to relate to technical, practical problems as well as to astronomical theory and pure mathematics; his evident respect for the sailors' work and his conviction that they should be paid more; his wide-ranging curiosity; and his love of language. Indeed, in addition to studying the sailors' dialect, Harriot spent time improving his Algonquian by conversing with Manteo and Wanchese.

AFTER MORE THAN A MONTH at sea, the *Tiger* arrived in what is today Puerto Rico. The captains of all seven ships had planned to rendezvous there before proceeding to Roanoke Island, but an unexpected storm meant that only one other ship managed to keep the appointment. It was captained by Thomas Cavendish, who would soon lead the third successful circumnavigation of the globe. One of the other ships had been forced onto a deserted island during the storm; some of her men died of hunger, others never returned from an expedition to find food, and only two were rescued, and interrogated, by Spaniards. The remaining four ships eventually made it to the Carolina Outer Banks, although the captain of one of them, which arrived before Grenville and Cavendish, simply dropped off his passengers on Croatoan Island (Manteo's homeland, not far from Roanoke Island) and went off to raid fishing fleets in Newfoundland.[9]

Cavendish had arrived in Puerto Rico twelve days after Grenville. The anonymous keeper of a journal of the *Tiger*'s voyage recorded that on first sighting the approaching ship, Grenville's company feared it was a French or Spanish warship; weighing anchor, they readied the *Tiger* for action. On recognizing the ship as one of their own, however, they fired their artillery for "joy at its coming," saluting the new arrival "according to the manner of the Seas." But the fact that none of the other ships turned up as planned was a portent of things to come.

A few days later, twenty Spanish horsemen appeared just across the river from the English encampment. After a tense meeting, the Spaniards agreed to trade food and water, but they did not return as promised. Grenville burnt the woods in revenge, and decided to continue making repairs and finding provisions elsewhere.

Sailing through the Indies, the two English ships captured a couple of small Spanish vessels, whose crew they planned to ransom in exchange for provisions. One of these prisoners later reported that Grenville had repeatedly assured them that he was under "the command of a great lord of England" (presumably Ralegh), who did not wish him to kill anyone, and who wished only for unity with the Spanish in the Indies. The prisoner also reported that the expedition included "men skilled in all trades," including about twenty "special people" whose food was served on silver and gold plates. Harriot was surely one of these; his status was such that he would be allocated his own house on Roanoke.[10]

After disembarking on Hispaniola, Grenville set about buying not only water and food for the voyage but also horses and livestock, as well as edible roots that could be planted to sustain the colony. He did not hide his colonial intentions from the Spanish, although he was cannily secretive about his colony's location. Nevertheless, such rapprochement was achieved between the English and the local Spanish settlers and officials that Grenville's men put on a "sumptuous banquet," complete with all the pomp their trumpeters and musicians could muster. But the *Tiger*'s astute journal-keeper was under no illusions that this would have been possible had the Spanish at Hispaniola had more troops on hand. He was quite right: the unnerved governor at Puerto Rico had already sent a detailed report to the Spanish king of all the doings of the English "corsairs," urging Philip to send warships to the Indies immediately.[11]

Leaving Hispaniola, Grenville and Cavendish's ships set out on the final leg of the journey to Roanoke Island. After two and a half more weeks of sailing—during which they were nearly shipwrecked by the shoals off the Florida coast—they reached the Outer Banks. It had taken nearly three months, and one can only imagine the relief they must all have felt. But the journal-keeper merely noted, "The 26th [June] we came to anchor at Wococon."[12]

Three days later, disaster struck. As Grenville and Fernandes attempted to steer the *Tiger* through the Banks, she ran aground. For two hours the crew tried to wrest control of the foundering ship, as the waves beat her against a hidden sandbar and the passengers feared for their lives. Eventually, the exhausted crew prevailed, and the ship was driven onto the shore. It seemed a miracle she had not split in two. But the miracle had come at a cost: the provisions in the *Tiger*'s capacious hold were spoiled by seawater.[13] This led to the loss of much of the colony's immediate food supply; it also meant that the salt-soaked stores of wheat were no good for planting, as Harriot noted.[14]

Perhaps even more damaging than the loss of these provisions was the fact that the accident ignited the disharmony among some of the men that had been smoldering during the voyage. For all Ralegh's careful planning—especially his attempt to forestall disunity by authorizing a council of all key personnel to advise Grenville—factions now began to emerge among some of the captains and governors. The *Tiger*'s journal-keeper was loyal to Grenville, noting the fact of the ship's grounding tersely, and blaming Fernandes, who was, after all, the chief pilot. But Ralph Lane blamed Grenville, whose leadership he had already begun to question. He claimed that Grenville was "tyrannous" and did not listen to advice.[15] Lane had also fallen out with Cavendish, who was allied with Grenville. While Lane excoriated Grenville's "intolerable pride and insatiable ambition," however, his own weaknesses would soon become apparent.

BEFORE HEADING TO ROANOKE ISLAND, Grenville wanted to investigate the unexplored country where they had landed. With four boats and several dozen men—including Harriot and John White, the artist—he set out to explore the area around Pamlico Sound and upstream of the mouth of the Pamlico River. The party came upon several mainland settlements, including Pomeioc, Aquascogoc, and Secotan (or Secoton).

The Englishmen were particularly well entertained by the local people at Secotan, where White made sketches of the village and its productive vegetable gardens and crops of corn. Soon after leaving Aquascogoc, however, it was discovered that a silver cup had been stolen from the English. Grenville wanted to continue exploring, with Harriot making preliminary maps of the coastline and rivers, so Vice Admiral Amadas was deputed to lead a party back to the village to reclaim the cup. He was not a good choice: his hot-headed behavior on the Thames was about to be repeated with a vengeance. Amadas demanded the return of the stolen item and was apparently told to wait until it was retrieved. But the cup failed to materialize, and the townsfolk fled the village. Enraged, Amadas and his men burned down all the houses and spoiled the corn crop.[16] Such a vindictive response was a far cry from the level-headed diplomacy evoked in Barlowe's report. Ralegh's rules of behavior had been broken before the new colony had even begun.

ON JULY 3, 1585, a small group (including Manteo and Wanchese) set out for Roanoke Island—about a hundred miles from where the *Tiger* was beached—in order to tell the chief, Wingina, that the fleet had arrived.[17] By the end of July, the *Tiger* had been repaired, and Manteo had arrived back from Roanoke Island with Granganimeo, who had cemented such friendly ties with the English during the Amadas-Barlowe visit the previous year. It seems that Wingina had sent him with an invitation to settle a little way from his village on Roanoke Island.[18] And it seemed to the colonists that things were looking up at last.

Once Grenville landed on the island, his men set about building what they called the fort, which probably meant a fortified village including barracks for the soldiers, accommodation for the settlers, a store, a jail, a forge and workshop for the metallurgist, and so on.[19] By mid-August, Lane was able to write to Walsingham—who was not only an investor in the colony but one of the most influential men at court—that the climate was so good, and the air so wholesome, that no one had been sick since their arrival in America. He added—ominously, as it would turn out—that all those who had been suffering from "lung diseases" during the voyage had already recovered.[20] The loss of the *Tiger*'s provisions weighed heavily, though, and Grenville decided to return to England for new supplies. It might have been more responsible for him to stay and govern, sending someone else for supplies, but perhaps he opted out because of the faction fighting. At any rate, Lane was left literally holding the fort.

Most of the ships that had carried the colonists and supplies to America eventually returned to England, carrying mail back to the motherland. In one such letter, Lane told Walsingham that he had agreed to stay on as governor of the colony, preferring "fish for my daily food, and water for my daily drink" to the "greatest plenty the [royal] Court could give me," because it meant he would be protecting the fledgling colony in the name of the queen and of "our most noble patron Sir Walter Ralegh." He spoke of the colony as a new English "possession," for which he would risk his life so that it would help free England from "the tyranny of Spain," that "sword" of the "Antichrist" of Rome. His own faith in Christ was so strong, he avowed, that he was sure that God would "command even the ravens to feed us, as he did [for] his servant, the Prophet Habakkuk."[21]

Lane's patriotic zeal and religious fervor set the tone for his relationship with the Algonquian people. Unlike Harriot, he invariably used the word "savage" in his writings, and his meaning of the term was clear in a letter he wrote to his friend Sir Philip Sidney (the famous poet, a friend of Ralegh, and Walsingham's son-in-law). Lane had been at Roanoke Island for less than a month, but he apologized for not writing sooner, saying he had had his hands full with the "infinite business" of governing "amongst savages," especially the "wild men of my own nation [presumably the soldiers], whose unruliness is such as not to give leisure to the governor to be almost at any time from them." His letters to various friends and backers at this time indicate that he had little interest in the Algonquian people. Apart from noting that they were "courteous," he was concerned primarily with the profits to be had from a country presently inhabited "only with savages."[22]

Lane did eventually develop good relations with several Algonquian leaders, especially Manteo, but his approach was necessarily different from Harriot's. Lane was a military man whose job was to keep order in the colony. Harriot, by contrast, had the leisure, the brief from Ralegh, and the temperament to quietly set about developing a relationship with Wingina and his people—a relationship that was based on mutual respect, and one that Harriot came to regard as friendship.[23] Such intimacy was fostered by his ability to communicate in Algonquian and his genuine interest in Algonquian language and culture.

White had already begun sketching the neat timber and wicker houses at Secotan, Pomeioc, and Roanoke, with their orderly gardens and crops, and now Harriot discussed with the people how they went about their agriculture, their fishing, their making of canoes and so on. He would later write up much of what he learned in his 1588 pamphlet, *A Brief and True Report of the New Found Land of Virginia*, and in his captions to White's drawings in the deluxe 1590 edition of his *Report*.[24]

In this report, Harriot noted that the Algonquians never "fatten[ed]" the earth with "muck, dung, or any other thing" but simply tilled the upper layer of soil and removed weeds and old corn stalks and roots. To do this, the men used long-handled wooden instruments almost in the shape of mattocks or hoes, while the women worked sitting down, so they used "short peckers or parers" about a foot long and five inches wide. Having prepared the soil, they left the weeds to dry in the sun for a day or two, and then they scraped them into many small heaps for ease of carrying them away for burning. Next Harriot gave details of how and where the seeds were placed in the newly tilled earth. He and his fellow Englishmen tried it out for themselves, too, tilling and sowing in the American way. He was delighted to find that their plot yielded a weight in corn that was five times the weight of wheat normally yielded by a similar-sized plot in England.

He went on to record all the edible native plants he could, using their Algonquian names—which he transliterated into Latin letters, since readers of his report would not be familiar with his unique phonetic script. It is telling that he did not simply anglicize the indigenous names as "native peas," "bush melons," and so on, as others in his position might have done.[25] Naturally enough, the other colonists did use English terms in their everyday communication, and Harriot listed these, too, as when noting that *wickonzówr*, which the English called peas, tasted similar to English peas but "far better."

He was especially fascinated by the way the Algonquians cooked their abundant produce. Their staple crop, corn—called *pagatowr* in their language and *maize* in the West Indies—yielded "a very white and sweet flour [that] makes a very good bread." The local peoples also roasted or boiled the corn for use in stews, for which purpose they used earthenware pots. Harriot commented, "Their women know how to make [these] vessels...so large and fine that our potters with their wheels can make no better." He listed a number of other indigenous foods, using their Algonquian names and giving details of the way they were cooked and how they tasted. A native melon called *tsinaw*, for example, made "a very good spoon meat [like] jelly, and is much better in taste if it be tempered with oil." There were abundant wild fruits and nuts, too, including a fruit "as red as cherries and very sweet; but whereas the cherry is sharp sweet, these are luscious sweet." Such attention to the details of cookery suggests that Harriot was not unfamiliar with the art himself. He certainly seems to have been a connoisseur when it came to the taste of food.

Next he listed the local species of fowl, fish, and game. He was particularly struck by the Algonquian way of building weirs out of plaited reeds and twigs in order to catch fish—a skill the English did not seem able to master. He also reported that the "bears of this country" made good meat and that in winter, when their stores of grain were running low, the Algonquians ate many of them. When Harriot and his party spent several months living rough while exploring and surveying the mainland over the winter of 1585–86, they went hunting for bears with the local peoples.[26]

All this detail suggests that Harriot frequently ate with the indigenous people—and clearly he worked closely with them as they sowed and reaped their crops and hunted and fished. He was impressed that amid all this abundant food the Algonquians ate moderately, while at the same time "making good cheer together." He concluded that their eating habits enabled them to live long, healthy lives: "I would to god we should follow their example." By contrast, the English suffered many diseases that he believed were due to "our insatiable appetite" and "sumptuous and unseasonable banquets." (To take an example, at an "informal supper" that Harriot and Ralegh attended in London, lamb, chicken, game, and fish were served—all at one meal. No doubt vegetables and various kinds of desserts were served, too.[27])

Harriot also praised the Algonquians' chief medicinal plant, sassafras, which they called *winauk*, a sweet-smelling wood that helped to cure "many diseases." He further recommended the health-giving

properties of a local herb called *uppówoc*. He described how the dried
leaves were pulverized into a powder and burned in clay pipes. The
Algonquians then sucked the smoke "into their stomach and head,
from whence it purged superfluous phlegm and other gross humors,
[and] opened all the pores and passages of the body...whereby
their bodies are notably preserved in health, and know not many
grievous diseases wherewithal we in England are oftentimes afflicted."
The ancient Hippocratic idea of bodily "humors" had been elaborated by the second-century Greco-Roman physician Galen, and it
still dominated European medical thinking in the late sixteenth
century.[28] There were four fundamental physiological humors, each
associated with an organ: phlegm from the lungs, yellow bile from
the gall bladder, black bile from the spleen, and blood from the liver.
(The role of the heart, and the circulation of the blood, would not
be understood until shortly after Harriot's death, with the work of
his countryman William Harvey.[29]) In the Hippocratic system, disease was caused by an imbalance of humors. Galen preferred other
causes, such as lesions and tumors, but he agreed that excess humors
or their waste products needed to be eliminated or "purged" from
the body, using appropriate drugs if necessary. Harriot apparently
believed that *uppówoc* was such a drug, and he noted that this health-
ful herb also grew in the West Indies: "The Spaniards generally call it
tobacco."

The Spanish had first visited the Indies nearly a century earlier,
so tobacco was well known in Europe. In England, people smoked in
inns and taverns, but it would be Ralegh, via Harriot, who made pipe
smoking fashionable among London's courtiers. In his report on
Virginia, Harriot told how the English at Roanoke had learned to
"suck" tobacco in the Algonquian way. Back in England, they adopted
and adapted the pipes brought back from America, and Harriot
claimed that since their return, many more people had begun to use
tobacco for its supposed medicinal virtues—including "men and
women of great calling."[30] Harriot himself would become addicted
to it, believing all the while that his habit was healthy.

A far more dramatic health issue than excessive smoking came
of the interaction between colonizers and native peoples: the deci-
mation of the indigenous population through European diseases.
The germ theory of disease and the phenomenon of acquired immunity would not be understood for another two centuries, which is
why Lane's comment that the wholesome American air had cleared
up the "lung diseases" of his men was so ominous. In a number of
villages, many local people began not only to fall ill but also to die

within a few days of the Englishmen's departure. The English themselves appeared to be relatively unaffected by the mysterious illness,
which fact seemed as miraculous to Harriot as it did to the Algonquians.
It also seemed that Wingina's people were unaffected, because Harriot
wrote, with uncharacteristic credulity, that the illness only afflicted villages that were less hospitable to the English.[31]

If Wingina's village did remain unscathed—Lane's men having
recovered from their "lung diseases" and other illnesses in the month
between their arrival at Wococon and their settling at Roanoke
Island—then perhaps the English parties that went off exploring the
country and "discovering" new villages became ill again when they
slept out in the open in the colder months, away from the shelter of
their houses on Roanoke Island. Under such conditions, they would
have been more prone to infectious flare-ups of respiratory illnesses
such as pulmonary tuberculosis, bronchitis, colds, and flu, any of
which could have been lethal to those with no immunity.[32]

Harriot was told about the tragedy by his Algonquian "friends
and especially the *Wiróans* [chief or leader] Wingina," who had visited the "four or five" stricken villages. The Roanoke people concluded that the Englishmen's God had avenged them against the
"wicked practices" of the neighboring villagers, and they asked
Harriot if he would entreat his God to punish likewise any of their
own enemies who had also abused the Englishmen. After all, they
said, it would be a win-win proposition. It would also be a validation
of "the friendship we profess them."

Wingina and many of his people already had a high opinion of
the English God, who, they felt, must be the source of the Englishmen's technological achievements. Harriot clearly took delight in
demonstrating the scientific tools he had with him, recording the
amazement with which the people observed such wonders as mag-
nets drawing pieces of iron, astrolabes and cross-staffs, magnifying
glasses that could focus the sun and start a fire, and a "perspective
glass whereby was showed many strange sights." He also noted the
Algonquians' awe at seeing for the first time "guns, books, writing
and reading, spring clocks that seem to go of themselves, and many
other things we had." The "perspective glass" in Harriot's list has
perplexed scholars, because the telescope was thought to be
unknown before the early seventeenth century. Harriot may have
been referring to an early, low-magnification telescope, such as
Leonard Digges had made, which may have been used for sighting
ships or land when at sea.[33] But some scholars believe Harriot's "perspective glass" was just a toy, consisting of an arrangement of mirrors

(as in a kaleidoscope) that distorted people's faces and bodies, like the curved mirrors at a fun house.[34]

Inspired by such technical wonders, and no doubt also because of their own love of ceremony, the local people often joined the English at their prayers and their singing of psalms. Wingina himself had sought the aid of Christian prayers when he was ill and his local priests could not cure him. So when he asked the English to show their "professed friendship" by asking their God to bring the mysterious illness to slay their enemies, Harriot was in a delicate position. Still, he was quick to say that "our God would not subject himself to any such prayers and requests of men," a rather Calvinist view. He added that God would want the people and their enemies to find peace and "to live together with us, and be made partakers of his truth." Harriot normally kept his religious views to himself, but in this particular passage of his *Brief and True Report* for Ralegh's future colonists, he seemed at pains to point out his sincere Protestantism.

When the Roanokes' enemies continued to die, Wingina's people still gave thanks to the English and their God. But even if the disease had not yet struck down anyone on Roanoke Island, unease began to fester, and Harriot described the various rumors that soon flew about the villages. Some believed the English were gods, or reincarnated avengers from the past. Others ascribed the illness to "invisible" English bullets manifesting as strings of blood sucked out of patients by local healers. Harriot was not impressed with such healers; he rarely made judgmental comments on Algonquian beliefs, but he noted that the healers were "not ashamed" to cover up their inability to cure the disease by making the more "simple" people believe in their invisible-bullets theory. Finally, yet others believed the mystery illness "was the special work of God for our sakes"—a belief that "we ourselves have cause in some sort to think no less." Harriot may have been a rationalist and a pioneer of early modern science, but faced with this seemingly miraculous natural event, his usual skepticism yielded to a religious explanation. This was in contrast to some of the "Astrologers," who suggested the disaster was due to the eclipse of the sun that Harriot had recorded on the voyage to America and "which unto them appeared very terrible." (It is not entirely clear whether Harriot was referring to English or Algonquian astrologers here, but superstition abounded in both cultures.) The astrologically minded also wondered about the malign influence of a recently sighted comet. Harriot himself discounted the eclipse and comet as causes of the strange plague, although he added cryptically that spelling out his reasons for this was not necessary in his brief report.

Harriot did not mention astronomy in *Brief and True Report*, although he intended to compose a much more extensive account of the Algonquians' culture later on. (Unfortunately, he never did—or his manuscript has not survived.) He did say that they had "no letters nor other such means as we to keep records of the particularities of times past, but only tradition from father to son." Of course, any traditional stories encoding detailed astronomical knowledge may have been secret, or too subtle even for Harriot's open mind and impressive grasp of the language. He didn't mention witnessing any stargazing activity, though; perhaps he thought it irrelevant to his report, or perhaps there was relatively little of it, because he noted that the people believed in a single creator-god, and they saw the moon, sun, and stars as only minor gods.

The Algonquians also focused their sense of spirituality on the earth, and they saw life force, or sacred power, in its natural features as well as its plants and animals. José de Acosta indicated a similar kind of worship when describing "a great mound of sand" in Caxamalca (in Peru), which was, he reported, "the chief idol of the Ancients." When he asked about its significance he was told that they worshipped it "for the wonder [of its] being a very high mount of sand in the midst of very thick mountains of stone."[35] Expressing wonder at the natural world's diversity seems human enough, but Acosta described this and other indigenous rituals as "foolish illusion" by a people in thrall to the devil in their "barbarous blindness." Arthur Barlowe, too, had equated the idols he observed with the devil, as noted. Harriot, by contrast, never mentions the devil; rather, he simply described what he saw and heard of their religion without passing any judgment at all—apart from emphasizing that the Algonquians did not worship the one "true," Christian God.[36]

He outlined as clearly as he could the Algonquians' pantheon of gods and the order of creation. He noted their belief in "heaven," described to them by a man from a neighboring village who reputedly came back from the dead while Harriot was at Roanoke. In his near-death experience, the man had walked with his dead father in a beautiful garden. The Algonquians also believed in some sort of hell, which they called *popogusso*, a place of perpetual burning, located beyond the sunset; this information, too, came from a man who had supposedly come back from the dead. Harriot describes all this in a remarkably dispassionate way, noting their idols, their temples, and their manner of keeping and displaying the corpses of dead kings.

(He did indicate sympathy for the priests whose job was to guard the bodies day and night, praying all the while and having "to reverence their princes even after their death.")

Harriot enjoyed what he called his "special familiarity with some of their priests," which is how he derived his knowledge of Algonquian religion. He especially enjoyed participating in their rituals. Whenever the community escaped some great danger such as a war or a storm, he reported, they all sat around a fire, men and women together, shaking gourd rattles, singing, and "making merry." Tobacco often featured in such rituals. The people would cast the powdered plant into "hallowed fires," giving thanks by making "strange gestures, stamping, sometimes dancing, clapping of hands, holding up of hands, and staring up into the heavens, uttering therewithal and chattering strange words and noises." Other rituals included casting tobacco into their fishing weirs in the hope of a good harvest.

Nevertheless, insofar as the official purpose of Harriot's writing *Brief and True Report* was the promotion of Ralegh's colonial ventures, his stated religious agenda did not stray too far from the Reverend Hakluyt's mission ("The people of America cry out to us...to come and help them, and bring unto them the glad tidings of the Gospel").[37] Harriot dutifully reported that he had taken every opportunity to preach—in Algonquian, "according as I was able"—from the Bible. The people immediately began to touch and kiss such an evidently sacred book, though Harriot tried to explain that it was the words that mattered, not the physical volume. The Algonquians probably saw his Bible as an object of power, the apparent source of the Englishmen's own wondrous technological powers, but Harriot naïvely interpreted their interest as a "hungry desire" for knowledge of Christianity and the "true and only God."[38]

Modern scholars have debated the impact of Harriot's mindset and his moral responsibility in the colonial program.[39] There is no doubt that in his report for Ralegh the primary purpose of Harriot's presentation of Algonquian culture and religion was to show that the people would welcome European knowledge of both technology and Christianity, and therefore not pose a threat to English settlers. His detachment about this seems chilling in hindsight. And although his emphasis on sharing knowledge was enlightened compared with the majority who emphasized religious conversion first and foremost, it was also paternalistic.[40] Yet there is no doubt that Harriot felt empathy and respect for many of the indigenous people, and his genuine interest in them shines through. His brief comments on

Algonquian culture, together with his little details of daily life—shared meals, festivals, making of canoes and pots, planting and hunting—bring to life a thriving and relatively harmonious community. His report also highlights the Roanokes' stable government, with a legal system in which the punishment suited the crime, and whose leaders were open-minded enough to engage cooperatively with such a vastly different people as the English. Harriot's account also reveals—especially in conjunction with his captions to White's drawings—that he himself took pleasure in the healthy, cooperative aspects of the Algonquian way of life.

HARRIOT REGARDED WINGINA IN PARTICULAR as a friend, and the chief sometimes accompanied him and his party on exploratory cartographic visits to neighboring villages. By March 1586, however, Wingina was losing faith in the English.[41] Their supply ship still had not arrived, and Lane had been harassing him daily to sell more food to the hungry Englishmen. (Lane generally exchanged copper for it.) Discord arose because Wingina's own stores were low, too; it was the end of winter, and the next harvest was months away. The Englishmen could not even manage to make adequate fishing weirs for themselves and badgered Wingina to allow his men to make some for them. Before this time of hunger, Wingina had prayed to the English God whenever he was sick, or when there was a drought or some other calamity, but now Lane's distress over food shortages suggested that this God was not so bountiful after all. (What neither he nor Lane knew was that one of the reasons for this shortage was that Amyas Preston and his partners, who had planned to bring the first new supply ship to Roanoke six months earlier, had been ordered by the government to travel to Newfoundland instead. There they were to inform English fishing fleets and traders of Philip II's abrupt ban on all English trade and to warn them not to try to sell their wares in Spain.)

The last straw was Granganimeo's death, after which Wingina suddenly changed his name to Pemisapan. No one seemed to know why. Perhaps Wingina felt the English God had failed to save his brother. Or perhaps Granganimeo had been so interested in trading for English goods that he had wanted to continue cooperating with the English, whereas Wingina was beginning to resent them. No doubt there were unprecedented divisions in the community, between those who supported the English and those who did not, and it is significant that Wanchese was now one of those who did not.

On the death of the revered and pro-English elder Ensenore, there was no holding back for Wingina-Pemisapan. He found it impossible to refuse Lane's demands, so he decided to avoid the issue altogether by moving his people away from Roanoke Island. At least this is what Lane believed, although Pemisapan's decision may have had more to do with moving away for the winter hunting season.[42] Either way, it would not be surprising if he intended to get rid of the pesky English, with their incessant demands and their strange powers of technology and disease, by starving them out as Lane assumed. Lane also accused the chief of fomenting fear of the English among neighboring villagers, telling them that the foreigners planned to attack them and advising them to hide their stores of grain and to refuse to trade.

Lane soon became convinced that Pemisapan was organizing a confederacy of thousands of fighters who would converge on Roanoke Island at a funeral feast for Ensenore. As he noted in his later report, he also believed that the Roanokes had tried to poison some of his men and that Pemisapan planned to burn down his house and "knock my brains out," as well as burn Harriot's house and those of all "the better sort."[43] (It is interesting that Lane singled out Harriot by name here: perhaps he was attempting to underscore Pemisapan's treachery.) Lane decided to weaken the supposed Algonquian alliance by making a surprise preemptive strike on a nearby village. But the villagers' scouts were watching, and after a skirmish in which shots rang out and several scouts were killed, the townsfolk fled into the hills.

Next morning, Lane sent word to Pemisapan—who had decamped to the mainland near Croatoan Island—that he wanted a parley. When he arrived, Lane found himself surrounded by seven or eight of the chief's key men plus their followers, and he panicked. Shouting "Christ Our Savior!"—the revealing, agreed-upon watchword in case of danger—Lane and his men fired on the group. Pemisapan was wounded and fled into the woods, but several of his supporters were killed. (Lane later reported that during the attack he had "looked watchfully for the saving of Manteo's friends": Manteo had remained loyal to the English, and his family still lived on Croatoan Island.) Two of Lane's men ran off in search of the wounded chief, and some time later, they returned with his severed head. And so it was that on June 1, 1586, the English killed their former host Wingina-Pemisapan. It was a disastrous end to what had begun as a remarkable coexistence between the two peoples.

News of Pemisapan's death quelled any hostile plans from his surviving supporters. This had been Lane's objective: he blamed Pemisapan for the unrest and had no intention of wiping out the rest of the Roanokes. But the only surviving details of Pemisapan's apparent treachery and Lane's response to it are in Lane's report to Ralegh. They are filtered through his eyes and presented as a justification for his actions. His tone sounds reasonable and honest, but he was a soldier, who kept what he called "a severely executed discipline" over his men. Given the savage reprisal led by Amadas in response to the theft of the silver cup back in July 1585, one wonders if Lane's discipline extended to the Algonquians. It is entirely possible that other such violent responses to petty pilfering had helped to turn Wingina against the English. It was Harriot who provided the hint that this was most likely the case.

Toward the end of his *Brief and True Report*, he concluded that "discreet dealing and government" would be necessary to win the Roanokes over to Christianity. This suggests a lack of such discretion under Lane's governorship. Harriot, ever the diplomat, seems to have been cautiously balancing his loyalties: to Ralegh, to Lane, and to the Algonquian people. Still, in one somewhat tortuous passage he does offer a sense of what had happened: "Although some of our company towards the end of the year showed themselves too fierce, in slaying some of the people, in some towns, upon causes that on our part might easily enough have been borne withal; yet notwithstanding because it was on their part justly deserved, the alteration of their opinions generally and for the most part concerning us is the less to be doubted. And whatsoever else they may be, by carefulness of ourselves need nothing at all to be feared." This at least makes it clear that some of the English did carry out deadly reprisals for petty transgressions by the Algonquians, but to avoid accusations of partisanship, Harriot acknowledges the Algonquians were in the wrong in their actions against the English. Unlike Lane, however, he is not surprised at their change of heart toward the English, given what they had suffered at the hands of the colonists. The only hope for future rapprochement now lay with "carefulness of ourselves"—that is, better behavior from the English.

It seems that Lane, whose hands were full with governing his unruly men, was simply not up to the task of working sensitively— "with carefulness"—with the people on whose land he was living and on whom he was dependent for the colony's survival. Many of the colonists, too, were disgruntled and troublesome, particularly once there proved to be no wondrous stores of gold and silver such as those found by the Spaniards in South America; the harsh, hungry conditions on Roanoke Island seemed all for nothing. Unlike Harriot, such men had not enjoyed the simple pleasure of living with the Algonquians, before it all went so very wrong.

ONE WEEK AFTER WINGINA'S DEATH, Lane's lookout, Captain
Edward Stafford, returned to the Roanoke settlement with a letter
from Francis Drake, whose fleet was on its way back to England after
looting its way through the West Indies. In the wake of Philip II's ban
on English trade with Spain, Drake had intended to establish an
English base in Florida, but Spanish resistance meant that all he
could do, in his characteristic fashion, was to weaken the Spaniards
by destroying and pillaging whatever of theirs he could. (Drake
hated the Spanish, but he got on well with the South Americans
themselves.) He had then detoured to Roanoke to offer reinforce-
ments; it seems he had heard about the fact that the Spanish were
mounting an armed expedition to locate and destroy the English
settlement.[44]

In his letter Drake offered to leave some provisions for the
colony, as well as a ship and some weapons and tools. The laborious
transfer of goods from Drake's fleet moored outside the Banks was in
full swing when on June 13 an unexpected storm blew up. It raged
for four days, and some of Drake's ships were cut adrift—including
the *Francis*, with the newly loaded provisions for Lane, along with
some of Lane's men, still on board. Drake conferred with his captains and generously agreed to provision another ship for the starving colony. But Lane conferred with those of his own "Captains and
Gentlemen of my company as then were at hand," and everyone, he
said, agreed with him on what to do next: abandon the colony and
go back to England with Drake. It seemed to Lane that "the very
hand of God" had first offered and then taken away the *Francis*, so
that now they should all leave, too.

What Lane did not say was that he had irreparably burned his
bridges with the Roanokes. Hakluyt would later say it for him, when
he published his own summary of the various reports from the
Roanoke voyages. He was in no doubt about the meaning of Harriot's
tortuous passage. Describing the colonists' hasty departure with
Drake, Hakluyt said it was "as if they had been chased from thence by
a mighty army, and no doubt they were, for the hand of God came
upon them for the cruelty and outrages committed by some of them
against the native inhabitants of that Country."[45] By June 19, 1586,
the English had left America.

A couple of calamities happened in the rush to board Drake's
ships. First, three of Lane's men were still away from camp, but with
another storm brewing Drake decided to leave without them.[46] No
one knows what happened to them.

Second, Drake's sailors had been at sea for ten months. They were sick and exhausted and were "aggrieved" at having to ferry Lane's men and their baggage through the choppy waters around Roanoke Island. To lighten their load, they threw overboard much of Harriot and White's work, along with books, maps, and a beautiful necklace of Algonquian pearls made for Queen Elizabeth by one of the company's skilled oyster catchers, according to Harriot's report.[47] Harriot must have been devastated, but all he said of the loss of his treasured, painstaking notes was that the pearls were lost "with many things else," due to "extremity of a storm" as they were leaving the country.

To add to the whole unhappy mess, just a day or two after Drake and Lane departed, Ralegh's supply ship arrived. Finding the settlement abandoned, the captain—or his men—chose not to stay long, despite the fact that Grenville was on his way with more colonists and supplies. In sending several hundred new settlers and soldiers, it is not entirely clear whether Ralegh supposed there would be plenty of "unused" land at the colonists' disposal or whether he assumed they would take what they needed by force if necessary.[48] The former seems more likely, given his emphasis on treating the "Indians" humanely. Also, Lane had been interested in reports about Chesapeake Bay from Harriot, White, and their party, who had established friendly relations with the local tribes in the area, so he may already have sent word that a new settlement might be made there. (In 1572, a Spanish-educated Powhatan who knew of Spanish colonial cruelty had led the massacre of a group of Spanish missionaries at Chesapeake. Any English settlers there would have to behave better than they had done on Roanoke.[49])

Grenville arrived two weeks after Lane's departure. He might have arrived before Drake if it weren't for his penchant for privateering. On his return from the colony the previous year, his piracy had recouped enough treasure to repay Ralegh and his principal investors. It also provided two skilled Spanish pilots, whom Grenville forced to help pilot his voyage back to Roanoke in 1586. It was common international practice to use captured seamen in this way, although according to a 1588 report to the Spanish government by two of his prisoners, Grenville apparently treated many of his captives "as slaves are treated in Algiers," forcing them to labor all day and chaining them up at night.[50] This time, though, by plundering on the outward journey instead of getting supplies to the starving settlement, Grenville had sealed the fate of the First Colony. He

CHAPTER 8

After Roanoke

HARRIOT ARRIVED BACK IN ENGLAND at the end of July 1586. Perhaps he had had the good fortune to sail home on the ship captained by his fellow mathematician, Edward Wright. Wright's work on navigation ran remarkably parallel to Harriot's, and soon they would both discover the secret of the fabled Mercator map. It is tempting to imagine them discussing the topic on the voyage home from America, although there is no evidence for it. Indirect evidence from the mid-1590s suggests that they knew each other then, so it is likely that they discussed navigation at some stage.[1] Despite the disastrous disintegration of the Roanoke Island settlements—English and Algonquian both—Manteo had returned to England, and Ralegh had not given up on America, so Harriot was still engaged in navigational research for him.

The Flemish cartographer Gerardus Mercator's 1569 map of the world was unique because of the way it represented "rhumb lines"—lines traveled over the curved surface of the earth when following a fixed compass bearing. When a sailor set such a fixed bearing, it seemed as if the ship kept sailing straight ahead, but in fact, should a ship continue to sail in the direction 45 degrees northeast, for example, it would follow a curved path that twisted its way toward the North Pole. For an ordinary seaman, this seemed counterintuitive, to say the least: How was it possible to sail northeast and end up almost due north of one's starting point, near the North Pole? Pedro Nunes—the pioneering Portuguese navigational theorist whom John

Dee had described in 1558 as the greatest living exponent of the "mathematical arts"—had shown that rhumb lines never actually meet the Pole but keep spiraling around it forever, in ever-decreasing "circles." To help keep the curvature of these paths in mind when plotting sea routes, navigators sometimes carried model globes, but these were cumbersome and expensive in an age before mass production. It would be much more helpful if curved rhumb lines could be plotted on flat maps as the straight lines they seemed to be—and this is just what Mercator achieved.[2]

Exploration was a commercial game, however, and explorers needed maps—so the savvy Mercator had refused to reveal his method for making his famous chart. Of course, it's possible that he simply thought it was obvious, although no one else seemed to think so, and soon after Harriot returned home he began to apply himself seriously to the "Mercator problem."[3] He had been thinking about it for some years. Nunes had given the first mathematical analysis of the strange behavior of rhumb lines, but he did not carry out the tedious calculations needed to complete the theory. He promised further work on the topic, but it never appeared. This may have been because he became worn down by critics—old-school scholastics who thought that mathematizing the art of sailing was an intellectual conceit. Sailing is a function of wind and sea, said one such Aristotelian (Diogo da Sà), and the sea knows nothing about constant angles and mathematical curves. In the face of such hostility, Nunes had vowed to "abandon mathematics, in the study of which, I have irretrievably lost my health."[4] Instead, it was Harriot who ultimately provided the first full mathematical treatment of this fascinating spiraling curve.

Like Dee before him, Harriot admired Pedro Nunes.[5] There may have been political tension between Catholic Spain and Spanish-ruled Portugal on the one hand and Protestant England on the other, but in science it was the exchange of knowledge that mattered. (At least to those whose main interest was science for its own sake: for the patrons of such people, commercial secrecy and nationalism were often paramount.) Harriot's goal was to take up where Nunes left off in his mathematical analysis of rhumb lines, which Harriot referred to as "helical lines." The English word "helix" was at that time of quite recent origin; this particular helix would be given its unique, modern name of "loxodrome" by the Dutch physicist Willebrord Snell, in his 1608 discussion of Nunes's rhumb lines.[6] Meantime, since no one knew just how Mercator had turned curved lines on the globe into straight lines on a map, finding a practical answer to this question was Harriot's goal in the late 1580s. His mathematical analysis of the loxodrome would come later.

If you look at a Mercator map—the type that used to adorn the walls of many a schoolroom—what jumps out is the fact that the lines of latitude and longitude are straight lines, not curved as they are on the globe. This means that the Arctic and Antarctic regions are spread across the top and bottom of the map, because the lines of longitude do not meet at the poles on the map. (This makes countries at higher latitudes look much bigger than they are, which is why Mercator's map has faded from classrooms; it is still an important tool for navigation.[7]) If you look more closely, you will notice that, unlike the vertical lines of longitude, the horizontal lines of latitude are not evenly spaced. This is the key to making straight rather than curved rhumb lines, so to solve the "Mercator mystery," the challenge was to work out how to determine this gradation in spacing, so that others could make maps using the same principle.

The first clue lies in the fact that since meridians of longitude actually do meet at the poles on the globe, the horizontal distance between any two meridians on the globe decreases as you move away from the equator. But if this distance is to remain constant on the map—so that the longitude lines can be drawn as evenly spaced straight vertical lines—then the actual horizontal distance between two given meridians, as measured on the globe, must be stretched more and more on the map as you move from the equator toward the poles. And as you move away from the equator, latitude increases, so the amount of horizontal stretching needed at any point on the map depends on the latitude at that point. In fact, this stretch factor is just the simple trigonometric quantity "sec φ," where "sec" stands for "secant"—which is defined, along with the sine, cosine, and tangent, in figure 15—and φ is the latitude of the line along which the distance between two meridians is stretched. A derivation of this stretch factor is given in figure 13.[8]

The second clue lies in the fact that by definition, any given rhumb line makes the same angle with each north-south meridian on the globe, because the compass bearing along a rhumb line is at a fixed angle from true north. So when the rhumb line is drawn on the map, it must make the same angle with each vertical (north-pointing) meridian (as in figures 14 and 15[9]). In other words—and in accord with intuition—the fixed bearing on the map must make a straight line. This means that on the map, the vertical distances between latitude lines must be increasingly stretched by the same factor as the horizontal distances between meridian lines, to keep the same angle-preserving symmetry as on the globe.

Local Mercator maps for specific voyages were not yet available, of course—and would not become available until the early seventeenth

century, when the stretch factor became common knowledge—so
Harriot's next task was to provide Ralegh's mariners with a set of
instructions and a table that substituted for such a map. First, he had
to work out how far a ship should sail in a given direction if it wanted
to change its latitude—or its longitude—by any particular number of
degrees.

To work out distances along a curved line such as a rhumb line,
it is easiest to imagine the line as composed of small straight segments; if these segments are tiny enough, then you can imagine
them meshing together to give the appearance of a single curved
line. This is one of the techniques used in computer graphics and
calculus today, but it was also the technique Harriot used to find the
formula for the change in latitude between two points on a rhumb
line of fixed compass bearing α degrees: change in latitude equals
distance traveled multiplied by the cosine of α. So, on rearranging
the terms in this formula (which is derived in figures 14 and 15), he
found that the required distance a ship must sail equals the required
change in latitude multiplied by the secant of α.

The change in longitude proved more complicated: it was equal
to the tangent of α times the "meridional part." For any particular
latitude, the meridional part is the stretched distance on a Mercator
map along a meridian from the equator to the given parallel. The
stretch factor for each latitude φ is sec φ, so finding the meridional
part at any latitude on a rhumb line broken up into tiny segments
required adding up the secants of the latitude of *each* segment (as
explained in figure 15b). For Harriot's formula for the calculation of
longitudes to be practical at sea, however, he would have to provide
Ralegh's navigators with a shortcut. They could not be expected to
add up these incremental distances every time, so he laboriously constructed a table of meridional parts. Today, the addition of infinitesimal quantities is done not step by step, the way Harriot had to do it
in the 1580s, but by the algorithms of "integral" calculus, discovered
by Newton and Gottfried Leibniz nearly a century later. Harriot's
ultimate analysis of the loxodrome would eventually yield an equivalent algorithm, but in his first approach to the problem, he ended up
adding by hand thousands of secants, as the latitudes of the segments
increased incrementally by just one angular minute at a time. Wright
did the same, when he produced a similar set of tables at around the
same time.[10]

At least Harriot and Wright had access to an updated table of
trigonometric values—of sines, cosines, tangents, and secants—to
help them construct their tables of meridional parts. The Frenchman

François Viète and the German Jesuit Christopher Clavius had put in the astonishing amount of work required to create tables of trigonometric values for angles increasing minute by minute. Their tables were published in 1579 and 1586, respectively.

Today, calculators have replaced trigonometric tables, but similar kinds of computational methods that helped table makers in the past are now used to program calculators to do the same job. The oldest extant trigonometric tables date back nearly two thousand years, to the *Almagest*. Ptolemy found trigonometric values for various angles by using theorems, proved by Euclid and others, about the geometry of a circle and its inscribed sectors and triangles. Mathematics students do this today when they learn to find trigonometric values for angles of 45°, and 30° and 60°, by applying Pythagoras's theorem to right-angled triangles containing these angles. In addition, various trigonometric formulae make it easy to generate additional values of sines, cosines, secants and so on. For example, if you derive the sine and cosine of 1°, then the "double-angle formulae" readily give the sine and cosine of 2°, and then of 4°, 8°, and so on by repeated application. This and other laborsaving fundamental trigonometric formulae were known to Ptolemy and proved more systematically by later mathematicians such as the tenth-century Arab Abu'l-Wefa. By the sixteenth century, new trigonometric formulae were being discovered. For instance, to help cut down the work in making his own tables Viète generalized the "double-angle" formula to a "multiple-angle" one.[11]

Ptolemy and his followers knew these trigonometric formulae in an equivalent geometric form rather than the algebraic functions used today. (And his "sine" magnitude was the length of the "chord" shown in figure 15a; it is double the modern sine, which derives from fifth-century India, via twelfth-century Latin translations of Arabic versions of Sanskrit works.) The mediaeval Indians and Arabs had also done trigonometry geometrically; these mathematicians include al-Khwarizmi, whose name gave us the word "algorithm," and who was a scholar at Caliph al-Mamun's ninth-century "House of Wisdom" in Baghdad. A work by al-Khwarizmi also gave us the word "algebra," a subject in which medieval Middle Eastern mathematicians made considerable progress. Some of them glimpsed the connection between geometry and algebra, although this was not made explicit until the seventeenth century, while trigonometry achieved its modern algebraic function form only in the eighteenth century.[12]

Algebraic (or analytic) geometry in general is usually held to have begun with René Descartes's *La géometrie* (*Geometry*), an appendix to his *Discourse on Method* of 1637. What is less well known is that Viète and especially Harriot had already begun the process. It is true that Descartes's work inspired later disciples to develop modern analytic geometry, in which the algebraic equations of curves are found from the coordinates (x, y, z) of points on the curves' geometrical graphs, and vice versa, whereas Viète and Harriot did not articulate a theory of curves and coordinates. Nevertheless, by the early 1600s—nearly four decades before Descartes's *Geometry*—Harriot would not only develop modern algebraic notation, as we'll see, but would draw diagrams such as those in plate 2, which indicate a connection between algebraic equations and coordinates on their locus curves. And in his work on spherical triangles, rhumb lines, and optics, he would be writing trigonometric relationships in a purely symbolic proto-function form a century before anyone else.[13]

In the meantime, the concepts and the notation of trigonometry were still evolving. When Harriot wrote up his early Mercator work for Ralegh's navigators—in a manuscript he called *Doctrine of Nautical Triangles Compendious*—he presented his formulae for changing latitude and longitude not in algebraic form but in an equivalent proportional form.[14] His use of the word "compendious" belies the unfinished state of the manuscript, but it perhaps indicates the huge number of calculations underlying it, and the new computational methods he was beginning to try out. In addition to the trigonometric theorems discussed above, "interpolation" is another way in which values of trigonometric or other functions can be generated. Ptolemy used a simple "linear" form of it, which had been known as far back as the Mesopotamians; in generating new trigonometric values using this method, the idea is to take two calculated or tabulated values, say $\sin 45°$ and $\sin 60°$, and to use the equation for the straight line between these two points in order to approximate the sine of any chosen angle between 45° and 60°. Harriot would later pioneer exact algebraic interpolation, but even in his *Doctrine of Nautical Triangles Compendious* he gives brief instructions on the use of a more accurate *non*linear algebraic formula, which was designed to give values halfway between two successive calculated or tabulated values; it could then be reapplied to give quarter intervals, and so on, and he presumably used it to help cut down the work of compiling his table of meridional parts.[15] (His Swiss contemporary Jost Bürgi also found an ingenious algorithm, although it was specifically for interpolating sine values. His unpublished manuscript was found only in 2013. In seventh-century India, Brahmagupta gave the first known algorithm for a nonlinear interpolation formula; it was an extraordinary but isolated and undeveloped result, expressed verbally and briefly.[16])

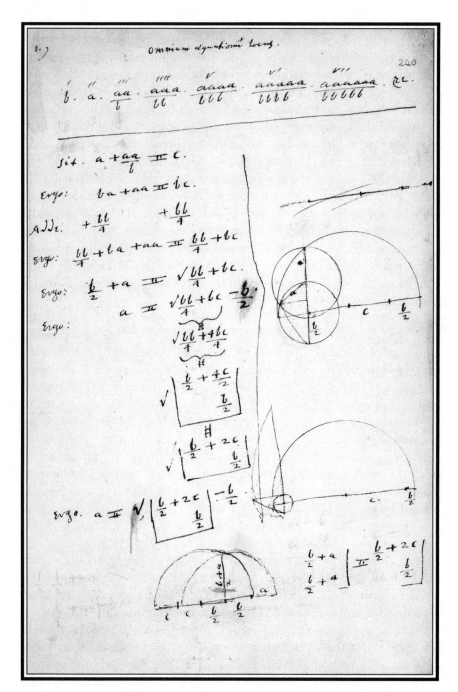

Plate 2: *Analytic geometry*: Harriot is using algebraic equations and corresponding geometric diagrams to explore what he calls "the locus of all equations." Note that his algebraic symbol for "equals" has two small vertical lines crossing the = sign, and his products are written one on top of the other and encased in a box.

Harriot may not have foreseen the development of calculators and computers, but his use of trigonometric and interpolation formulae show that he understood much of the mathematics that underlies their process of computation. In the late 1580s and early 1590s, however, he and Wright had no option but to solve the Mercator problem by making their extensive, laborious tables of meridional parts, with a little help from old and new computational methods. Two less accurate, less detailed, unpublished sets of tables had been produced earlier, and the better of them was by none other than the mystical John Dee.[17] Dee's work dates from around 1558—a decade before Mercator's map—when he was corresponding with Nunes and Mercator. In the 1580s, he was on the Continent, and Harriot apparently knew nothing of his early attempt. It is likely, though, that he and Dee discussed the problem sometime after Dee returned to England in 1589: Dee and Harriot would become friends when Dee became involved with Ralegh's Durham House circle in the early 1590s.[18]

Unlike Mercator, Wright, and Dee, Harriot never achieved the fame he deserved for his discoveries—the practical navigational tables and methods in *Arcticon*, his algebraic geometry and interpolation, his table of meridional parts, and his discoveries still to come. He was a perfectionist, which helps explain both his achievements and his failure to publish and secure his reputation. Even as he wrote up his work on the Mercator problem for Ralegh's navigators, he had his sights on a more sophisticated, more accurate method for calculating meridional parts than all those tiresome additions of secants. So, although the title and content of his *Doctrine of Nautical Triangles* manuscript suggest a potentially publishable treatise, he moved on before it was even finished.

IN HIS QUEST FOR A perfect solution to the Mercator problem via a mathematical analysis of the loxodrome, as well as because of the need to protect his patron's commercial advantage, Harriot inadvertently left the way open to Wright, who has gone down in history as the first to publish a complete, relatively accurate table of meridional parts. Not that Wright had intended to go public quite so soon. Like Harriot, he was working for a patron: the Earl of Cumberland, a naval commander and privateer who would have preferred to keep Wright's work for the advantage of his own maritime investments. But Wright had shown his tables to a friend, who published an extract in 1594—with an acknowledgment of Wright but without Wright's prior permission. In 1597, another colleague published the entire method, saying only that he obtained it "from a friend of mine."

Most upsetting of all to Wright were the actions of Jodocus Hondius (who had engraved the Molyneux globes in 1592). Hondius had borrowed Wright's manuscript, promising not to publish any part of it without Wright's permission; instead, Hondius studied Wright's tables and used them to design and print his own map of the world, using Mercator's "stretched" grid. Wright was incensed that he had lost his priority. Elizabethan critics could be unforgiving, and Wright now feared that if he did publish his work, critics would dismiss it as derivative of Mercator and Hondius. On the contrary, he lamented, "I learned [Mercator's secret] neither of Mercator nor any man else."[19] Wright finally published his work—in his 1599 *Certain Errors in Navigation*—only when he discovered that he was about to be plagiarized completely. Another book was going to press, purporting to be by Cumberland's former chief navigator Abraham Kendall (who in 1596 had died at sea, along with his new captain, Sir Francis Drake); when Cumberland asked Wright to check the manuscript before printing, Wright "found it everywhere to agree with mine, and to be a copy of the same book word for word, which I made and presented to his Lordship [Cumberland] seven years before." In the end, though, Wright's book made him famous throughout Europe.

Yet from the distance of centuries it has become clear that Harriot was the better mathematician. As would happen so often, however, Ralegh's schemes and dreams took up Harriot's time, and throughout the 1580s and '90s he would repeatedly have to put his mathematical research aside. In fact, soon after his return from America in July 1586, he had felt compelled to help his friend and patron.

The failure of the Roanoke colony, and subsequent reports from disgruntled colonists, had added fuel to a fire of public opinion that was destroying Ralegh's reputation. Highborn courtiers had long been jealous of the place of the mere son of a tenant farmer in the queen's affections. Such jealousy, together with tales of his lavish lifestyle and supposedly arrogant manner, and the continuing perception that he was exploiting consumers and businessmen through his monopolies in the woolen and other trades, were turning him into the man nearly everyone loved to hate.

Of course, Ralegh was no more to blame for social inequality in England than any other successful courtier. As a Member of Parliament, he worked hard for his West Country constituents, among whom he remained extremely popular. And no matter how high he rose at court, he kept his "hick" Devonshire accent.[20] As for his colonialist ambitions, it is easy to frame them in hindsight, but for his time, he approached the issue with more intelligence and humanity than most. (His anti-Catholic attitude to colonizing Ireland was another matter. He saw Catholics as enemies, whereas Algonquians were potential Protestants: his humanity went only so far.) Despite the failure at Roanoke, Ralegh immediately set about the arduous and expensive process of preparing for a new, nonmilitary colony, to be governed by John White—the artist and veteran of the First Colony— with the help of twelve assistants or councilors. Tellingly, the new colony was not intended to displace the native inhabitants but to live on vacant land they had told Harriot about, near Chesapeake Bay.[21]

Harriot had of course spent much time with White mapping and recording America, and he wished his venture well. He did not intend to join him, though; he had his mathematical research, and Wingina's death no doubt weighed heavily on him. But he wanted to do what he could to help promote Ralegh's new project and to salvage Ralegh's reputation. So he had begun writing his *Brief and True Report of the New Found land of Virginia,* opening with a broadside against the rumormongers who had spoken maliciously of Ralegh and his original American settlement: "There have been diverse and variable reports, with some slanderous and shameful speeches bruited abroad by many that returned from [Virginia]." To counter these naysayers, he wanted to set forth a "true" report of the colony—written by one who had been "specially employed" in "dealing with the natural inhabitants, and having therefore seen and known more than the ordinary"—so that investors or prospective colonists who had become doubtful about participating in the new venture might reconsider. Moreover, quite uncharacteristically of him, given his natural reticence, he wanted to set the record straight "in public manner."

Acknowledging that people might be confused by the diversity of reports about the First Colony, he wanted to begin by setting down a few words on what he thought had motivated the negative rumors. It was true that a few members of the colony had been sent home early "for their misdemeanor and ill dealing in the country" and had been "worthily punished" (an allusion to Lane's "unruly" men, the "wild men of my own nation.") However, "by reason of their bad natures," they had slandered not only Lane but Roanoke Island, too. Other adverse reports were made by men who "had little understanding, less discretion, and more tongue than was needful or requisite" and who pretended to have more knowledge of the area than they did. Such boasters claimed to know all the faults of the country, although, unlike Harriot and White, few of the First Colonists had been as far as Chesapeake Bay, where White planned to settle.

Some of the disgruntled colonists had had gold and silver as their main aim, and when these ores were not found, such men, wrote Harriot, "had little or no care of any other thing but to pamper their bellies." He did not mean to tar with this brush those with legitimate economic disappointments. Many small traders who had invested in the colony had lost money because Barlowe's land of milk and honey had been overhyped.[22] (Later in his report, Harriot took great care to set out in detail a realistic assessment of the potential for "merchantable" resources in Virginia.) And finally, for some colonists the rigors of the adventure were simply too much. You could not find in Virginia the amenities of life to be found in English cities, so there was no "dainty food, nor any soft beds of down or feathers." This had not bothered Harriot: for all the luxury of Durham House, he had not forgotten his lowly Oxfordshire roots. Life in the colony must have been challenging, though—especially in the last few months when food supplies were running perilously low.

Printed in 1588, his *Report* is the only work of Harriot's to be published in his lifetime. It was then republished in 1589—along with Barlowe's and Lane's reports and other firsthand material—in Hakluyt's *Principal Navigations*.[23] It seems that the manuscript version of Harriot's *Report* did help Ralegh's cause, as by January 1587 enough investors had been found to form a corporation, headed by White, to found the City of Ralegh at Chesapeake.[24] Then, on May 8, 117 colonists—91 men, 17 women, and 9 children—set sail for Virginia. Manteo was with them, and so was another Roanoke, Towaye. Both had returned to England with Drake the previous year, and now they were returning home.[25] Manteo had been appointed, in effect, Ralegh's chief of Indian affairs in the Roanoke area.

Harriot had very likely taught White and the other captains some navigational theory before they departed. In addition, he and White had produced the first surveyed maps ever made in North America. They would be used for the next half century and thus helped the colonial enterprise, for better or worse. From a scientific point of view, though, Harriot's care in quantitatively surveying the landscape is yet another example of his eye for detail and his ability to overcome technical and scientific problems. The coastline has changed in four hundred years, but satellite images have confirmed just how accurate these maps were.[26]

IN AUGUST 1587, JOHN WHITE'S granddaughter became the first English baby born in America. She was named Virginia, after her new country, but her birth was not an auspicious omen.[27] During the voyage, White had fallen out with his chief pilot, Simon Fernandes. One of their points of disagreement was Fernandes's penchant for piracy. White quite rightly disapproved of taking colonists into potentially deadly encounters. Unfortunately, Fernandes had taken charge of the voyage right from the start, sidelining White and playing by his own rules—although White's report is the only evidence we have with which to judge Fernandes's apparent faults. As White tells it, when they arrived at the Carolina Banks, they learned from Manteo's people on Croatoan Island that Wingina's men—including Wanchese— had avenged their leader's death by attacking and driving away the sixteen soldiers Lane had left manning the fort on Roanoke Island. These men presumably died of starvation; they were never found. Yet Fernandes, who supposedly wanted to engage in more privateering before the summer ended, refused to take the colonists on to Chesapeake Bay. Instead, he deposited his passengers on Roanoke Island, where the houses of the First Colonists were overgrown, the fortifying embankments razed, and the prospects ominous.

White still had hope. The Croatoans had promised to come to the settlement with other local leaders to negotiate peace between the two peoples. But then an unarmed Englishmen, George Howe, was brutally killed by Wanchese and his friends as Howe was fishing alone for oysters. It must have been terrifying for the colonists, finding Howe's battered body and knowing that Lane's soldiers had likely been driven to their deaths. White had learned nothing from Lane's mistakes, however. When the Croatoan peace party failed to turn up as promised, fear led him to avenge his countrymen's deaths in a nighttime attack on a hostile neighboring village. The first (and apparently the only) casualty was a friendly Croatoan, who had come with his people to gather crops abandoned by Wanchese's allies when they fled the village after killing Howe. Harriot had specifically noted the Algonquian tactic of fleeing after a skirmish, but White had not taken sufficient notice.

Manteo—who had recently been baptized a Christian—was furious. Undeterred, White eventually persuaded him that his people had precipitated the attack by failing to keep their promised rendezvous. Harriot would have acknowledged there was fault on both sides, but White was already out of his depth.

Three days after his granddaughter's birth, he set off for England to reprovision the colony. Despite the disaster on Croatoan Island, he had been extremely reluctant to leave his post, but so eager were the colonists for speedy supplies that they had pressured him to go. They were worried that any supply ships dispatched from England might go straight to Chesapeake Bay, the colonists' original destination; with his knowledge of the coastline around Roanoke Island and with his family depending on supplies, White seemed the best person for the job. He would never see his daughter and granddaughter again.

CHAPTER 9

War, and a New Calendar

IN NOVEMBER 1587, White arrived in England to find that the government had embargoed all English shipping: no ship was to leave England—not even supply ships for Ralegh's colony.[1] Instead, all were to ready themselves for an invasion by Spain. Unlike her father, Elizabeth had not maintained England's navy; her treasury was nearly empty—thanks partly to years of military aid to Protestants in France and the Low Countries—and she now had to augment her fighting fleet with private ships.

The Spanish had long been looking for an excuse to attack the English, whose corsairs had been so successfully raiding their ships and New World settlements. After the execution of the Catholic Mary Queen of Scots back in February, the king of Spain felt he had his justification and began preparing his Armada. Mary's final downfall had been her apparent role in yet another Catholic plot to assassinate Elizabeth, the Babington Plot, named after one of the chief planners, Anthony Babington. Walsingham had sniffed out the conspirators in late 1586. They died gruesome traitor's deaths, and their property was confiscated. The queen gave Babington's houses and land to Ralegh; this tainted gift was the first property Ralegh had ever owned.

Mary, too, had been tried and convicted, although an agonized Elizabeth had stayed away from the proceedings and then put off signing the death warrant for three months. All the while she had been hoping desperately for a way to save Mary. After all, the Queen of Scots was her cousin and a fellow monarch; like most European royalty at the time, Elizabeth believed in the divine right of kings. In the end, she saw no way out of her awful impasse. She signed the warrant, though she did not yet give approval for carrying it out. Her councilors went ahead and executed Mary anyway. This betrayal, and the royal blood on her hands, left Elizabeth utterly traumatized. Her people, on the other hand, were relieved that the traitorous Queen of Scots was dead at last: bonfires were lit, "and all the bells in the city of London rung for joy," as a contemporary observer put it.[2]

To Mary's twenty-year-old son, the Protestant James VI of Scotland (and future James I of England), Elizabeth eventually managed to write a letter protesting her innocence in an execution forced upon her by her councilors and lawyers and carried out without her final approval. James had earlier tried to plead for his mother's life, but he responded to Elizabeth with grace. He acknowledged her "long-professed good will" toward his mother and hoped that she would now work with him to "strengthen and unite this isle, establish and maintain the true religion." In earlier correspondence, the Virgin Queen had alluded to the possibility that James might be her successor; now he was ready to forgive Elizabeth her regicide if she honored this understanding.[3]

As for war with the Spanish, Elizabeth had been trying to avoid it for years. But now, in the wake of the execution of the Queen of Scots, she felt she was losing the propaganda war between the two nations. In a new papal bull—thousands of copies of which were to be distributed among English Catholics—the pope had reaffirmed the 1570 excommunication of the Protestant queen. Cardinal Allen, a vociferous expatriate critic of Elizabethan religious policies, had also weighed in, publishing a diatribe in which he reinvigorated the long-standing Catholic catch-cry against the queen's legitimacy. Ever since her ascension to the throne in 1558, Elizabeth had been dogged by Catholic denial of the legality of her father's divorce from his first wife, and now Allen went so far as to refer to the queen's mother, Anne Boleyn, as "an infamous courtesan."

Elizabeth had long known she was ruling on a religious knife edge. In fact, her attempt to steer a middle path between Catholicism and Puritanism was upsetting not only Catholics but also some Protestants. Several years earlier she had told senior Anglican clergy that radical Protestants complained she was "of no religion—neither hot nor cold, but such a one as one day would give God the vomit."[4] She had been resolute enough then, but now Allen's personal attacks devastated her, and she finally yielded to the advice of her admirals and accepted the inevitability of a war with Spain. Which was just as well: English spies reported that a fleet of Spanish ships was ready to invade England if and when Philip II gave the order. And what a fleet it was, with its intimidatingly huge galleons. Philip could well afford the best, given the gold and silver that Spain had plundered from South America over the past century.

A twenty-four-hour coast watch was put in place, with a system of beacons of burning resin whose light would flash inland from beacon to beacon when Spanish ships were sighted. Finally, on July 19, 1588, the moment arrived, and the dreaded Armada was spotted off the English coast, making its way through the English Channel from Plymouth toward the mouth of the Thames. Philip had ordered that his smaller ships should sail upriver to capture London, "that rich and flourishing City." He hoped to rely on English Catholics to turn to his aid, but he had underestimated the extent of English unity in such a crisis. (Such spiritedness would famously help save the country nearly four centuries later during another dramatic gathering of private English ships in the Channel: the rescue from Dunkirk in 1940.) The Venetian ambassador to France sensed this spirit and privately warned that despite Philip's superior navy, the English were "men of another mettle from the Spaniards." They were, he added, "expert and active in all naval operations, and great sea dogs." The pope's ambassador concurred: Philip "goes trifling with this Armada of his, but the Queen acts in earnest. Were she only a Catholic she would be our best beloved."[5]

Ralegh had counseled that in the shallow coastal channels the smaller, swifter English ships would be more effective than the Spanish galleons, and indeed the English commanders outsmarted the Spaniards, rapidly mobilizing their nimble vessels and taking advantage of a favorable wind that enabled them to steal in behind the Spanish ships and hem them in. The ensuing battle in the Channel raged for many days. Then, on the night of Sunday, July 28, the English unleashed a ghostly fleet, empty of men but packed with gunpowder. Drake, Hawkins, and others had offered some of their own older vessels for these eerie, sacrificial fire ships that slipped silently toward the Spanish flotilla, lighting up the night until the heat was such that the ships' guns automatically exploded amid the helpless galleons. Surviving Spanish ships fled north, only to be wrecked by storms and rocky coasts. But the Armada was huge, and many more ships waited ominously in the Channel. In early August, Elizabeth made her famous speech to the troops at Tilbury on the Thames: "I know I have the body of a weak and feeble woman, but I have the heart and stomach of a king, and of a king of England too, and think foul scorn that . . . any Prince of Europe should dare invade the borders of my Realm." She promised to take up arms herself to fight with her courageous people. Fortunately there was no need. Ralegh was right, and the agile English ships prevailed, with the help of the fire ships, local English knowledge of the tides and winds around the coast, and some fortuitous storms. By early September, the mighty Armada was defeated.[6]

FAR LESS OF A TRIUMPH was White's attempt to return with supplies to Roanoke Island. Back in April, he had managed to get permission to take two of Grenville's ships that were deemed too small for England's defense (they were of twenty-five and thirty tons; the smallest ships in the fleet Harriot had sailed on displaced fifty tons). Unfortunately, the chief captain, Arthur Facy, was more interested in privateering than in taking White to America—and he had some success, plundering a Scottish and a Breton ship. Attacking ships of nominally friendly countries was sheer piracy, of course, presumably initiated by Facy with willing support from his sailors. Lowly seamen did the hardest work in sailing vast, dangerous oceans, yet they endured appalling shipboard conditions, suffering often-brutal discipline, and they could be a rough lot whose sole goal was to find treasure.[7]

In May, White's ship was on the receiving end of piracy when it was attacked by a French warship. (White later reported it as God's punishment for the earlier "thievery by our evilly disposed mariners.") A fierce and bloody battle ensued, in which dozens were killed or severely injured; White and several prospective colonists were also wounded. In the end, the English were forced to surrender and hand over the booty they had seized en route. After hours of unloading, the French discovered they had been "over greedy," as White put it in his later report: one of the French boats sank under the weight of the Englishmen's loot. The Frenchmen then returned most of the weapons and sails they had taken from the beleaguered English ship and "left us two cables and anchors." They had taken everything else, though, including the provisions for White's colony. So the English had little option but to repair their damaged rigging and sails and return to England.[8]

After the defeat of the Armada, Ralegh set about organizing yet another supply ship and finding new investors for the City of Ralegh enterprise. But not until 1590 did White manage a successful return to Roanoke Island. We can only imagine how the colonists had fared in hostile territory during the three years since White had left for supplies. Equally, we can only imagine the shock White must have felt on returning to find that all the colonists had vanished. No one knows what happened to them, although research into their fate is currently ongoing. Some scholars believe they merged peacefully with Manteo's Croatoan people, or with the Weapemeoc tribe fifty miles to the northwest of Roanoke Island. Others believe they moved to their original destination, Chesapeake Bay, and lived peacefully with the Chesapeakes until they and their hosts were massacred by Powhatans (a claim that appears in Jamestown documents but is still being debated).[9] It certainly seems that the colonists had left Roanoke Island willingly enough, leaving behind a carved sign saying "Croatoan." According to White's report, however, when he set out for Croatoan Island, the ship's cable broke, and she almost ran aground; in the ensuing efforts to right her, she lost all but one anchor. Winter was approaching, and they were short on supplies—some of White's fleet had failed to turn up, having gone privateering instead—and once again White felt he had no option but to return to England.[10]

In an irony of timing, 1590 was also the year that Harriot's *Brief and True Report of Virginia,* accompanied by engravings of White's drawings and paintings made by Theodor de Bry, was published in deluxe editions in four languages: Latin, English, French, and German. Hakluyt had been the conduit between de Bry—a Frankfurt-based Flemish printer—and Harriot, who wrote additional captions to explain White's drawings of Algonquian life.[11] The book, *America, Part I,* was an international hit, and Harriot made a modest note to himself, recording the publication of "my discourse of Virginia in 4 languages."[12] But just as *America* was making Harriot, White, and Ralegh famous in Europe, Ralegh's interest in North America appeared to be waning. He had spent such enormous effort, and so much money, only to be repaid with disaster upon disaster. He had already begun to set his sights closer to home—Ireland, where the queen had given him forty-two thousand stolen acres in Waterford-Cork, on which he planned to establish an English colony. Gossips gleefully whispered that Ralegh was in a kind of exile there, sidelined at court by his new rival: the handsome, twenty-two-year-old Robert Devereux, Earl of Essex.[13] The hot-headed son of the queen's cousin Lettice Knollys, and the stepson of Elizabeth's youthful love Robert Dudley, Earl of Leicester, young Essex seemed bent on bringing Ralegh down. In 1588, he had even gone so far as to challenge him to a duel, but cooler heads in the Privy Council had intervened.

Of course, the facts were different from the spiteful rumors. Ralegh had been in Ireland at the end of 1588 on government business:

ferrying soldiers and fortifying defenses against a possible attack by
stray Armada ships scattered along the Irish coast after fleeing the
fighting in the Channel. Once free of these duties, he sojourned at
his new Irish property, where it seems he began work on his *Ocean's
Love to Cynthia*.[14] This was love poetry to the queen—his remote,
adored, ever-renewed moon goddess Cynthia. He was the ocean (the
"Wa'ter," a play on the Devonshire pronunciation of Walter), ever
moved by the moon's power.[15]

Both Ralegh and Elizabeth had long played their parts in their
courtly romance. It was a theater piece charged with sexuality—a
drama of romantic love that both players knew was an elaborate
game that symbolized her power and his dependence on her. Ralegh,
with his dry humor and his gift with words, had privately jested at the
charade.[16] But the queen, with her wit and her power, was surely
having the last laugh. In a man's world riven by dangerous factions
and religious divides, she knew she needed every trick in the book to
hold on to her place as one of Europe's greatest "princes."

While Ralegh was poetically romancing Cynthia, his Ireland-
based friend Edmund Spenser was writing his own allegorical mas-
terwork on Elizabeth and her court, *The Faerie Queene,* in which Ralegh
reputedly played a part.[17] Ralegh was so impressed with Spenser's
genius that he took him back to court and introduced him to the
queen. She, in turn, was so impressed that she gave Spenser a pension so that he could continue to work on his epic. (She evidently
missed the ironic—and sexist—criticism of her reign that some
scholars see in parts of *The Faerie Queene*.)[18]

In 1589, Harriot had spent time as a technical advisor for
Ralegh's colony in Ireland (Ralegh had leased some of his lands to
English settlers) and had been rewarded for his work there and in
the New World with a gift of Irish property: the isolated medieval
Abbey of Molana, near Youghal.[19] Accepting such a gift constitutes a
Faustian bargain by today's standards. Ralegh was still active in supporting the Elizabethan oppression and forced colonization of the
Irish. Four centuries ago, however, few, if any, of Harriot's contemporaries would have seen the analogy. When Marlowe's play *The Tragical
History of Doctor Faustus* opened in London that very same year, many
would have seen not an allegory but a literal choice in Faustus's decision to turn away from the rigors of science as a source of knowledge
and to sell his soul for the devil's gift of two dozen years of magical
power.[20] Ireland was not seen as a moral bargain in the same sense;
instead, the Protestant conquest of Irish Catholic "rebels" was framed
as a religious duty.

Harriot, however, also showed an interest in indigenous Irish ways, likening their manner of fishing by shooting darts to one of the Algonquian methods.[21] Not that he spent much time in Ireland after 1589. He and Ralegh were soon back at Durham House, where Harriot began taking astronomical observations to update his table of solar declinations for Ralegh's future navigators. (Ralegh had not yet given up on North America completely, but he had also begun thinking about South America.) To make sure these declinations were as accurate as possible, he also needed to measure the obliquity of the ecliptic—the angle between the apparent yearly path of the sun and its apparent daily path, or equivalently, the angle of the tilt of the earth's axis. This obliquity, as noted earlier, is "about" 23.5 degrees. The qualification is because the angle changes very slightly over the centuries, because of the strange phenomena of "nutation" and "precession." By modern reckoning, in Ptolemy's time the correct angle was 23 degrees 40 minutes and 50 seconds, while 1,450 years later in Harriot's time it was 23 degrees 29 minutes and 30 seconds. A minute is 1/60th of a degree, and a second is 1/60th of a minute, so these values are 23.68 degrees and 23.49 degrees, respectively. Today the obliquity is 23.44 degrees. Harriot's measurement was spot-on for his era, even more accurate than that of Tycho Brahe, considered the most accurate naked-eye astronomer in history.[22]

THE PHENOMENON OF PRECESSION WAS discovered more than two thousand years ago, by Hipparchus. He called it the "precession of the equinoxes." Other early peoples, notably the Mayans, may also have discovered it, although evidence is opaque and controversial.[23] Thanks to Ptolemy, however, we know that when Hipparchus compared his own observations with those of Timocharis and Aristyllos 150 years earlier, he noticed that all the longitudes of the stars had changed slightly, so that all the stars appeared to be moving, in the same direction, along the ecliptic. In a virtuoso feat of mathematical imagination, he realized that geometrically, this drift in the position of the stars corresponded to a small rotation of the celestial sphere's north-south axis, as if it were pivoting around its own center. Such a rotation meant that the direction of the North Celestial Pole itself was actually changing.

To see what this has to do with equinoxes, it helps to imagine the north-south axis as a vertical drinking straw that passes through the middle of a horizontal circular disc of cardboard, so that the disc is halfway along the straw. (Or stick a pencil through a small piece of paper, through the middle of a circle drawn on the paper.) The disc represents the plane of the celestial equator, and its edge represents the equator itself. Rotate the straw so that it pivots around its middle, both its ends tracing out small circles in the air, so that the whole straw traces out an imaginary double cone. The orientation of the disc will change as the straw rotates, so that the outer rim of the disc (or the circle on the paper) oscillates. On the celestial sphere, if the position of the equator oscillated like this, then so did the point where the equator crossed the fixed ecliptic—the point that defined the equinoxes (as in figure 9). Hipparchus calculated that this oscillation was so slow that the equinoxes were moving by just a tiny fraction of a degree each year.[24] He also figured out that as the axis inched its way around, the point of intersection between ecliptic and equator was changing in such a way that each spring equinox came just a little earlier than it did the year before. In this sense, it *preceded* the equinox of the previous year, and hence the term "precession of the equinoxes."[25]

It would take Newton to explain *why* precession was happening. Eighteen centuries after Hipparchus (and a century after Harriot was telling his students about precession), Newton realized that it is the earth's north-south axis that is circling around, not the celestial sphere's as Hipparchus had supposed. Newton's explanation hinges on the earth's everyday rotation about this tilted axis, and on the gravitational pull of the sun and moon on the spinning earth. This pull makes the earth's axis itself rotate, just like a spinning top or gyroscope that has been pushed out of equilibrium. Instead of tipping over under the gravitational pull of the ground, the axis of a slightly tilted spinning top rotates around in a small circle. You can readily see the way the "north pole" of a top makes such a circle as the top spins, but the slow circle or "wobble" described by the earth's axis was measurable only by the most careful astronomers.[26]

Of course, when Harriot was taking new measurements at Durham House, no one knew about the role of gravity in planetary motion. Nevertheless, the ancients had bequeathed a great deal to the late sixteenth-century pioneers who, over the coming decades, would lay foundations for Newtonian physics and astronomy. For instance, Harriot and his peers knew that it takes about seventy-two years for the earth's axis to rotate through 1 degree—so that the celestial sphere of background stars also appears to move by 1 degree.[27] This means it will take nearly twenty-six thousand years to make one 360-degree circular turn. During this cycle, the direction of "north" changes ever so slowly as the axis rotates.[28] Today, as in Harriot's time, the earth's North Pole points approximately toward the star Polaris. (Or, as the pre-Copernicans had it, Polaris is near the North Celestial Pole.) But as Harriot would tell his next batch of pupils, precession was causing Polaris to move even closer to the north, because the angular distance between Polaris and the Pole was decreasing. Quoting Ptolemy, Harriot noted that Hipparchus's estimate of this distance had been 12 degrees and 24 minutes, but that now, according to Harriot's own observations, it was 2 degrees and 56 minutes. He calculated that this angle would continue to decrease until Polaris was only half a degree from the Pole. (Today, it is about 1 degree.) Then, he said, it would slowly move away again, ending up 48 degrees away "if the world do last so long."[29]

What this means is that in about thirteen thousand years' time the earth's axis will have rotated so that "north" is in the direction of the star Vega, not Polaris. When that happens, the seasons will have reversed, so that the northern summer will begin in December—although for a number of years now, scientists have been aware that additional factors are perturbing the axis's wobble, changing the direction of "north" and "south" in unexpected ways. These factors include gravitational changes on earth due to melting polar ice sheets and changing amounts of continental groundwater.[30]

Precession was important for astronomers and navigators, but it had also caused heated debate in Harriot's time because of its surprising effect on the timing of religious festivals. With 360 degrees in a circle and approximately 365 days in a year, a precession of 1 degree every seventy-two years corresponds to about a day every seventy-two years—and this meant that the humble, everyday calendar was drifting slowly out of line with the seasons. A day or so every century is neither here nor there in the short term, but it does mount up—and the calendar in use in Harriot's time was already sixteen centuries old. To be more precise about the accumulated number of days of seasonal drift, astronomers needed to carefully define what they meant by a "year."

A 365-day year had been used since antiquity. Five or six thousand years ago, the ancient Egyptians had noted that when the bright star Sirius rose just before sunrise, the life-giving waters of the Nile would soon flood the water-starved fields.[31] They also recorded that there were 365 days between this beneficent dawn conjunction of sun and star. The Central American Mayans and the South American Incas, among others, also had a 365-day calendar. In his *Natural and Moral History of the East and West Indies,* José de Acosta gave a detailed, firsthand account of the

Mexican calendar. It had eighteen months of twenty days each, and the remaining five days were called "days of nothing." On these days, "the people did not any thing, neither went they to their Temples, but occupied themselves only in visiting one another." Even the priests rested, "ceasing their sacrifices" on these five days that were needed to make a 365-day year. All in all, he said, the Mexican calendar was "ingenious enough, and certain, for men that had no learning."

Of course, the Mayans were highly sophisticated, building pyramids and observatories and developing hieroglyphic writing— although Acosta dismissed this as pictorial rather than a grammatical written language. Nevertheless, he did take to task those Spaniards who assumed all the "Indians" were lazy and stupid and who destroyed native artifacts for being symbols of superstition. Acosta appreciated the achievements behind these artifacts—including the Mayan pictorial calendars and the Guatemalan books made of leaves—and lamented their destruction due to his countrymen's "foolish and ignorant zeal."[32] But he considered the Mayans to be poor astronomers in comparison with the Peruvian Incas. He had been a Jesuit missionary in Peru since 1571, and he believed the Peruvian calendar was superior to the Mexican because it was based on the cycles of both the sun and the moon, like the European calendar. The moon's cycles delineated twelve rather than eighteen months, with no additional days needed to fit with the solar year.

It was one thing to measure a year by the rising of a star such as Sirius, but the precession of the equinoxes meant that such a star-based ("sidereal") year did not quite match the time between successive spring equinoxes, which defined the seasonal ("tropical") year. It was this latter, season-based year that was the basis of the calendar used in Europe in the early 1580s. (Acosta notwithstanding, some modern scholars think the Mayans may have been good enough astronomers to have determined both a sidereal and a tropical year.[33]) The European calendar dated from 46 BCE, when Julius Caesar instituted a 365-day year *plus* an extra day every four years, so that the average calendar year was 365¼ days. This so-called Julian calendar was an attempt to account for the precession of the equinoxes, but still there was a problem, because the time from one spring equinox to the next was a fraction less than Caesar's calendar year: 365.2422 days as opposed to 365.25, a difference of 0.0078 days.[34] After 128 years, this difference amounted to a whole day (because $1/0.0078 = 128.21$).

The effect of precession (and the need for leap years) means that the date and time of the equinox is not exactly the same each year. Today the March equinox usually occurs sometime on the twentieth or twenty-first. In fact, March 21 had been the official Christian date of the spring equinox since 325 CE. By Harriot's time 1,260 years later, however, the seasons had moved ahead of the Julian calendar by about ten days (1,260 divided by 128 is approximately 10). In other words, in Harriot's time the date of the actual spring equinox was March 10 or 11 by the Julian calendar, according to Harriot's written lessons for Ralegh's captains, while the official equinox was ten or eleven days later.[35] This was problematic for some. Easter Sunday was timed to occur on the first Sunday after the first full moon after the northern spring equinox—but how could people feel secure celebrating Easter ten days too late?

Although Caesar's calendar had been an attempt to fix the problem of seasonal drift, scholars had continued to debate the issue—including Ptolemy in 150 CE, the eighth-century English monk the Venerable Bede, the ninth-century Arab astronomer al-Battani, the eleventh-century Persian mathematician-poet Omar Khayyam, the thirteenth-century English philosopher-scientist Roger Bacon, and Copernicus in the early 1500s. By the 1570s, the timing of Easter had become so pressing to Catholics that Pope Gregory XIII had commissioned his best mathematicians and astronomers to make accurate astronomical measurements of the seasonal year and to construct a new calendar that kept better pace with precession. As a result of their deliberations, Gregory had issued a papal bull in 1582 proclaiming a brand-new calendar for Catholics.

It was not a popular move, since it meant that ten days appeared to be "lost" from people's lives before the new calendar could begin: the pope decreed that on the day after October 4, 1582, the calendar would read not October 5 but October 15. Not surprisingly, many uneducated folk were terrified at having their days "stolen" from them and suspected all sorts of evil intent by the pope and his "conjuring" mathematicians.[36] But this ten-day jump would bring the northern spring equinox from around March 11 to around March 21, the official Christian date for the equinox.

In response to Gregory's initiative, in 1582 Queen Elizabeth had consulted with *her* best astronomers, notably Dee and Thomas Digges, who approved the pope's decision. The new "Gregorian" calendar had a 365-day year plus leap years, but the 366-day years no longer occurred every four years; rather, leap years were those whose date was divisible by four (as in the Julian calendar) *except* for century years (1600, 1700, and so on), which must be divisible by four hundred to qualify as a leap year. It sounds convoluted, but with an average calendar-year length of 365.2425 days, it would take three thousand years for the Gregorian calendar to run ahead of the seasons by just one day. (The difference between a calendar year of 365.2425 days and the seasonal year of 365.2422 days is now just 0.0003 days, and 1/0.0003 = 3,000. Omar Khayyam had devised an even more accurate calendar than the Gregorian, although its leap years were harder to remember and its extra accuracy not worth the practical problems.)

Despite the efficacy of the new calendar, public outcry stayed Elizabeth's hand—an outcry due more to Protestant prejudice than to superstition. In particular, the archbishop of Canterbury, Edmund Grindal, baulked at the idea of accepting a reform by "the Antichrist" (the pope). Protestant regions on the Continent also rejected the change, so that although the pope prevailed in most Catholic countries within a few years, most of the Protestant jurisdictions took a century or more to accept the Gregorian calendar. It was not adopted in Britain until 1752. Today, it is virtually universally accepted for everyday secular use (although critics are seeking further reform for geopolitical and mercantile reasons[37]).

IN THE EARLY 1590S, Harriot's interest in precession was its effect on the angle between the ecliptic and the celestial equator, as mentioned, and he needed to measure the correct angle for his era in order to accurately pinpoint the altitudes and declinations of the sun and stars (as in figure 10). According to modern calculations, Harriot's new table of solar declinations was correct to within a couple of minutes of arc—far more accurate than any contemporary table—and it meant that latitudes calculated from his table were accurate to within a couple of nautical miles.[38]

At the same time that Harriot was determining the obliquity of the ecliptic "by my own experiment with an instrument of 12 foot long upon [the] Durham House leads," Ralegh and White made further, ultimately fruitless, efforts to send a fleet to Croatoan Island.[39] By 1593, White had retired to Ireland, finally abandoning his Lost Colony to "the merciful help of the Almighty."[40] Ralegh would keep searching for the colonists, if in a desultory way, mounting a rescue expedition as late as 1602.[41] For now, though, he had other, more personal things on his mind.

CHAPTER 10

New Chances

D ESPITE THE RIGORS OF PREPARING for the defenses
against the Spanish Armada and searching for the Lost
Colonists—and despite Essex and the court gossips—Ralegh
found time for relaxation. In particular, in recent years he had enjoyed
many a pleasant social evening with a new friend: Henry Percy, the
9th Earl of Northumberland.

Four years younger than Harriot and eleven years younger than
Ralegh, the Anglican Northumberland was also heir to a distinguished Catholic family. His uncle and his father—the seventh and
eighth earls—had allegedly participated in plots aimed at replacing
Elizabeth with Mary Queen of Scots: his uncle had been executed in
1572, and his father, who had been implicated in the 1583 Throckmorton
Plot, was sent to the Tower, where he committed suicide. According
to a contemporary account, he did this to forestall the seizure of his
property and title that a conviction of treason would have mandated.[1]
This meant that now, for all that Essex and other courtiers sneered
at Ralegh, regarding him as a parvenu, Ralegh had been befriended
by a man who was not only an earl but was also—via his kinship with
an earlier king—eighth in line to the throne.[2]

Northumberland was a rather dissolute young man, spending
his time at "hawks, hounds, horses, dice, cards, apparel, mistresses,"
according to his own later testimony.[3] His account books from the
late 1580s tell the same tale—and they also give a glimpse into
Northumberland's growing friendship with Ralegh.[4] For instance,
one day, Northumberland lost £10—about $4,000 today—to Ralegh

at cards. Another day he had his blood let by Ralegh's surgeon. Yet another day, Ralegh's servant delivered some goods for him; and on various occasions he and Ralegh exchanged elaborate gifts.

Despite his penchant for luxuries, gambling, and women, Northumberland wanted to be taken more seriously than a wealthy playboy, and he was drawn not only to Ralegh's glamor but to his intellectual pursuits. By 1591, when he was twenty-seven, Northumberland had made such progress that he was becoming renowned for his interest in science and alchemy and would soon earn the nickname "Wizard Earl."[5] This interest was further kindled by his growing friendship with Harriot, whose mathematical ability and knowledge of astronomy made a deep impression on him. By January 1592, he was buying tobacco sourced by Harriot, who was still convinced of its health-giving properties, and who had become a regular guest at the supper parties of the Ralegh-Northumberland circle. Apparently Harriot reveled in more than the conversation. He was drawn to the gambling, too, and inevitably to thinking about the mathematics of chance. How could he not, in such a circle? A typical entry in Northumberland's accounts records that the earl won £53 2s. "at play," while losing £74 5s. 4d. Another entry shows he lost a shilling tossing coins with Harriot, who evidently bet only small sums.[6]

The mathematics of probability was an almost untouched field in Harriot's age. People had long been playing games of chance, of course; these games likely developed from methods of divination, such as the throwing of sheep knucklebones in ancient Greece and yarrow stalks in China, and ancient dice have been found that are several thousand years old.[7] People had also long been interested in the kind of mathematics that underlies the analysis of chance, perhaps beginning with the Indians some two thousand years ago. Until the sixteenth century, however, no one seems to have put the two together in any systematic way. Before then, the mathematical focus was on more "worthy" topics than gambling. For instance, instead of asking how many ways two dice add to 7, or how many ways you can choose a hand of nine cards containing three of a kind, third-century Mediterranean mathematicians had asked such things as how many combinations of Aristotle's four elements were possible, or how many combinations of "malignant humors." Ancient and medieval Indian mathematicians had asked how many combinations of the six tastes—sweet, sour, bitter, salty, pungent, astringent—were possible, or how many ways a group of long and short syllables could be combined in poetry. Number mysticism featured, too, as in medieval Jewish scholars' interest in how many words of two, three, four, and more letters can be made from the twenty-two letters of the Hebrew alphabet, and in sixteenth-century Christian numerological attempts to decode the Book of Revelation.[8] These and other early mathematicians discovered algorithms for their combinatorial problems, but they do not appear to have developed a mathematical concept of probability.[9]

The probability of getting a 7 on tossing two dice is found by working out the number of combinations in which the dice scores add to 7, and comparing it with the total number of possible scores. Harriot listed all the ways two dice can land, as follows: suppose the first dice (or "die" as the singular is sometimes spelled) lands with a 1 face up; then the other dice can land with a 1, 2, 3, 4, 5, or 6. Harriot listed these six possibilities as 1,1; 1,2; 1,3; 1,4; 1,5; and 1,6. If the first dice lands with a 2 face up, the other dice can still land with a 1, 2, 3, 4, 5, or 6—and so on for all six possibilities for the first dice, giving $6 \times 6 = 36$ possible pairings for two dice. He also painstakingly listed all $6 \times 6 \times 6$ (= 216) triples for three dice, and similarly for four and five dice. Clearly he was feeling his way here. Like a handful of others before him—notably the early sixteenth-century Italian Niccolò Tartaglia, who, with his compatriot Girolamo Cardano, was a pioneer in the field—and like others after him, Harriot needed to make his laborious lists in order to discern the general patterns that define the mathematics of chance.

Beside each of his possible pairs, triples, and so on, Harriot listed the sum of each set of digits. In the case of a pair of dice, the pairing 1,1 adds to 2; 1,2 adds to 3; through to 1,6, which adds to 7. Similarly, 2,1 adds to 3; 2,2 adds to 4; ...; 2,5 adds to 7; and 2,6 adds to 8. And so on. He noted that six of these pairings add to 7 (which, he also realized, was the most likely sum on the toss of two dice: there are fewer than six pairs for every other possible sum). He did not speak, as we do now, of the probability of getting a 7 on tossing two dice equaling the number of ways two dice add to 7 divided by the total number of ways two dice can fall—namely, $6/36 = 1/6$—but he had calculated the chances nonetheless.

At some stage (there is no date on the manuscript) he estimated how much money could be wagered per hour in throwing dice. He reckoned on ten throws per minute and a rate of a shilling a throw. He also wrote down a brief calculation suggesting that if you threw six dice at a time, you would need to play 46,656 times in the one game in order to have a reasonable likelihood of winning—one for each of the $6 \times 6 \times 6 \times 6 \times 6 \times 6$ possible ways six dice can land. Even if you played 46,656 times, however, many of your scores would be duplicates, so you could not be certain of winning. Whatever Harriot's reason behind his calculation, he showed that at a shilling a throw, and with twenty shillings in a pound, 46,656 throws came to £2,333 16s.— the equivalent of nearly a million dollars today.[10] Perhaps he was trying to make concrete for Northumberland the fact that the odds were stacked against him. Few people then understood the counter-intuitive laws of chance. Or perhaps he was simply creatively "doo-dling." Another calculation on the same page calculates the odds of throwing six dice so that they all land with a different number uppermost.

Although Harriot was not the first to do these kinds of calcula-tion, his exploration of chance—like his early work on the Mercator problem—would turn out to be a springboard into more advanced mathematics, where he would make original discoveries. To take an example, his work on the mathematics of dice would help him to pioneer the general "binomial theorem." That research would take many years, given the usual interruptions, including not only his ongoing teaching and research on navigation—Ralegh still had his sights set on South America—but also the work involved in keeping Ralegh's accounts. Nevertheless, it is this period of the early to mid-1590s, when he began to think about such nonnavigational topics as probability, that marks the beginning of a scientific change of direc-tion for Harriot: one in which the focus was on mathematics and physics for their own sake. It was a change of direction that would benefit hugely from his meeting with the scientifically inclined gam-bler Northumberland.

IF NORTHUMBERLAND'S ACCOUNT BOOKS' references to gambling point to the likely motivation for Harriot's work on combinations and chance, an intriguing entry intimates something significant about Ralegh's love life. Northumberland's list of expenses for New Year's gifts in 1587 notes that he had given the queen an embroi-dered kirtle—a long dress worn under a coat and costing the princely sum of £18—and that he had also spent nearly £14 on fifteen yards of velvet "for Mistress Throckmorton." Velvet was an aristocratic gift. Hierarchy was so important in Elizabethan England that a 1574 Proclamation on Apparel declared who could and who could not wear such a regal fabric.[11] As for Elizabeth Throckmorton (or Bess, as she was known), she was a cousin of Francis Throckmorton of the infamous plot, and she was one of the ladies of the queen's privy chamber. It is not at all clear why Northumberland was giving her such a lavish present. Perhaps she had done him a favor at court; she certainly had intimate access to the queen. Regardless of the reason, the entry in Northumberland's account book is important because it shows that Bess Throckmorton was known to the Northumberland-Ralegh circle as early as 1587. This was also the year that Ralegh was made Captain of the Queen's Guard, which meant he was responsible for the queen's personal security—and this meant he had unprecedented access to her private chamber and to the ladies who attended her. Bess then disappears from the record until 1591, when she turned Ralegh's life upside down—and therefore Harriot's, too.

Bess's Protestant father, Sir Nicholas Throckmorton, was a younger son in an aristocratic family whose titled heir was Catholic. Like the Percys, and the Tudors themselves, the Throckmortons were a family divided between these two faiths. Nicholas had been England's ambassador to Scotland and France. He had worked hard to save Mary Queen of Scots, by counseling her to abdicate her claim to the English throne and by mediating her case with Elizabeth. In France, he had served England well for the most part and had been respected as "a very wise and expert man," as one contemporary put it. He was also forthright. There is a story that Elizabeth had once vented her fury at his insistence on some diplomatic course of action that she was reluctant to take, shouting, "God's death, Villain, I will have thy head!" To which Nicholas retorted, "In that case, you will do best to consider, madam, how you will keep your own on its shoulders."[12]

Bess inherited her father's feisty determination, but little else. She was literate but not learned (the literacy rate for Elizabethan women was about 10 percent), she was not conventionally beautiful, and she did not have an appealing dowry.[13] After Nicholas died when she was six, her mother had loaned Bess's £500 inheritance to the Earl of Huntingdon. This was not unusual in an era without savings banks—Northumberland's accounts contain numerous mentions of loans received and given—but in this case Huntingdon defaulted. When her mother died in 1587, the twenty-two-year-old Bess was left with some pearls, some fine clothes and household linen, and a lavishly canopied feather bed. She also had a small income from a property in Mitcham, but it was not a dowry that would attract a covetous suitor.

Ralegh, of course, had no need of money. He saw something special in Bess, something that made her catch his eye as no one else had done: her frank and passionate personality. In this she was a match for Ralegh's other love—the unreachable Elizabeth. Indeed,

Bess was no ordinary woman. To take just one example, she would never give up fighting to retrieve her £500 from Huntingdon and his heirs. It would become a symbolic fight for justice, but the measure of this determined woman is that in 1622, fifty years after she lost her inheritance, she won her case: the Privy Council ordered it repaid, with 50 percent interest.[14]

In the late summer of 1591, the twenty-six-year-old Bess had discovered she was pregnant with Ralegh's child. The Elizabethans nominally adhered to the stricture of celibacy outside marriage, but premarital conceptions were not uncommon. In fact, the community did not hide the issue under the proverbial carpet but expected men to support their illegitimate offspring. Bess probably did not know it, but Ralegh himself had already fathered a daughter, following a dalliance with young Alice Goold when he was in Ireland in 1589. She was the daughter of a judge, but nothing else is known about her, and Ralegh had left Ireland before the pregnancy was discovered. He denied any intimacy when Alice's father wrote to him some months later, but he would eventually acknowledge and make provision for this "reputed" daughter in his will of 1597.[15] Of course, he was far from alone in such behavior. His nemesis Essex continued womanizing and creating illegitimate babies even after his marriage in 1590.[16]

Illegitimate offspring from secret dalliances may have been tolerated as part of Elizabethan life, but the queen reacted furiously if any of her favorite courtiers, or any of her female attendants, were caught misbehaving. They were expected to act decorously. Even more importantly, they were expected to be honest with the queen. In particular, her permission was needed before her ladies could marry, and generally—when they conducted themselves via a respectable courtship—she happily granted it, giving them presents and attending their weddings. They would then leave her service. Her attendants were privy to the most intimate details of her life and court, and she needed them to have undivided loyalties and responsibilities.[17] But when her ladies carried on secret courtships behind her back, she felt personally betrayed, and the fate of such a woman was not happy. She would be banished from court in disgrace and would find it difficult to find a husband if the father of her child did not marry her. She might even spend time in the Tower if the queen felt aggrieved enough. Knowing all this, any joy Bess felt at her pregnancy must have been undermined by considerable fear.

Ralegh did not abandon his mistress as many men had abandoned theirs—and as he had abandoned Alice Goold. For him, everything was different with the tall, charismatic Bess, and many years later, he would remind her in a famous letter that she was his chosen one—the one he truly loved, and whom he had chosen to marry, albeit secretly.[18] So secretly, in fact, that no one knows the date. It may even have been before the pregnancy, although if Bess had taken anyone into her confidence, it would have been her adored older brother Arthur, but according to his diary, he only found out about the marriage on November 19, 1591.

Ralegh brazenly denied the eventual inevitable rumors, hoping he could keep up his role as the queen's devoted "lover." But now he could not afford to be smug. He had witnessed the queen's fury when Essex lost favor the previous year for secretly marrying Walsingham's daughter. Bess played her part, too, hiding her condition from her companions in the privy chamber for as long as she could, then taking leave from court to steal away to the sanctuary of Arthur's home. Her baby was born there, on March 29, 1592; a month later, she left him with his nurse and went back to her duties at court.

The baby had been baptized Damerei, after an apparent Plantagenet forebear that Ralegh had dug out. Sick of the sneers at his "low-born" status, he had engaged a genealogist a few years earlier. Present at the baptism—in the role of godfather, no less—was the Earl of Essex. Arthur, the other godfather, had recently moved in to Essex's circle, drawn partly by his saber-rattling Protestantism. Ever since his cousin Francis's treason, Arthur had been trying even harder to prove himself a useful and faithful Protestant servant to the queen.[19]

The Essex connection would grow even more incestuous when Northumberland married Essex's sister in 1594. Meantime—and for the moment, at least—Essex kept Ralegh and Bess's secret as they went about their business as usual.[20] For Ralegh, this meant organizing a privateering raid on Spanish treasure ships, because the English victory in 1588 had not been the end of Spanish-English hostilities. The Spaniards were rebuilding their Armada, and the English sought to cut it off before it matured by destroying Spanish ships at harbor and keeping up their attacks on the source of Spain's wealth: its trade and treasure fleets from South America.

Ralegh was anxiously awaiting favorable winds for the fleet's departure when the queen decided, as usual, that she could not risk his life in such a dangerous venture. He was too enterprising and hardworking to lose. Besides, she would sorely miss his company if anything happened to him. He was by far the most brilliant of her young courtiers. He eventually managed to persuade her to let him escort the fleet past Spain, so that he could direct the captains and ensure that his planned routes were followed. Perhaps, too, he wanted to be away from court to ease his guilty conscience.

Almost as soon as he set sail, his secret broke. The queen was devastated. She had trusted Ralegh completely—and she had trusted Bess, too. She had never suspected for a moment the secret assignations that must have been going on under her very nose. She commanded Martin Frobisher to set sail with orders for Ralegh to return from sea immediately. At the end of May, he was placed under house arrest at Durham House. Bess was not allowed to stay with him. The woman who had publicly unmasked the fairy-tale "romance" with the royal favorite could not be allowed to live in the queen's own house, and Bess was bundled off to some unknown location. Arthur and his wife were interrogated, and everyone waited fearfully for the queen's next move. And while they waited, Essex moved inexorably back into the sun. Elizabeth allowed this other "lover" to bask once again in her courtly favor, just so long as his wife remained safely out of view, banished to the country.

Toward the end of June, Ralegh was hopeful that the storm would pass over him, because not only did the queen allow him to keep his right to live at Durham House, she confirmed the coveted lease she had made over to him six months earlier: the lease to Sherborne, a choice Church property in Dorset with a run-down castle set amid spacious and fertile grounds. It seems that she was waiting for Bess and Ralegh to show some sort of remorse for their betrayal of her friendship and patronage.

But that was not their style. Bess was proud of her love and her husband, while Ralegh felt he had worked so long and hard on behalf of the queen that *he* deserved *her* apology for arresting him. One can admire their self-esteem, perhaps, but not their insensitivity or their strategy. On August 7, 1592, Ralegh and Bess were sent to the Tower.[21] Durham House's gloomy portent had become manifest.

Bess must have been terrified, but her letters from prison show her marshaling her formidable courage and making the best of it. She wrote to one of her friends that she and Ralegh "are true within ourselves, I can assure you." She even had the nerve and insouciance to sign herself "E.R." After all, she was no longer Bess Throckmorton but Elizabeth Ralegh—*Lady* Ralegh if you please. Nevertheless, she could not hide her fear. It was a particularly hot summer, and the plague was so rife in London that all the theaters were closed. Bess was clearly both sick and anxious, wondering aloud: "Who knows what will become of me when I am out [of the Tower]?"[22]

As for Ralegh, he had already tried out the histrionics of the heartbroken lover denied sight of his goddess-queen. Now, in the Tower, he poured out his heart in a frenzy of poetic wounded pride, adding a new dimension to his portrait of Cynthia: she was no longer solely the eternal, bountiful goddess but also the heartless lover, the fickle goddess with the power first to bestow joy and freedom and then to cruelly take it away.[23]

> Despair bolts up my doors, and I alone
> Speak to dead walls, but those hear not my moan.
> …
> The blossoms fallen, the sap gone from the tree.
> The broken monuments of my great desires.
> From these so lost what may th'affections be,
> What heat in cinders of extinguished fires?

After many more poetic images of his anguish at losing Cynthia's love, he alludes to her power to renew and create—and yet, he laments, she "leaves us only woe, which like the moss, having compassion of unburied bones, cleaves to mischance, and unrepaired loss."

The queen apparently never saw these tormented verses.[24] But it was not Ralegh's gift with words that led to his delivery from the Tower; rather, it was his talent for strategy and organization. The raid he had organized on Spain's treasure fleet, and from which he had been so unceremoniously recalled, had arrived back in England with the biggest "prize" of the privateering war: the 1,600-ton *Madre de Dios,* which was carrying over 500 tons of valuable spices, as well as rubies, diamonds, pearls, amber, and musk. Ralegh had written from his prison that he would be needed to sort out problems associated with the venture, including paying the mariners, and he was quite right: the sailors had already begun looting the spoils and selling them cheaply to the eager buyers who flocked to see the colossal treasure ship. (The captured galleon was ten times the size of the queen's *Tiger,* in which Harriot had sailed to America. According to Harriot's calculations, it was 1,457 tons—not the official 1,600 tons—compared with the *Tiger*'s 149 tons.[25]) After a week of chaos, the queen's share of the booty had diminished by some £20,000. By September 15, there was nothing for it but to release Ralegh.[26]

He arrived in Devon to a hero's welcome. He may have been unpopular elsewhere in the country because of his monopolies, wealth, and power, but this was his home territory, and he was greeted with shouts of joy. The mariners, too, cheered at seeing him.[27] He quickly saw to it that the seamen were paid and that their stolen goods were, where possible, recovered. He was also the only one who knew just which investor was owed what percentage. In light of the mess he had been called upon to clean up, his wounded outrage at the imprisonment of such a tireless and loyal servant as he can perhaps be seen as more sober fact than hubris.

While the queen may not have revoked Ralegh's leases and monopolies, in approving the division of the spoils she took her revenge, making sure that Ralegh, the brains and the energy behind the whole thing, came out of it with a slight loss while the other investors made handsome profits. These others included Edward Wright's patron, the Earl of Cumberland, who had joined the raid late in the piece. Yet Ralegh's personal backers suffered the loss of their investments along with his, and he railed against the injustice of it.[28] And all the while, poor, sick, lonely Bess lingered in the Tower. Her baby Damerei's noble name could not undo the misfortune of his star-crossed conception, and he died—likely succumbing to the plague—before he was six months old. While Ralegh pretended that his marriage was unimportant to him, busying himself by "toiling terribly," as Robert Cecil put it, in order to win back the queen's favor, Bess wrote urgent letters to friends who might be able to secure her release.[29] Arthur, too, worked tirelessly on his sister's behalf during the long, slow months that followed.

Finally, on December 22, the effort paid off and she was released. Like Lady Essex, Lady Ralegh was permanently banished from court. Unlike Essex, Ralegh was banished, too. We can infer how furious the queen must have been, determined as she now was to bring down this once-incandescent Elizabethan star. He had paid a devastating price for his love—and love it must have been, for Ralegh knew the risks he ran by marrying Bess secretly and incurring the wrath of his sovereign. With banishment came the free time for a honeymoon of sorts, at Sherborne, but it did not last long. Once Parliament resumed in February 1593, Ralegh spent most of his time at Durham House, desperately trying to show the queen how useful he still could be, and generally feeling very sorry for himself.[30] Harriot never wavered in his loyalty to Ralegh, now that he had been brought to his knees. But he had also been drawing ever closer to Northumberland.

In keeping with his continuing ambition to be more than simply a nobleman and profligate gambler, Northumberland was building a superb library. In 1591, he had employed Walter Warner as a live-in librarian, with a pension of £20 a year. A bold scientific thinker in his own right, Warner soon became Harriot's colleague and loyal friend.[31]

Northumberland was keen to mentor dispassionate, original scholars—those who loved science and mathematics for their own sake, rather than those whose motivation was personal gain, or dilettantes "who have more wares laid out than they have to sell," as he put it.[32] Consequently, he recognized not only Warner's talent but also Harriot's exceptional brilliance, which he acknowledged in an extraordinary way. In 1593, on being admitted to the prestigious Order of the Garter at Windsor Castle, he celebrated by giving Harriot a gift of £80!

Northumberland's generous gift, together with access to his extensive library, was a promise of things to come, and Harriot was beginning to spread his intellectual wings. He was not yet free to fly, though. He was still acting as Ralegh's assistant, and he also gave private lessons on various subjects to occasional students who had heard of his reputation as a teacher.[33] But then, as if Ralegh's troubles were not enough, a new drama unfolded. This time, Harriot was personally caught up in the fray, and all hope of focusing on original scholarship evaporated.

CHAPTER 11

Setback

A T THE END OF 1591, Elizabeth had ratcheted up her attempts to control the illegal entry of Jesuit missionaries, ordering new searches to be made for hidden priests and for recusants who failed to attend regular Anglican services. English Catholics had long resented the loss of their power. Durham House itself symbolized the Protestant appropriation of the Catholic Church's property that had followed the dissolution of the monasteries under Henry VIII. Nevertheless, most resident English Catholics had been prepared to get on with life, holding their masses quietly and attending Anglican services just to keep the peace. This new wave of persecution, however, led radical expatriate polemicists to renew their attacks, conveniently forgetting that until the pope stopped advising Catholics to show loyalty to their Church rather than to their queen, then rightly or wrongly she was going to conflate Jesuit proselytizing and overt dissent with treason.[1] Using the pseudonym Andreas Philopater, one of these angry writers penned a *Responsio* to Elizabeth's anti-Catholic edicts, in which he took a sideswipe at Ralegh along the way.

During 1592 and 1593, the *Responsio* was a bestseller in four languages.[2] An English pamphlet reinforced Philopater's attack on "Sir Walter Ralegh's School of Atheism, and of the Conjuror that is Master thereof." In particular, the author worried that "Ralegh's teacher" would urge him to influence the queen to promulgate an edict denying the immortality of the soul. He also claimed that scholars at this "school" were taught to jest at "both Moses and our Savior, [and at] the old and new Testaments," and that "the scholars [were] taught among other things, to spell God backwards." Aside from the perceived insult of turning "God" into "dog," this last accusation may have had its origin in the cabbalistic number mysticism associated with forming words of power, or secret names for God, from different combinations of letters of the Hebrew alphabet—the kind of "learned conjuring" that Marlowe had evoked in his recent play, *Doctor Faustus*.[3]

The intriguing question is who "Ralegh's teacher" might be. Philopater likely had Harriot in mind as the Conjuror and teacher of Durham House, but Harriot was more interested in applying combination theory to the mathematics of gambling than to number mysticism. John Dee was the other possibility. In 1584, Catholic radicals had targeted him as "an atheist" given to "figuring and conjuring." Not surprisingly, he now feared that he was the supposed Conjuror in Ralegh's "School of Atheism," although Harriot's notes show that he regarded himself as the target. After all, at the time of these attacks he was well known as the live-in teacher at Durham House. Protestant polemicists, too, attacked Ralegh's "school," and Harriot also saw himself as one of the "Mathematicians...harbored in high places, who will maintain it to the death, that there are no devils," as the Anglican writer and pamphleteer Thomas Nashe had put it in his 1592 *Pierce Penniless's Supplication to the Devil*.[4]

The charge of atheism was a handy one to wield against anyone you disagreed with. It was easy to tar with this brush anyone who doubted the literal truth of even one word of the Bible. Neither Ralegh nor Harriot was literally an atheist. On the contrary, judging from the brief written record, Harriot lived his life according to the principle of calm submission to the will of God. But he was more interested in religious injunctions about neighborly love than in fire and brimstone, while, as we've seen, his refusal to characterize the Algonquian religion as devil worship contrasted with the attitudes of more orthodox travelers such as Arthur Barlowe and Father Acosta.[5] There also seems little doubt that Harriot had enjoyed freethinking conversations with Ralegh and his circle—conversations along the lines outlined in the testimony of the Anglican vicar Ralph Ironside. In mid-1593, Ironside and Ralegh, along with Ralegh's brother Carew, were among the guests at a dinner party not far from Sherborne. Toward the end of the evening, Carew was feeling so relaxed by the wine that he abandoned all caution and began asking provocative questions of the two vicars present. When he challenged them on the nature of the soul, Ironside could not hold back: "Better it were...that we would be careful how the souls might be saved than to be curious in finding out their essence."[6] This was too much for Ralegh, who requested a proper answer from Ironside.

Of course, like everyone else Ironside did not have a clue about the composition of such a numinous, hypothetical construct as the "soul." Instead of admitting this, he infuriated Ralegh by giving the standard definition that "the reasonable soul is a spiritual and immortal substance breathed into man by God, whereby he lives and moves and understands, and so is distinguished from other creatures." Ralegh persisted by asking what *was* that "immortal substance breathed into men?"—and Ironside had no recourse but to give a circular answer. A frustrated Ralegh asked for a better argument, to which Ironside retorted that "such demonstrations were against the nature of a man's soul." In other words, he added, the senses detect substances that are "sensible," while "man's soul, being insensible, [is] discerned by the spirit." Ralegh actually agreed with this time-honored argument, as his writings show. It was Ironside's repeated use of circular logic that infuriated him, as Ironside's testimony reveals when examined in the calm of hindsight.[7]

In the early 1590s, however, times were anything but calm. Rumors about the dinner-party debate flew around the countryside, and Ralegh's supposed atheism and irreverence grew more extreme with each telling. Although his writings suggest that his attitude toward religion was quite orthodox by today's standards, what Ralegh loved to do was to debate. Many Elizabethan intellectuals enjoyed speculating about metaphysics, discussing unknowable topics such as the nature of God and the soul. It sharpened their theology as well as their logic.

Another scholarly debate about religion, between two unnamed participants, was recorded anonymously the following year. The skeptical interlocutor has been plausibly but not certainly identified as Harriot.[8] Either way, the discussion gives insight into the kinds of genuine questions being asked by intellectual freethinkers in the early 1590s. For instance, how could God be *both* the First Cause—the immaterial instigator of everything in the material universe—*and* an intelligent, omnipotent being? Surely intelligence is a material quality, so God cannot be both First Cause and intelligent. He cannot be omnipotent, either, if he is intelligent and therefore limited by materiality—and if he *were* omnipotent, then why did he allow the war in heaven between the angels? Furthermore, "how can angels read, or see, or reason, or talk if they do not have the same organs as men?" The questioner also posited the notion that there were no biblical miracles witnessed these days, and indeed why believe in them to begin with? After all, he asked, "Isn't all true knowledge physical or experimental?"

Harriot generally preferred mathematics and experimentation to speculation, which both supports and undermines the possibility that he is the questioner here. The emphasis on experiment supports it, but generally he was not given to speculation about things he accepted as self-evident—including the existence of God, made manifest, it seemed to almost everyone at the time, in the very existence of the world. On the other hand, it is quite possible that he did ponder the nature of God, angels, and "separated souls" in conversations with friends such as Dee. He certainly read treatises by supporters of angels and demons, notably the German Benedictine abbott and magus Johannes Trithemius, who died in 1516. Trithemius had also been a collector of old Roman manuscripts containing classical shorthand scripts; apparently he wanted to use them for his secret spells and invocations, although Harriot was more likely to have been interested in these codes as a model for his phonetic alphabet for representing the Algonquian language.[9]

Tantalizing as it is, to posit Harriot as the unknown skeptical interlocutor in the conversation on angels, more certain evidence of his freethinking on religious matters lies in some brief notes he made on the ancient Skeptical philosophy of "doubt" and in his unorthodox consideration of the doctrine of atomism. His friend Nathaniel Torporley would make notes a decade or so later on his and Harriot's debates on this latter topic.[10] (No one knows how or when Torporley met Harriot, but he was one of the better mathematicians in the Northumberland circle.) In these later conversations, and in his own later working manuscripts, Harriot appeared to support the ancient doctrine *ex nihilo nihil fit:* "from nothing, nothing comes." This saying was widely, although not uniquely, associated with the ancient atomists Democritus, Epicurus, and Lucretius, who believed that matter was created from primordial, eternal atoms (that is, from something rather than nothing).[11] Not surprisingly, this "pagan," pre-Christian doctrine was anathema to Christian Creationists—including the Reverend Torporley.

In fact, in 1590, the Anglican theologian Richard Harvey had attacked those who denied the immortality of the soul and who believed the world was eternal rather than divinely created in six days. Harriot would indeed consider the idea of an eternal universe, in the sense that its atoms were eternal; meantime, he believed he was one of those "jugglers" attacked by Harvey. (Harriot was relieved to note, though, that Harvey's brother Gabriel, an academic and pamphleteer, had defended "mathematical practitioners" in his response to Nashe's attack on "mathematicians who maintain there are no devils"; Gabriel named Thomas Digges, Harriot, and Dee as "profound mathematicians.")[12]

While Harriot would later consider the idea of eternal atoms, he would also dabble in biblical chronology (as would Newton more than half a century later).[13] Some of this was research for Ralegh, who was by then composing his *History of the World*, his unfinished and magisterial work, but perhaps Harriot had not made up his own mind about the Creation idea. In one of his manuscripts he doodled "in the beginning God made heaven and" in his Algonquian-inspired phonetic script.[14] This was most likely just idle play, although the omission of "earth" could suggest Harriot's belief in an eternal world. Of course, he might simply have been interrupted before finishing, but regardless of whether he believed in a moment of Creation or not, Harriot saw himself as a faithful Christian. As for *ex nihilo nihil fit*, he wrote this in his work on the nature of matter and the mathematics of infinity (of which more later)—it had nothing to do with the kind of theological argumentation engaged in by Ralegh at the dinner with Ironside or by the unknown and freethinking interlocutor. But it does show that Harriot was not constrained by religious orthodoxy in his thinking about the nature of reality. Although he did not publish these ideas, he did discuss them with others, as Torporley's notes show, and word got around, so that despite his personal faith he would be forever tagged as an atheist espousing the heretical *ex nihilo* doctrine against the biblical version of Creation.

Harriot's reputation for atheism was such that even the astrologer Simon Forman attacked him. Forman claimed to have devised a secret method for calculating longitude at sea. A reliable method for longitude would have been a great boon, but Forman gave no such account. What he did say was that he had been encouraged to publish a defense of his mystical method (in 1591) in order to silence those "counting themselves great and cunning [clever] Clarkes." These included "the unbelieving S[t] Thomas," quite likely referring to Harriot, who included Forman's book, along with Harvey and Nashe's, in a list of published references to himself. His mathematical colleague Thomas Hood may also have been Forman's "doubting Thomas": Harriot did not record his own doubts about Forman's longitude claim, but Hood publicly dismissed it.[15]

While Forman's attack was motivated simply by anger at Hood and Harriot's rejection of his longitude method, many religious folk found mathematics and experimental science so mysterious that they linked them with magical codes and spells. Perhaps it didn't help that Dee engaged in both kinds of pursuit, but Harriot had become a special target in the murky world of heretic hunting. His intellectual standing meant that he was regarded as a bad influence on Ralegh, rather than vice versa.[16] He was even dragged into an investigation into Christopher Marlowe's religious beliefs. In his testimony of May 12, 1593, a certain Richard Baines claimed that Marlowe had said that "Moses was but a juggler, and that one Heriot, being Sir W. Raleigh's man, can do more than he." (Machiavelli had claimed that Moses's miracles were just "juggling tricks" designed to keep the superstitious Hebrew masses in his power. Presumably Marlowe thought that Harriot's command of mathematical science was equally awesome to the uneducated people of his own day.) Baines also claimed that Marlowe had said that Christ was a bastard and "that all those that love not tobacco and boies [boys] was fools."[17] Ralegh and Harriot had certainly helped make tobacco fashionable, so it was hard not to read into Baines's statement a link between Marlowe and the Durham House circle. (Scholars are still debating the extent of the connection.[18])

Then, on May 30, Marlowe was killed—in a dining-house brawl whose cause is still a mystery—before his case could proceed to trial.[19] But the rumormongers' tongues continued to wag, and by March 1594, "atheist" stories about Ralegh and Harriot reached such a crescendo that a new enquiry was held at Cerne Abbas in Dorset (the county in which Sherborne was located). This was an enquiry into heresies espoused by Dorset denizens, including the "impious opinions concerning God...of Sir Walter Rawleigh, Mr Carew Rawley, and one Heryott [Harriot] of Sir Walter Rawleigh's House."[20]

The evidence presented was all hearsay. Several witnesses singled out Harriot in particular, saying they had "heard" that he held such "atheistical" opinions as denying the resurrection.[21] Ironside chose not to give evidence in person, but his notes against Ralegh were tendered. With no firm evidence, the investigation was dropped. But Harriot would never again be free of the slurs, or the shadows surrounding Durham House.

THE STRESS CAUSED BY THE atheist rumors and the Cerne Abbas enquiry sidetracked Harriot from his mathematical research on combination theory and navigation. He had earlier advised his friend Robert Hues on the mathematics of rhumb lines, which Hues included in his bestselling 1594 book (*Tractatus de globis*) on the uses and underlying mathematics of the Molyneux globes. In acknowledging his friend, Hues looked forward to the publication of Harriot's own "expected tract" on rhumb lines, but it was not only Harriot's quest for a perfect loxodrome solution that led him to lay aside his "expected tract" and his *Doctrine of Nautical Triangles Compendious*: the enquiry also played a part.[22] The increasingly insecure Ralegh responded to the drama by desperately trying to prove his orthodoxy, joining in the local search for Jesuits secreted in noble Catholic households. In questioning one such priest, Father Cornelius, Ralegh famously became so impressed with the man's sincerity that he tried to save him from execution. He could not have it both ways, though: he had helped to capture the priest, but he no longer had power in London, and now, in Dorset, he was forced to oversee Cornelius's eviscerating traitor's death.[23]

Still tormented by his banishment from court, Ralegh soon thought up a better way of redeeming himself with the queen. He had long been inspired by Spanish stories of an undiscovered El Dorado in the remote, mysterious land of Guiana, and now he decided it was time for him to stake a claim for England in gold-rich South America.[24] With administrative help from Harriot—and from Lady Ralegh, who hated to see her husband so restless and unhappy in his exile from court—he managed to obtain investment for his new venture from Lord Admiral Charles Howard and also from Sir Robert Cecil, who had recently become the youngest-ever privy councilor. Cecil was the son of Elizabeth's trusted chief minister and Lord Treasurer, William Cecil (Lord Burghley). Two years before Burghley died in 1598, Cecil, at the age of thirty-three, would take on the post of secretary of state.[25] He was a man on the rise, but the Raleghs trusted him, and Howard, too.

Ralegh sent out a reconnaissance party to ascertain the lie of the land, especially any Spanish defenses near the mouth of the Orinoco River on the northeast coast of South America. This was his proposed route into Guiana (which today consists of Suriname, Guyana, French Guiana, and the Guiana Highlands of southeast Venezuela and northern Brazil). In September 1594, in a letter keeping Robert Cecil informed of his efforts, Ralegh added a poignant postscript: "I had a post this morning from Sherborne. The plague in the town is very hot. My Bess is one way sent, her son another way, and I am [greatly troubled]." (The Raleghs now had another son, eleven-month-old Walter, or Wat, as his parents called him).[26]

In London, the theaters had been closed for two years because of the plague, but the emergency eventually eased, and they had reopened by November, when Marlowe's *Tamburlaine* and *Doctor*

Faustus were revived. The Lord Mayor declaimed against the plays' "lascivious devices, shifts of cozenage and matters of like sort." He was worried that theatergoers would "imitate" the "said lewd offences." His counterpart the following year would be equally appalled when Shakespeare's *Romeo and Juliet* and *Richard II* were performed. These were also deemed to be "nothing but profane fables, lascivious matters, cozening devices, and other unseemly and scurrilous behaviors." This Lord Mayor, too, believed that drama should encourage the "avoiding of those vices which they represent."[27]

Ralegh and Harriot had little time for the moralizing of Puritans, and for the moment little time to spare for the theater, either. As Ralegh's exhilaration grew at the prospect of sailing into the New World at last—now that the queen no longer wanted him by her side—Harriot had been updating his *Arcticon* for Ralegh's new captains.[28] The fleet set sail on February 6, 1595. There were four ships, two boats, and some three hundred men. Captaining one of the boats was the former Oxford fellow Lawrence Keymis, who had abandoned mathematics for the excitement of Durham House and the thrill of "discovering" the mystical Guiana.

Harriot, by contrast, was becoming ever more interested in mathematics and physics for their own sake. He was back in form after the anxiety of Cerne Abbas and had so many new ideas to pursue, as well as research projects to finish—including his analysis of the loxodrome, his work on interpolation, algebra, and the mathematics of chance. With a life as interrupted as his, it was hard to focus on one topic at a time, let alone publish his findings.[29] Yet while his and Ralegh's interests were diverging, their friendship remained undimmed. Harriot had seen Ralegh off as the fleet left for Guiana, and he also oversaw the business side of the venture while Ralegh was away.

This latter responsibility came at a cost. For some years the wealthy merchant William Sanderson had overseen Ralegh's maritime investments. In particular, he had acted as Ralegh's representative in the formal negotiations over the distribution of profits from the *Madre de Dios*. Sanderson was a few years older than Ralegh, but he had married Ralegh's niece in the early 1580s and had supported Ralegh's North American ventures and those of his half brothers Humphrey and Adrian Gilbert. He had also financed Molyneux and his globes. Sanderson had raised nearly £45,000 for Ralegh's Guiana venture, and a couple of weeks before the fleet departed he had collected all the relevant accounts of moneys raised and spent and given them to Ralegh and Harriot to audit. As was usual in business at that time, some of this money was Sanderson's, some of it was borrowed in his name, and some he had raised from others, for which he and Ralegh stood bonded. With so much credit at stake, and knowing the risks of sailing into uncharted territory, Sanderson wanted to protect himself against financial liability in case Ralegh didn't return. In turn, Lady Ralegh wanted to ensure her own protection from claims by Sanderson. Suitable documents were drawn up, but just as he was about to set sail, Ralegh ordered that the document indemnifying Sanderson be given to Harriot for safekeeping. Sanderson was incensed at this lack of trust, and at the fact that his insurance, together with all his accounts, was in Harriot's hands rather than his own. He turned his back on Ralegh without so much as a "bon voyage," and Harriot was left with even more responsibility for Ralegh's affairs.[30]

At the same time, Harriot began to audit Northumberland's accounts and help with his various lawsuits.[31] (Northumberland's household accounts show that he was often in court over such things as tenants encroaching on his woods or enclosing commons, as well as pursuing—or being pursued over—loans.[32]) In return, Northumberland gave Harriot a lifetime grant of land in the county of Durham, which meant that Harriot could keep the rent from the land. Northumberland had estates in eight counties in England and Wales, which brought in an average of more than £6,500 a year, making him one of the wealthiest men in England.[33] He also gave Harriot the tools needed for original research: the run of his library, and the chance to converse with like-minded mathematicians such as Warner—as well as with Northumberland himself, whose interest in science continued unabated. Over the next few years the "Wizard Earl" would do his own experiments in order to find out what happens when two billiard balls collide and to investigate the "impetus" of water and other falling bodies—and no doubt he was still interested in the practice and mathematics of gambling, too.[34] Harriot would make original contributions to all these topics and more in the coming years.

His commitment to mathematics and physics did not waver even when Ralegh returned from Guiana with further fantastic tales of El Dorado and the promise of gold hidden so deep in the jungle that no European had been able to claim it. Indeed, Guiana still had "her maidenhood," as Ralegh put it in his 1596 account of his adventure, *The Discovery of the Large, Rich and Beautiful Empire of Guiana.* The country was "never sacked, turned, nor wrought, the face of the earth has not been torn,…the graves have not been opened for gold, the mines not broken with sledges, nor their images pulled down out of their temples. It has never been entered by any army of strength, and never conquered by any Christian prince."[35]

Despite the sexual metaphor "maidenhood"—and despite the hundreds of beautiful young women who "came among us without deceit, stark naked"—Ralegh was quick to point out that, just as in Virginia, he did not tolerate his men behaving like those Spaniards who had raped both women and land in South America. (Recent research suggests that it is almost certain that Columbus's men brought syphilis to Europe.)[36] Ralegh did allow himself to muse on the beauty of one particular woman he met. Not only was she the "best favored" woman he had ever seen, but also "she stood not in awe of her husband, as the rest [did], for she spoke and discoursed, and drank among the gentlemen and captains, and was very pleasant, knowing her own comeliness, and taking great pride therein." Aside from the color of her hair, she reminded him "of a lady in England"—presumably his Bess: she loved him, but she was too independent-minded to be in awe of him. If anything, it was the other way around, as Robert Cecil and his friend Lord Henry Howard would soon delight in pointing out.[37]

For all the romance of El Dorado—*Discovery of the Large, Rich and Beautiful Empire of Guiana* would become a bestseller and be translated into several languages—Ralegh had not yet found any gold. He had a bigger plan than mere quick plunder, though: not conquest but cooperation between the English and the Guianans. This was partly a savvy plan to beat the Spanish to the legendary gold-rich territory by forming an alliance in which the Guianans would be on an equal level with the English, who would protect the country from other invaders in return for limited access to land for colonization and mining. The plan also reflected Ralegh's determination to win access to new resources—and new Christian converts—in a relatively humanitarian way.[38]

In the end, though, it was his gift for naval strategy rather than his dreams of empire that finally brought Ralegh back into the queen's favor and back into court. First, he played a brilliant role in the sacking of Cadiz in 1596—the latest bout in the ongoing, half-declared war with Spain. Essex and Lord Admiral Howard had led the fleet, but Ralegh's advice on where to land safely had proved crucial in the victory. The following year, Ralegh, Robert Cecil, and Essex organized another raid. The venture, led by Essex, was a failure, due to the fleet being scattered by storms near the Azores and to subsequent miscommunication. Stranded at Fayal under sustained Spanish fire, Ralegh successfully attacked and sacked the town, in direct contravention of Essex's earlier orders. It was the major "success" of the campaign, but Essex tried to have Ralegh court-martialed; instead it was his own reputation that suffered. With his decline, Ralegh's exile ceased. In June 1597 he returned to his old post as captain of the Queen's Guard. And yet, although he was soon given access to the queen in her privy chamber, Ralegh was never again the brightest star for Elizabeth. Still, she made over the lease to Sherborne, the home he and Bess had rebuilt and had come to love, so that now he owned it outright.

Harriot had no interest in the shifting sands at court and was not seduced by the lure of South American gold, although he certainly wished Ralegh well. To help shore up support for Ralegh's further exploration of Guiana, Harriot wrote to Cecil, telling him what he had gleaned about the Orinoco River from reading Acosta's *History of the Indies.*[39] But his days of teaching for Ralegh were coming to an end. He would continue to take care of Ralegh's accounts, and he would still do the occasional navigational work for him, such as drawing up a scale map from geographical details brought back from Guiana. (In telling Cecil about the map, Harriot added, "I think it is of great importance to keep that which is done as secretly as we may, lest the Spanish learn to know those harbors and entrances, and work to prevent us."[40]) Although he retained an interest in the New World, Harriot yearned to go deeper into the foundations of pure mathematics and physics—to discover "new worlds" in numbers and the laws of nature rather than through geographical exploration.

And so, by 1597, he had sold the Irish abbey Ralegh had given him and, with Ralegh's blessing, had taken rooms in Syon House at Northumberland's estate on the Thames, eight miles west of London. Harriot still kept his room at Durham House; he and Ralegh would always remain close. He seems to have got on well with Bess, too, and with her brother. Arthur recorded a number of visits from Harriot, and at one stage, he gave Harriot a gift of "three yards of black satin, at 12s a yard." Harriot was so fond of black—the mark of the intellectual— that when Ralegh made a will in 1597, he bequeathed to Harriot all his own black suits. (In addition he left him a gift of £200 and an annuity of £100, along with all his books and some of his furniture.) The will also stipulated that most of Ralegh's estate was to be bequeathed not to his wife but to their son, Wat: Elizabethan women could not own property. They could inherit and use the income from property (and Ralegh's will provided for Bess in this way), but they were not free to own or dispose of it without their husbands' consent. Until Wat came of age, Ralegh's will specified, his estate should be managed by Bess's brother Arthur, two cousins, and Harriot, whom the Raleghs evidently trusted as if he were family.[41]

Still, Northumberland was impossible to resist. The following year, he freed Harriot from administrative duties by offering him a lifetime pension of £80 a year (four times Warner's pension). In doing so, Northumberland gave Harriot free rein to follow his thoughts, and he took full advantage of this freedom.

CHAPTER 12

Royal Refraction

THE FIRST TOPIC HARRIOT CHOSE to pursue at Syon was optics, particularly the nature of the reflection and refraction of light, which interested him solely out of scientific curiosity. It was also a topic with a long history. Most ancients spoke not of light rays but of rays of vision—it was thought that the eye sent out rays that illuminated objects. This idea was decisively overturned by the eleventh-century Arab scholar Abu Ali ibn al-Haytham, one of the greatest of all medieval scientific thinkers, who argued that light must be emitted or reflected from an object in order for our eyes to see it. The *nature* of light, however—and whether it resided within each visible substance—was still a mystery. Nevertheless, the ancients had known that when a ray, be it a ray of vision or of light, hit a reflective surface at an oblique angle, it was reflected back at the same angle (but in the symmetrically opposite direction, as in figure 16a).

In his *Optics*, Ptolemy had made a serious attempt at finding a similar law of refraction—a quantitative law describing the way light is bent at the surface of a transparent medium, such as when you step into ankle-deep water and notice that your foot is not quite in line with your leg. It is as though the light rays have slightly changed direction in the water, although why this should happen is much less apparent. In fact, for two thousand years scholars had been unable to quantify the amount of bending in this intriguing everyday phenomenon, let alone explain its cause. Ptolemy had taken some of the earliest steps when he measured various angles of incidence and refraction (defined in figure 16a), and from these measurements, together with a number of geometrical diagrams, concluded that there was "a definite quantitative relationship" between the angle of incidence and the corresponding angle of refraction. By slightly tweaking some of his experimental data, he even managed to fit a reasonably accurate regression formula to successive angles of refraction. In the end, though, the only correct relationship he deduced was qualitative: light is bent more sharply in an optically denser medium.[1] Eight centuries later, not even al-Haytham—whose name was latinized as Alhacen or Alhazen in Harriot's time—had been able to uncover the elusive law.[2]

Recently discovered manuscripts by al-Haytham's predecessor the tenth-century Baghdad scholar Ibn Sahl, show that in his analysis of lenses he used a geometric construction that gives, in hindsight, the correct law of refraction. It was a brilliant analysis in "geometric optics"—the drawing of diagrams showing incident rays and the estimated subsequent paths of reflected and refracted rays—and it was the first mathematical study of the way lenses reflect and refract light. Perhaps somewhere there is another, undiscovered manuscript detailing how Ibn Sahl arrived at his construction. In the extant manuscripts, however, his "law" arose as a purely geometric exercise: he was trying to show how to construct the best geometric lenses for focusing light for burning—he was not looking for a law of refraction, and he did not refer to experimental measurements of angles of incidence and refraction or generalize his result to all refractive media. In other words, on the available evidence, he did not deduce from observation a general law of refraction, and nor was this his goal.[3]

By Harriot's time, still no one had been able to abstract the amount of bending into a satisfactory universal law of nature. In fact, there were very few such laws known at that time. The ancient laws of light reflection and of levers and floating bodies (the last two due to Archimedes), Euclid's geometrical laws of flat space, and the fact that the intensity of light diminishes according to the square of the distance from the source were about the sum total.[4] Of course, there were hypotheses about the way nature works, such as Ptolemy's and Copernicus's models of planetary motion, but these were not precise enough to provide general mathematical laws. Kepler would change this situation in 1609 when he formulated his quantitative laws of planetary orbits, but meantime he was still sifting through Tycho's astronomical data.

Kepler was also exploring optics. In particular, like Tycho and almost everyone working in optics, he wanted to understand refraction so that he could solve the problem of (what we now call) atmospheric refraction.[5] Harriot had mentioned this problem to his navigation students, too. Just as refraction displaces your foot in the water, so the line of sight to a star is displaced by atmospheric refraction. One way of noticing this was that astronomical measurements taken during different weather conditions differed from one another. Today, we know that this happens because light is bent when it passes through different layers of the atmosphere, each layer or region having a different density. We also know that air temperature, humidity, pressure, and temperature gradient all play a part in the way atmospheric density changes with height above the earth. Consequently, each time the light from a star moves into a denser region, its direction of travel is bent a little more. But we judge a star's position by a direct line of sight—our eyes are not aware of all these subtle bendings—and so we misjudge its true direction. In the late sixteenth century, however, the nature of the atmosphere was a mystery, though it had long been surmised that light was being refracted somehow during its passage from the stars to observers on earth.

In the 1590s, Tycho was making new estimates of the actual amount of this refraction, which he did by comparing measured positions of stars with those calculated from the geometry of the celestial sphere.[6] Some scholars, including Tycho, also realized that atmospheric refraction means that we see the sun setting after it has actually set: what we see on the horizon at sunset is refracted rays bending up from below the horizon. Similarly, we see the sun rise earlier than it actually does. It turns out that atmospheric refraction also explains the twinkling of distant stars. Some early peoples, notably Torres Strait Islanders, had observed that changes in the weather could be predicted by changes in this twinkling.[7] It was a remarkable observation, and we now know that refraction is the key to why it happens. Changes in the weather bring changes in moisture and air pressure, and therefore in density, which lead to changes in atmospheric refraction. During turbulent conditions associated with dramatic weather changes, the amount of refraction oscillates as the different layers of atmosphere swirl about, so the apparent direction of the star's light oscillates, too. This means that the light itself appears to zigzag subtly, giving the appearance of twinkling. When the air moves strongly enough, twinkling stars can appear colored—just as a dewdrop refracts different colors when you look at it from slightly different directions.

Harriot would later discover why this change in color happens. In the meantime, in the absence of detailed knowledge of the atmosphere, a number of suggestions had been made to explain why this celestial refraction was happening at all. In the thirteenth century, the Polish scholar Witelo had conjectured that "vapors" rising from the ground caused refraction near the horizon. More generally, though, atmospheric refraction was considered to arise when light moved from the crystalline heavenly orbs to the rarefied heavenly "ether" that was presumed to fill space, and finally to the dense air and "vapors" above the earth. Aristotle had posited this hypothetical "ether" as a fifth "element," in addition to fire, earth, water, and air; for Christians, such a pure celestial substance fitted nicely with the idea that the heavens were "divine" and eternal, as opposed to the generation and decay that characterized earthly matter. So when one or two intrepid astronomers suggested that refraction, being an earthly phenomenon, "proved" that the "heavens" were made of the same stuff as the earth's air and terrestrial "vapors," it was a controversial topic.[8] Tycho, for one, disagreed, as would Kepler, and as had al-Haytham. They thought astronomical refraction simply "proved" that Aristotle's heavenly "ether" was of a different substance from the earthly atmosphere.[9] Neither of these opposing conjectures was correct in detail, but one thing that everyone agreed upon was that refraction occurs when light passes from one transparent medium to another of different density.

This, then, is what Harriot focused on. Earlier, he had advised his navigation students that in a relatively short voyage across the Atlantic, the astronomical inaccuracies due to refraction were not worth worrying about and that he would discuss the topic elsewhere.[10] It seems that he never got around to it. Rather, while everyone else was speculating wildly on the underlying cause of atmospheric refraction, he focused on unraveling the nature of refraction itself. In the scholarly atmosphere at Syon, he began his research by reading what al-Haytham and Witelo had had to say on the subject. Al-Haytham was still the master of optics; he had absorbed and gone beyond Ptolemy's *Optics*, while Witelo had closely followed al-Haytham. Harriot read Friedrich Risner's 1572 *Opticae thesaurus*, a hugely influential volume combining both al-Haytham's *Book of Optics* (*Kitab-al-manazir*, translated into Latin as *De aspectibus*) and Witelo's *Perspectiva*. Unbeknown to Harriot, Kepler would soon begin reading Risner's edition, too, while a decade or so later, Willebrord Snell would read Risner's 1606 *Opticae libri quatuor*. All three men were about to make breakthroughs in the field of optics.[11]

Harriot also read the most up-to-date treatise on refraction, Giambattista della Porta's 1593 *De refractione*. Kepler wanted to read it but could not find a copy.[12] Della Porta, whose translator had dedicated an earlier work to Northumberland, was already famous for his encyclopedia of natural magic—and, as Dee's story shows, key tools of the "natural magician" were curved "magic" mirrors and crystal balls, which della Porta discussed in both his encyclopedia and his *De refractione*. Modern scholars disparage most of his work on refraction, but Harriot did not have the advantage of hindsight. Instead, he set out to test the unsubstantiated conjectures in *De refractione*. He did this by extracting a relationship between angles of incidence and refraction from della Porta's geometric drawings of his hypothetical paths of refraction in a spherical crystal ball (just as modern scholars have extracted the law from Ibn Sahl's diagrams). Then he compared its predictions with his own measurements of refraction between air and glass. He found della Porta to be wrong— just as he found Witelo's tables of refraction to be inaccurate. (In fact, Witelo had just copied Ptolemy's tables.[13])

Harriot's first measurements were taken at Syon on August 11, 1597, but in modern scientific style, he also took them the next day, and again on August 21. He would repeat them over the next few months and years. He carried out his earliest observations by immersing a staff in water and marking off the position of its image when viewed through the water. You can get the idea by immersing a straw in a flat-bottomed glass of water. Hold the straw at an angle, and if you look side-on at the top of the water, you see the immersed part of the straw bent upward toward the water surface. Put your finger on the outside of the glass, level with where the bottom end of the refracted image of the straw appears to meet the side of the glass: this point is marked *f* in Harriot's diagram, shown in figure 16b (adapted from his manuscript BL Add MS 6789 f406r). Then, looking from the outside of the glass, you will see that your finger points to a spot that is higher than the actual lower end of the straw, which is marked *d* in Harriot's diagram. Harriot marked off these points carefully and repeated the process as he changed the angle of the staff.

After Harriot had taken measurements of the positions of his refracted staff, he used them to calculate, from trigonometric tables, the various pairs of angles of incidence and refraction. In other words, he did not measure his angles of incidence (i) and refraction (r) directly but calculated them from his measurements of the image of his staff, via the trigonometric ratios $\sin r = \dfrac{bc}{bf}$ and $\tan i = \dfrac{bc}{cd}$ (as in figure 16b).[14] It was an elegantly simple experimental design. Nevertheless, as you will notice by experimenting with a straw, it takes a skilled and patient experimenter to take reasonably accurate measurements giving one pair of angles, let alone for the large number needed to uncover the law of refraction. That is likely why few, if any, of Harriot's peers had studied refraction in this meticulous way.

This may seem surprising, because empirical evidence is the foundation of modern science, and people had been aware of its importance for many centuries. In particular, al-Haytham had suggested sophisticated experimental designs for measuring reflection and refraction. However, like other so-called fathers of the experimental method, including Roger Bacon and his later namesake Francis Bacon, it appears that al-Haytham did not carry out rigorous, repeated experiments in a systematic way. Not that these experimental "fathers" didn't try: al-Haytham did manage to experimentally rule out Ptolemy's idea that the angle of refraction has a direct, essentially proportional relationship to the angle of incidence. He also explored lenses and mirrors, the nature of the eye, the optical illusion that makes the moon appear larger near the horizon, and more, while 250 years later Roger Bacon wrote on lenses, gunpowder, astronomy, and anatomy. On the surviving evidence, though, most of these and other pioneering "experimental" authors generally discussed "thought experiments" rather than rigorous, repeated practical ones.[15]

Religious philosophy no doubt played a part in this. In the fifth century St. Augustine had popularized a Neoplatonic Christian philosophy in which experimentally tinkering with nature was frowned upon. Divine inspiration alone was regarded as the way to "true" knowledge. Plato himself had preferred intellectual speculation to experiment not because of God but because he felt that the senses deceive us as to the true nature of things. Of course, he was half right. His student Aristotle had disagreed, although he spoke of everyday empirical observation rather than rigorous experiment. Sixteen hundred years later, the Franciscan friar Roger Bacon's attempts to discern the truth from nature rather than scripture had led to his imprisonment and the banning of his books in 1277. Medieval Islamic science, too, had often tended to focus on contemplation of the divine, via mathematics and a search for conceptual unity, rather than reductionist experimental quantification.[16] So, although known if isolated attempts at "modern" experimental deductions date back at least to Archimedes, Harriot's contemporary and countryman Francis Bacon was going against tradition when he publicly espoused an experimentally based philosophy in the early 1600s.

Bacon was genuinely interested in such scientific questions as the nature of matter and the cosmos and wrote a number of works on these and other topics; nevertheless, because he was a lawyer and Member of Parliament by day, his focus was political rather than strictly scientific. He saw science as a way of literally wresting control over nature and using such knowledge in the service of the state: "knowledge is power," as he said.[17] To facilitate this goal, he believed that the foundation of science needed an overhaul to free it from the Aristotelian and Neoplatonic past. It seems that Harriot did not read his work, although in 1608, Bacon would note privately that he was inspired by what he had heard of Harriot, Northumberland, and Ralegh, who were, he said, "already inclined to experiments."[18]

Bacon's "inclined" perhaps sums up the situation in Harriot's time. It is not that people didn't experiment. It is more a question of timing and purpose. Harriot's "inclination" was first and foremost to experiment, rigorously and repetitively, then to describe his results mathematically. In other words, he wanted to understand the *what* before asking *why*. For many of his forerunners and contemporaries, the process was reversed: they felt such a need to understand why a phenomenon happened that they speculated rather than experimented.

This relative "lack of inclination" to experiment among Harriot's peers and forerunners meant that as far as refraction went, Ptolemy's fifteen-hundred-year-old tables were still authoritative. Even Kepler relied on Witelo's version of them when he began searching for a law of refraction in the early 1600s. Some of Ptolemy's data was inaccurate, as noted, but Kepler thought it good enough to lead him to an idealized law. He, too, was influenced by Neoplatonic Christian philosophy.[19]

As for Harriot, it took much more than a few sets of careful experimental results to find the law of refraction, as his hundreds of pages of notes on the topic testify. He rarely dated his manuscripts, which were riffled through and left in disarray after his death; his pages and pages of diagrams, calculations, and tables—some rough, some beautifully drawn up—are all jumbled together, so that it is virtually impossible to know exactly how he proceeded from his first measurements in 1597. Key to his process, however, was finding many pairs of angles i and r, including a new set of measurements that he took sometime around 1600 or 1601 using a new, more accurate method. Instead of immersing a staff in water, he immersed his astrolabe.[20] This was a circular apparatus for taking astronomical observations: a scale was marked around the outer edge of the circle, as on a protractor, and when the star was viewed through sighting holes on a movable inner circle, its angle above the horizon could be measured from the scale. Harriot turned this instrument to a very different use in his study of refraction. When he marked off the images of the scale marks on the immersed part of the astrolabe, he found that they appeared higher in the water than they actually were—just as the immersed end of a staff or straw appears bent upward, so that its refracted image is higher in the water than the actual straw.

When he plotted his observed refracted images of these scale marks, he found that they lay on a smaller, concentric circle inside the circle of the actual astrolabe scale. In fact, the same construction is embodied implicitly in his immersed staff diagram (figure 16b). Imagine that the bottom end of the staff lies on a circle of radius bd, while the bottom of the refracted image lies on a circle of radius bf. Harriot's astrolabe experiments had shown that for any angle of incidence, all the refracted images lay approximately on the same circle, which would be the one with radius bf in figure 16b. In other words, the ratio of the radii bf and bd would be the same for every possible pair of angles i and r. Harriot would surely have realized immediately that this ratio is equal to $\dfrac{\sin i}{\sin r}$ (as you can see in the endnote[21]).

Whatever his reasoning, Harriot had found the law of refraction by 1601. What it means is that if i is the angle of incidence in water and r the angle of the same light ray refracted into the air (as in figure 16b), then for any choice of i, you can find the corresponding r without the need for measuring but simply by solving the equation $\dfrac{\sin i}{\sin r} = k$. In the case of Harriot's measurement of light moving from water and being refracted into the air (and into the observer's eyes), the constant, k, is called today "the index of refraction" of air with respect to water. It needs to be determined from experiment, so a number of pairs of both i and r have to be measured accurately first, before the law can be used. Harriot's law was completely general. He experimented with refraction in different media—such as light refracting through a glass prism rather than water—and in each case, the ratio of sines was constant, although this constant value was different for each different medium. In other words, Harriot found that each different pair of refracting substances has its own refractive index.

He wrote up his experimental results in a number of calculations and diagrams based on his astrolabe experiments. Chief among the diagrams was his impressively titled *Regium*, indicating Harriot knew that he had found something special, something royal.[22] He scaled his drawing so that his larger circle, the astrolabe scale in air, had a radius of 1,000,000, while the radius of the smaller refracted image circle was 748,955. Nowadays we use decimal point notation, so today (following Leonhard Euler in the eighteenth century), Harriot's larger radius would be scaled to 1, and his smaller one would be 0.748955. This scaling means that the index of refraction k can be read directly from the inner circle, and Harriot's value of 0.748955 is an excellent determination of the refractive index of air with respect to water.

Harriot found refractive indices for other pairs of media, as mentioned, but his *Regium* stands out from most of his papers: it has turned gray from the fingerprints of all the scholars who have eagerly studied it from Harriot's time to our own. After all, the law of refraction, in all its generality, was a superb discovery. It had eluded scholars for two thousand years, and even Kepler—who was about to publish a seminal work on geometric optics and the nature of vision, the most substantial advance since al-Haytham—could not find the correct law. In May 1603, he would write to a friend: "Measuring refractions, here I get stuck. Good God, what a hidden ratio!"[23] Kepler's seminal work was his commentary on and extension of Witelo (*Ad vitellionem paralipomena*), which he published in 1604, by which point he still had not found the correct law of refraction. Nor had he found it when he published his study of refracted image formation in his *Dioptrics* in 1611. Unlike the painstakingly thorough Harriot, Kepler would throw even his unfinished ideas into the public arena, in order to stimulate research, and also to ensure his priority.

Harrriot, on the other hand, did not seem to want to be famous.[24] His law of refraction should have given him instant and enduring fame. Instead, the first to publish it was Descartes, in 1637. The Dutch physicist Snell had also written this law, in about 1621 (two decades after Harriot), in an unpublished manuscript.[25] His countryman Christiaan Huygens—who in 1678 would develop the first successful model of refraction, using a wave theory of light—implied that Descartes deduced the law of refraction only after having seen Snell's work while he was in Holland. Be that as it may, Snell's representation of his law was geometric, and therefore it was implicit, as in Harriot's *Regium*. Descartes, too, presented his law geometrically, although he noted that the law of refraction was trigonometric, which is why he has long been considered the first to publish the sine law of refraction, as it is known today. After all, virtually everyone at that time expressed trigonometry geometrically.

But there is more to Harriot's discovery. He was arguably the best algebraist of the early seventeenth century, in the decades between Viète's death in 1603 and the rise of Descartes in the late 1630s, and in his optical work in the early 1600s he had begun to express trigonometric relationships in essentially their modern algebraic form.[26] A closer look at his carefully drawn *Regium* shows some additional trigonometric calculations at the bottom of the page, but that was not all: elsewhere he wrote his law of refraction explicitly in terms of sines, when making computations for his tables of refraction.

These neatly drawn tables list angles of incidence increasing degree by degree from 1 degree to 90 degrees, along with the corresponding angles of refraction; there are several such tables, each for different media, and they, too, show the traces of many scholarly fingerprints. Harriot labeled his tables so that it was clear that the direction of the light ray is reversible (as Ibn Sahl had also apparently realized, although al-Haytham had not[27]). That is, when changing your perspective from "water to air" to "air to water," you simply interchange the labels of the angles i and r. For Harriot's law of refraction, this means, in effect, that $\dfrac{\sin i}{\sin r}$ becomes $\dfrac{\sin r}{\sin i}$, so that if the refractive index of air with respect to water is known to be 0.748955, then the refractive index of water with respect to air is $1/0.748955 = 1.335194$.[28]

Harriot's were by far the most accurate tables of refractions in existence at that time, and his working sheets show how he did it. On July 20, 1601 (in a manuscript, BL Add MS 6789 f266, dated at Syon), he recorded a dozen pairs of angles found from experiment and compared his results with values obtained "by calculation." A further search of his papers shows that by "by calculation" he meant the sine law of refraction. In making tables for refraction from water to air, for example, he already knew that the refractive index was 0.7489552, so all he had to do now was to take successive values, degree by degree, for the angle r in air (shown in figure 16b), find their sines, and then calculate the sines of the corresponding angles in water by multiplying $\sin r$ by 0.7489552. In other words, he used the sine law in the form $\sin i = k \sin r$.[29] Finally, he used trigonometric tables (and interpolation) to find each angle of incidence from its sine value.[30] Then, to make separate tables of refraction from air into water, he just had to interchange his i and r (and invert his k).

There are pages and pages of such calculations, all done by long multiplication to six-figure accuracy and systematically set out around the explicit sine law. (Almost all of Harriot's arithmetical calculations in his various projects were correct; he was careful in everything he did. He was also human, and very occasionally he made a slip, such as misreading his own handwriting.[31]) Harriot's discovery, if made public, would have been momentous not only because

v.55. 748955.52
       81918528 (8
         1497910
7.50.38 3744775
         748955
         6740595
         748955   7—7
         5991640
        121322322
       6135079861160

v.67.   748955.52
         920804.2
43.35.0   2995820
          3744775
          1497910
          6740595   4—4
         01222111
        0894160733320

v.65.   748955.52
         906807.5
42.44.55 5242685
          2246865
          4493730   5—5
          6740595
         600122101
        678783159185

v.63.   748955 E2
         818006
41.52.40 4493730
          748955
          6740595   3—3
          5991640
         1111221
        667323398730

v.61.   748955 E8
         814609
40.55.25 6740595
          748955
          4493730   7—7
          2995820
          5242685
          5991640
         112243131
        655050273145

v.59.   748955 G2
         887167
39.56.20 5242685
          4493730
          748955   5—5
          5242685
          3744775
          5991640
         120223321
        641979510485

v.57.   748955 E5
         838640
38.54.42 5242850
          4493730
          5991640   1—1
          2246865
          5991640
         1123221
        678126089850

v.79.   748955
         987677
47.19.25 5242685
          1497910   161
          4493730
          748955   3—3
          5991640
         11214321
        735194449785

v.77.   748955 E
         874370
46.52.0   5242685 0
          2246865
          2995820   6—6
          5242685
          6740595
         10122311
        729789283350

v.75.   748955 E2
         865878
46.20.20 3744775
          1497910
          6740595
          3744775
          4493730   0—0
          6740595
         11223321
        723434358375

v.73.   748955 E0
         858304
45.44.38 2995820
          2246865
          4493730   0—0
          3744775
          6740595
         11223211
        716228662320

v.71.   748955 E2
         945818
45.5.5   5991640
          748955
          3744775   1—1
          2995820
          6740595
         11224321
        708150433690

v.69.   748955 E1
         933880
44.21.50 59916400
          3744775
          2246865   2—2
          6740595
         0133321
        699209408900

people had been looking for the law of refraction for two millennia but because of its importance in the construction of lenses. Spectacles had been in use for several centuries, but the process of lens grinding would require a whole new level of accuracy with the invention of the telescope in 1608. This is hindsight, though. To be fair to Harriot, it is also hindsight to expect him to have published his law of refraction. There were no scientific journals at that time. People published books, and even pamphlets, but not short research papers of the kind that would have suited Harriot's concise style.

The first British scientific society, one of the first in the modern world, was the Royal Society of London, founded in 1660 and chartered in 1662. This was largely in response to Francis Bacon's writings on the importance of sharing scientific knowledge, and had it existed in Harriot's time, he might have been tempted to present his law of refraction to the society and to see it published in the Society's *Philosophical Transactions*. Instead, in 1601 he did not rush into print with his stunning discovery. He simply moved on to the next creative challenge. He had not yet finished with optics: the rainbow beckoned.

(Image Opposite) *Plate 3: Using the sine law of refraction to tabulate angles of incidence and refraction*: For example, at the top of the middle column, Harriot's 748955 (i.e., 0.748955) is his experimentally determined refractive index for air with respect to water; taking an angle of refraction of 67°, he multiplies the refractive index by sin 67° = 0.920504 (which he has scaled as 920504) and gets the equivalent of 0.689416. (The u in front of his 67° is his symbol for "sine;" he denotes degrees by a bar over the 67) He then finds the inverse sine of this to be 43°35'0", which is the corresponding angle of incidence (cf. Figure 16b). He has many pages of such calculations, enabling him to tabulate angles of refraction corresponding to angles of incidence from 1° to 90°.

CHAPTER 13

Spirals and Turmoil

L IKE MANY BUSY AND CREATIVE people, Harriot usually had several projects on the go. He no doubt needed a change of pace every now and then from the laborious calculations and painstaking experiments that had led him to his law of refraction and would soon lead him to explain the rainbow and its colors. In the opening years of the seventeenth century, he continued his research into optics, but he also began seriously exploring pure mathematics and doing experiments on projectiles and falling motion. He was still mulling over his earlier unfinished projects, too: the loxodrome and the mathematics of combinations. His manuscripts at this stage show many pages of tinkering—rough diagrams, calculations that seem to go nowhere, and experiments that peter out, as well as promising new work and the occasional gems of discovery and resolution, such as those hidden in his tables of refraction and their construction.

Even in the comfort and freedom of Syon, however, life often interrupted Harriot's study. Some of his papers contain shopping lists, scribbled in the margins of his research work, including such items as "a scarf," "a riding cloth," shoestrings, ginger, "galls for ink," candles, books, "a ribbon for my dagger," and the ever-present tobacco and pipes. There were occasional additional little "notes to self," as we might call them today—such as "ask my apothecary if he has any good sarsaparilla yet." (Sarsaparilla was a tropical American plant used as a cure-all.) In addition to the mundane things, one scrawled note poignantly testifies to the tumultuous times in which

Harriot lived and points to yet another drama surrounding Ralegh.[1] It refers to "L. of E.," Ralegh's nemesis, the Lord of Essex (earls were often referred to as lords).

In the closing years of the sixteenth century, Essex had lost favor with the queen, not only on account of his poor leadership in the failed 1597 Azores venture with Ralegh but also because of his unsanctioned actions as commander of an unsuccessful English campaign in Ireland. Essex felt that his decisions were misunderstood at court—that it was one thing to issue orders from London and quite another to decide tactics on the ground. Still, he had no sensitivity to Elizabeth's role as a sovereign who should be seen to be in charge, and she wrote to remind him sharply that "the eyes of foreign Princes" were watching the Irish campaign waged in her name.[2] He had no sensitivity to her personal vulnerability, either—her awareness that at sixty-five, she was too old to play the virgin goddess role by which she had maintained stable rule in a man's world for forty years. Unlike Ralegh, whose loyalty to the queen was steadfast, and who still played his role in the courtly theater by professing undying love for his sovereign, Essex was openly contemptuous of her.

To make matters worse for Essex, Ralegh was in the ascendant. Although he had not yet realized his dream of discovering El Dorado, he had befriended the Guianans to such an extent that a few of them had returned to England with him.[3] Like Manteo and others before them, they lived with Ralegh as friends or wards, later returning home and offering loyal assistance to the English in South America. Life in general had been fairly good to Ralegh during the previous few years. He was back in favor with the queen, as we've seen, and he and Bess had managed to spend some happy times at Sherborne and in London, socializing with friends such as Bess of Hardwick, and the Sackvilles at Knole (ancestral home of the twentieth-century writer Vita Sackville-West).[4] In 1600, he had taken up a post as governor of Jersey, where it seems he even managed to spend time with his daughter by Alice Goold.[5]

Life had not been so good for the country as a whole. For four years unseasonal summer rain had ruined the harvests, and by the late 1590s there was widespread starvation. The government did what it could, offering food handouts to the needy and banning corn exports and enclosures of commons, but discontent was festering.[6] And then, as now, scapegoats were found. As early as 1596, the queen had responded to these "hard times of dearth" by issuing a proclamation calling for the expulsion of the "diverse blackamoors" who had recently "crept into this realm," where they were "fostered and relieved" by noble families, whereas starving English people needed "the relief that these people consume." The Raleghs and the Percys had apparently "fostered" some "blackamoors," former African slaves freed from the Spanish by the English during their raids, who had found their way into various noble English households, where they lived as wards, servants, or, perhaps, as curiosities.[7]

Then as now, too, hard times bred political instability, especially as Elizabeth still refused to formally name the Scottish King James VI as her heir. The ever-present factions at court grew more divisive as courtiers formed alliances with each other and with Elizabeth's potential successors. (James was the most likely, although other claimants included his cousin Arbella Stuart, who was also Bess of Hardwick's granddaughter). In this feverish, fearful climate, Essex began to foment lies against Ralegh and other rivals, and from 1599 onward he had been working on various ways of becoming kingmaker. He already had the people's support on account of his promises of reform, which included tackling the abuse of monopolies that had made his rivals rich while the poor were starving. It must be said that by now Ralegh even had the patent on playing cards.[8] Essex had had his own share of monopolies in happier times as queen's favorite, but now he was aligned with those who decried this kind of privatization of wealth through indirect taxes from the people. The queen promised to give consideration to the issue (and on November 28, 1601, would issue a Proclamation Reforming Patent Abuses), but Essex would not wait. He had been secretly courting the Scottish king, and one of his early plans had been to enforce regime change with the help of soldiers who had served under him in Ireland. His closest allies had dissuaded him from such overt treason, and in the end, he and his inner circle formed a less violent plan. They would take Ralegh and other key courtiers prisoner so that Essex could gain an audience with the queen. He would then "humbly" make his demands about a new power structure under James. Of course, such a structure would see Essex and his allies promoted and his enemies destroyed.

On February 8, 1601, he made his stand.[9] He took to the streets with two or three hundred armed men, but it all went terribly wrong for him. This was partly due to chance events that he could not have foreseen and partly because the rebellion was ill conceived, having been made with a few supportive earls only the night before. This precipitate action was possibly the result of (false) rumors that Ralegh was planning to have Essex killed.[10] Essex had also counted on wider public support, but there was no mass rallying of Londoners.

Instead, confusion reigned as the plan unraveled and his chief co-conspirators fled. By evening, Essex had surrendered.

Prosecutor Edward Coke, encouraged by Judge Francis Bacon, cobbled together a case for treason that relied too much upon rumors and supposition.[11] The inevitable aftermath followed quickly, and on February 25, 1601, Essex prepared for his execution. As Captain of the Guard, Ralegh was required to oversee the deaths of traitors, no matter how personal the circumstances, and this time, apparently Harriot was with him. It seems that it was while waiting with the "Lord of Essex" that Harriot scrawled his note: *L of E ready. Lord forgive him this horrible sin this bloody and crying shame.*[12]

From the scaffold, Essex repented the lies he had told about Ralegh and other rivals. And he had not intended to kill the queen, who was horrified by Essex's death despite the fact that she had seen no alternative but to sanction the decision of her appointed judge and prosecutor.

Harriot was distressed by the whole Essex drama, and some years later, he apparently read Bacon's 1604 apologia for the controversial trial.[13] Northumberland must have been ambivalent, too. Essex was his brother-in-law, after all. And like Essex, Northumberland had been courting James as Elizabeth's likely successor. Unlike Essex, Northumberland tried to put in a good word for Ralegh, too. But it was a rather tepid word; Northumberland was even-handed in his account of Ralegh's strengths and weaknesses.[14]

HARRIOT HAD BEEN THROWN OFF course by all this turmoil and tragedy, but he was soon enough back at work at Syon. Rough work on the pages grouped with his condemnation of Essex's "horrible sin" reveals that, along with optics, algebra, and other topics, he was thinking in a different, more mathematical way about the loxodrome, the spiraling curve made by a rhumb line—a line of constant compass bearing—along the curved surface of the globe.

Harriot's work in this batch of notes deals with "spherical triangles"—triangles made from three intersecting "great circle" arcs on a sphere.[15] Because they have curved sides, they are more complex than ordinary triangles, although the ancients had formulated some significant results in spherical trigonometry. More recently, Pedro Nunes had attempted to use the geometry of spherical triangles to help navigators steer a course not quite along a rhumb line but along successive arcs of great circles. He'd done this because the shortest route between two relatively nearby points on a sphere lies along a great circle (or a geodesic, in today's terminology), and so he considered this to be the most efficient method of sailing.[16] Harriot's friend Robert Hues, like Nunes's critic Diogo da Sà, was dismissive of Nunes's method, which he thought too unwieldy to be practical.[17] Harriot's goal, by contrast, was purely mathematical. His work on spherical triangles may have begun with his reading of Nunes, but he ended up exploring the subject for its own sake.

It was the same with all the mathematics underlying Harriot's exploration of the loxodrome. He created new results, some of which would ultimately lead him to a more elegant solution of the Mercator problem than adding up thousands of secants by hand; in the meantime, he explored each of these results for its own sake, only putting them all together later. In doing this he exemplifies the historical interconnection between practical motivations and pure, curiosity-driven research.

In thinking about how to analyze the loxodrome—his nautical "helical line"—Harriot's first purely mathematical breakthrough concerned mapping curves in three-dimensional space to two dimensions. When angles between lines on a curved surface such as a globe are mapped to the same angles on a flat surface, the map is called conformal (or angle-preserving) in modern mathematical terminology. For instance, a rhumb line makes a constant angle with each meridian on the globe, and so does its straight line equivalent on a Mercator map. However, Harriot now showed that Mercator's projection is not the only way to keep the same angles when going from a sphere to a plane: "stereographic projection" is conformal, too.

Stereographic projection itself had been known since Ptolemy. It maps each point on a sphere to a unique point on a flat surface such as a map. Actually, it maps each point uniquely except the North or South Pole, which is used to determine the correspondence. It is somewhat analogous to shining a light onto the top of a transparent ball set on a table and drawing on the table the shadows of lines painted on the ball.[18] Harriot chose the South Pole and drew lines from this pole through other points on the surface of the sphere. The intersection of each line with the equatorial plane gave the corresponding image on the plane of the point on the sphere, as in figure 17a. Showing that this projection preserves angles was significant enough on its own: it seems that no one before Harriot had proved the general conformal nature of pointwise stereographic projection.[19] (By "general," I mean that Harriot proved that the angle between *any* two lines on the sphere is the same when those lines are stereographically projected onto a plane.)

Harriot did not stop at this result, significant as it was, but went on to make use of it in his loxodrome analysis. Since a rhumb line spiraling around the globe makes an equal angle with each meridian of longitude it crosses, he showed that when this loxodrome is stereographically projected onto the equatorial plane, it forms another spiral with the same "equiangular" property (as in figure 17b). Not surprisingly, this second spiral is now called the equiangular spiral. It is seen in the cross-sections of nautilus shells, although Harriot would not have known it. No one had studied the equiangular spiral before: it is usually held that Descartes discovered it in 1638.

Harriot had this result by the early 1600s, so yet again his friends had reason to lament that he did not publish his work.[20] Instead, he proceeded to explore his new curve further. In doing so, he found its length, from the "beginning" of the spiral (at the center of the concentric circles in figure 17b) to any given point on it, thereby notching up yet another first. He was not only the first to find the length of the equiangular spiral—a result generally accredited to Evangelista Torricelli, who published it in 1645—but the first to find the exact length of *any* curve. In fact, in 1637, Descartes would echo Aristotle by maintaining that it was not possible for "human minds" to find such a length.[21] The closest anyone had come to knowing the length of a curve was the approximate circumference of a circle, known since ancient times using what can be thought of as approximations of the number pi (π) in the modern formula $C = 2\pi r$. (Pi has no exact value. Its name and symbol date from the 1700s.)

Finding the length of a curve whose equation is known is now done using the algorithms of integral calculus. Harriot had neither these algorithms nor an equation for his equiangular spiral, which is represented today as $r = e^{k\theta}$, where θ is defined in the endnote, and e is an "irrational," "transcendental" number like pi and is approximately 2.718.[22] (A number is irrational if it cannot be expressed as a ratio of integers or, equivalently, it has no finite, or no repeating, decimal expansion, and it is transcendental if it is not "algebraic"—if it cannot be found as the root of a polynomial equation.) Harriot used a mnemonic to help him memorize pi to seven decimal places: "cadaeibf," which becomes, when letters are given digits corresponding to their alphabetical order, 3.1415926. He was an early adopter of decimal fractions, having been inspired by Simon Stevin's 1585 book on the topic, although as noted earlier he expressed these without a decimal point. (Decimal point notation became popular only toward the end of Harriot's life.) But he did not know about the number e, which was first studied explicitly by Leonhard Euler a century and a half later.[23] Nevertheless, Harriot was one of the first to glimpse e implicitly, as was the Scottish laird John Napier, in his 1614 publication on his discovery of logarithms; the connection is shown in the fact that the equiangular spiral is now also called the logarithmic spiral. (A logarithm is a power: the logarithm "to the base 10" of the number 100 is 2, because $10^2 = 100$. The logarithm "to the base e" of the radius r of the equiangular spiral is $k\theta$.)

These glimpses of concepts that would later become fundamental to mathematics illustrate the process of mathematical and scientific discovery, and studying Harriot's work—that chaotic bunch of long-lost manuscripts—helps us understand how modern mathematics and science began to emerge, and how it continues to develop. Ideas are often generated in a haphazard way as researchers experiment with various methods to solve specific problems, and it is only later that someone is able to unify the intuitive steps of his or her forerunners into a satisfying, rigorous whole.[24] Newton is the most famous such unifier, but his work, too, required later development by a succession of disciples who continued the distillation process. It is thus with all the great discoveries, particularly, perhaps, in mathematics and mathematical physics. Because most of Harriot's work long remained unpublished, few later researchers were able to refine his methods and to build upon his ideas. Nonetheless, others rediscovered his results—independent codiscovery or rediscovery being another feature of the mathematical and scientific process—so his work illustrates the way such discoveries begin.

For instance, his discovery of the equiangular spiral offers not only a hint at the underlying nature of the number e but another glimpse of the foundations of calculus. Just as he had broken up his rhumb line into segments in order to calculate changes in latitude and longitude, so he had broken up his spiral into tiny segments in order to find its length, as in figure 17c. The smaller the segments, the closer the figure comes to the spiral, until in the "limit," when there are virtually an infinite number of segments, the figure fits the spiral exactly. Proof of this fit would not be complete until this type of infinite limit was made mathematically rigorous, 150 years after Newton and Leibniz found the integral calculus algorithms for "adding up" infinitesimal quantities—which illustrates the point about the process of mathematical discovery and its later refinements.

The idea of approximating curves with straight-line segments is an obvious one, of course, and the Egyptians and Babylonians had approximated circles by polygons—triangles, squares, pentagons, hexagons, octagons, and so on. This seems to be how they worked out their approximations for the area and circumference of a circle. The Greeks, notably Eudoxus, Euclid, and Archimedes, developed this method more rigorously, by giving the polygons an increasing number of ever-smaller sides. Then, in the method now known as the method of exhaustion, they showed that the error in the approximation of the polygon to the circle could be made as small as you like, simply by choosing more sides for the polygon. Yet most Greek mathematicians were wary of infinity and did not presume to say that an infinite-sided polygon would give the exact circle.

Harriot, however, did presume this. While a few of his near-contemporary forerunners also presumed it to varying degrees, Harriot is perhaps the earliest pioneer of the relatively sophisticated *algebraic* precalculus of the early seventeenth century.[25] In the case of his segmented spiral, he realized that the lengths of the successive segments—those marked *bd, de, ef,*...in figure 17c—formed a "geometric progression," a series of numbers in which the ratio of each successive pair is the same. Euclid had derived a rhetorical algorithm for the sum of an arbitrary number of terms in a geometric progression, a result taught today in high school. But unlike Harriot—who was the first to generalize such algorithms by expressing them algebraically in terms of an arbitrary n^{th}-term—Euclid did not consider what happens if the series goes on forever, so that it has an infinite number of terms (or, in modern language, so that n "tends toward" infinity). Harriot took that step and used the result in his calculations of the length of the spiral.[26] He then went on to find the length of the corresponding loxodrome.

The key thing here is Harriot's process of discovery. He used a mix of trial and error, intuition, and rigorous mathematical inference and deduction. He tried several approaches: drawing diagrams, writing out algebraic equations, and trying specific numerical examples. He was feeling his way, picking out ever more concrete examples. Only when he had satisfied himself with numerical examples did he write out the general equations for the lengths of his spirals.[27] This was a significant breakthrough: he had managed to find formulae for the sum of the lengths of the infinitesimal segments in both his equiangular spiral and loxodrome, without having to add up each segment by hand. So he thought about applying his results to the calculation of a meridional part, which requires the sum of the secants of incrementally increasing latitudes along the segmented rhumb line. He wanted to find a way to use his work on spirals to help him "add up" all these secants using clever algebra instead of brute-force arithmetic. The complete solution still eluded him for the moment, though.

2.)

$v.i.$ 2,908,882,045,634²

$v.i.$ 2,908,887,045,634²

```
1 1 6 3 5 5 2 8 1 8 2 5 3 6
  8 7 2 6 6 4 6 1 3 6 9 0 2
1 7 4 5 3 2 9 2 2 7 3 8 0 4
1 4 5 4 4 4 1 0 2 2 8 1 7 0
1 1 6 3 5 5 2 8 1 8 2 5 3 6
  5 8 1 7 7 6 4 0 9 1 2 6 8
    2 3 2 7 1 0 5 6 3 6 5 0 7 2
    2 3 2 7 1 0 5 6 3 6 5 0 7 2
    2 3 2 7 1 0 5 6 3 6 5 0 7 2
2 6 1 7 9 9 3 8 4 1 0 7 0 6
5 8 1 7 7 6 4 0 9 1 2 6 8
```

argiter b:
2t. bc.
per lineas solas.

$$\overline{1 0 2 2 2 2 2 2 3 4 6 5 3 4 3 3 2 3 1 3 1}$$
8 4 6 1, 5 9 4, 7 5 5, 4 1, 1 8 4 4 4 5 8 4 6 1 9 3 6

$v.i.$ $v.i.$ basse.   8,4 6 1,5 9 4,7 5 5,4 1 1
      2,9 0 8 8
      2,9 0 8 8

$v.i.$ $v.i.$ gnat.   8,4 6 1,5 9 4,7 5 5,4 1 7,6 6 2, aaa aaa aaa
100,000,000,000,000,000,000,000,000,000,000

$v.84.7.$ 99,999,99 7,5 3 8,4 0 5,2 4 4,5 8 8,1 5 5, aaa aaa aaa

$v.2.77.$ 99,999,99 7,5 3 8,4 0 5,2 4 4,5 8 2,3 3 7, aaa aaa aaa

$v.i.$ $v.i.$ 4 9,9 9 9,9 9 3,7 6 9,2 0 2,6 2 2,2 9 4,0 7 7, aaa aaa aaa

     4 9,9 9 9,9 9 3,7 6 9,2 0 2,6 2 2,2 9 1,1 6 8 aaa aaa aaa

$v.i.$ $v.i.$ 4 9,9 9 9,9 9 5,7 6 9,2 0 2,6 2 2,2 9 1,1 6 8 aaa aaa aaa

$v.i.$ $v.i.$ 5 0,0 0 0,0 0 4,2 3 0,7 9 7,3 7 7,7 0 5,9 2 2, aaa aaa aaa

     5 0,0 0 0,0 0 4,2 3 0,7 9 7,3 7 7,7 0 8,5 3 1 aaa aaa aaa

vbr 7 7 7 1 0 6 5 1 1 1 2 8 0 2 0 4 7

$v.45.$ 7, 0 7 1, 0 6 7, 8 1 1 8 6 5, 4 7 5, 2.

$v.46.$ 99,999,995,769,636,467. nog.

$v.89.54.$ 99,999,995,769,202,532. nor.3.

$v.46.$ 99,999,995,769,202,890,8. to much.
       890,7. to little.

$v.b.$ 7,071,068,111,028,020,47. non. 48.

$v.ab.$ 4,113,780,266,233,8390. non.1.)

$v.bc.$ 2,908,881,922,565,348,3. to little.
          348,5. to much.

$v.fz.$ 2,056,890,220,139,771,45. to little.
     2,056,890,220,139,771,53. to big.

$v.dz.$ 4,113,780,440,279,54290. to little.
     4,113,780,440,279,54310. to big.

$v.d.$ 7,071,067,811,865,475,2. to little.
     7,071,067,811,865,477, to much.

$v.i.bc.$ 2,908,882,045,634,245,96374.

$v.d.$ 7,071,067,961,446,755,770. non.1.)

b. ac. c. ab.
c. ab. a. bc.

$v.90.$ $v.i.$ $v.45.$ $v.fz.$
$v.dz.$ $v.90.$ $v.i.$ $v.d.$

$vzt:$

Instead, by 1603 he had used stereographic projection to find a significant property of spherical triangles.

The ancients had known that the area of a plane triangle is half its base times its height (because a triangle fits into half a rectangle, whose area is "base times height"). But they did not know the area of a spherical triangle, whose sides are made of curved lines on a sphere rather than straight lines on a page, as noted. Harriot's papers show him trying out various approaches to find a solution to this question: fitting triangles into circles, measuring angles in selected spherical triangles, and, ultimately, splitting a sphere into groups of pairs of equal spherical triangles and stereographically projecting them onto a plane.[28] In this way, he found the formula for the area of a spherical triangle, and he wrote up his proof on September 18, 1603. He must have been proud of it, because he rarely dated his manuscripts. (He usually only dated his final theoretical and mathematical results, and especially his experimental results, in order to show his replications of the experiment over different days.) And it was indeed worthy of pride: his contemporary Henry Briggs—the first Savilian Professor of Geometry at Oxford—would later praise it as a "special contribution by a modern mathematician to an old problem."[29]

Once again, he did not publish (the formula would be rediscovered and first published nearly three decades later); this time, he certainly had a pressing excuse. It seems he had begun his work on spherical trigonometry around the time of Essex's execution, but he finished it at the dawn of a far more personal tragedy. Harriot was not a person who sought solace in work in the face of emotional dramas: he would produce no further theoretical results, nor do any experiments, for the next nine months.[30] A clue to the drama appears

(Image opposite) **Plate 4**: *Trial-and-error calculations on spherical triangles*: This appears to be early rough work on the problem. The top diagram shows a spherical triangle with a known angle of 45° and a side of arc length 1´. Harriot then uses trial and error to estimate the other angles and sides in the spherical triangle, using or checking the sine rule for spherical triangles (known in plane form to Ptolemy, and stated for spherical triangles by Abu'l-Wefa in the tenth century). The numbers on this page are sine values to more than fifteen decimal places, all multiplied out by hand. Note Harriot's comments "too little" and "too much" or "too big" as he tries to find the right value. The lower diagram suggests he may then be exploring the area of the triangle by analogy with the method for plane triangles. The interesting aspect in this rough work is his trial-and-error approach, requiring an inordinate amount of work to discover new results, especially in an era before calculators.

(Petworth House/Leconfield, HMC 240 V (240 f404); reproduced with the kind permission of Lord Egremont and the West Sussex Record Office.)

in another of Harriot's asides—a list of names on an otherwise blank page: S. W. R [Sir Walter Ralegh]; L. Grey; L. Cobham; Mr. G. Brookes; S. Ar. Gorges; S. Ar. Sanaher; S. Griffith Marcum.[31]

Arthur Gorges was Ralegh's cousin, who had sailed with Ralegh in Essex's failed expedition against the Spanish in 1597, and was as well an MP, a poet, and a translator. Lord Grey was the son of the Lord Grey who had ordered the notorious 1580 massacre in Ireland. I can find no relevant record of Ar. (Arthur, presumably) Sanaher, although Harriot's writing is difficult to decipher and it may be another name altogether. But Sir Griffin Markham was arrested for treason in July 1603, as were Ralegh, Lords Grey and Cobham, and Cobham's brother George Brookes.

CHAPTER 14

Changing of the Guard

THE TROUBLE HAD BEGUN FOUR months earlier. On March 24, 1603, Queen Elizabeth died at the age of sixty-nine. The last years of her reign had not been happy ones: famine and plague had decimated the population, while Essex's execution had left a deep wound. As Arthur Throckmorton had put it after his friend Essex's death, "It seems the state was sick. I hope this letting blood will do it good."[1] It didn't: in the final months of Elizabeth's reign, everyone had been nervous about who would become the new monarch, and ambitious courtiers had stepped up their efforts at maneuvering their way into positions of power.

In those tense times, all the key players at court behaved badly. Ralegh was his usual witty, bombastic, tactless self in Parliament, and he blamed young Cecil, probably unfairly, for blocking his appointment to the Privy Council. Cecil would retaliate in spades. Essex had already tried to turn James VI of Scotland against Ralegh, and soon Cecil's friend Lord Henry Howard was secretly whispering his own treacherous words to the likely new king of England. The former Guiana investor Lord Admiral Charles Howard had turned against Ralegh, too—he had been an Essex supporter. As for Cecil, he was growing weary of Ralegh's histrionic ways, and to ensure his own favor with James, he, too, began quietly disassociating himself from Ralegh, tarnishing his former friend's reputation in order the better to burnish his own.

Tellingly, after almost forty-five years of rule by a queen, Henry Howard and Cecil both disdained Lady Ralegh's spiritedness—the very quality that Ralegh loved (most of the time[2]). Cecil, and especially Howard, saw this as a weakness in Ralegh, and they made capital of it by adding rumors of Bess's supposed influence over the Durham House coterie to their lies about its supposed opposition to James's succession. To underscore the latter point, Howard told James that Ralegh, Northumberland, and Lord Cobham made a "diabolical triplicity, that denies the Trinity." Cecil played the same card, though more smoothly. He admitted to James his former friendship for Ralegh; after all, everyone knew that aside from his support of Ralegh's expeditions, there had been a closer bond. After Cecil's wife died, Bess and Ralegh had encouraged the Cecils' son Will to spend as much time with them and their own son Wat as he liked, and young Will had grown up adoring Ralegh. Now, though, Cecil told James that it was only from "private affection" that he had supported "a person whom the most religious men do hold in anathema."[3] The old atheism rumors against Ralegh, Harriot, and their friends had an insidious new lease of life.

Northumberland, however, had been having his own correspondence with James, offering him advice about the best way to make a peaceful transition to the role of king of England. James did, in the end, inherit Elizabeth's throne, and he was most grateful for Northumberland's "sincerity and love," as he put it. When he made his first royal progress through London, Northumberland rode at his right hand. Ralegh's former investor Lord Admiral Howard rode on the left.[4]

If most people initially welcomed this changing of the guard, hoping that James would be more tolerant and even-handed, the new king would soon show a pronounced autocratic streak. In his early speeches to Parliament he used the paternalistic analogy between a king and a father, emphasizing the absolute authority and divine right of kings.[5] He and Ralegh were bound to clash, even if his mind had not already been poisoned by the viperous whispers of those who still pretended to be Ralegh's friends. Besides, James hated tobacco. In 1604, he would issue his *Counterblast to Tobacco*, in which he found it a miracle that such a vile custom, "brought by a father so generally hated"—presumably Ralegh—should have become so popular in England.

James immediately, and understandably, placed his own man in Ralegh's position of captain of the Guard. He also continued the reform of monopolies, so that Ralegh lost a key source of his income. (James awarded some of these monopolies back to his new favorites. Lord Admiral Howard took over Ralegh's wine patent, for instance.[6])

Then, to add insult to injury, not two months after Elizabeth's death, James acceded to the bishop of Durham's request to return Durham House, ordering Ralegh to vacate within two weeks. Ralegh was heartbroken. He had lived here for twenty years and had spent £2,000 in improvements. Now, to be ordered to move a household of forty, as well as nearly twenty horses, in just two weeks seemed the utmost cruelty.[7] Worse was still to come.

James had told Northumberland he intended to extend more tolerance to Catholics, but once in London, he prudently moved cautiously. It did not take long for a couple of Catholic priests to hatch a plot to capture the king and force him to accept their demands for tolerance. Markham, Grey, and Brooke joined in, and when the rumor broke and the plotters were arrested, Brooke not only confessed to the plot (the Bye Plot) but also mentioned seditious plans of his brother, Lord Cobham. According to the eventual prosecution reports, Cobham had been involved in a scheme—the so-called Main Plot—to funnel funds from the Continent to England through the ambassador of the Spanish Netherlands; some of the money supposedly was to be used to help organize a coup that would put the English-born Arbella Stuart on the Scottish King James's new English throne. The idea was that she would institute religious tolerance, though apparently she disagreed with the plan and alerted James. Cobham also reputedly organized a Spanish pension for himself, in return for supporting a pro-Spanish English foreign policy.

Brooke also claimed, on hearsay, that Ralegh was involved in his brother's plan; Cobham and Ralegh were good friends. When Ralegh was questioned on the matter, his first response was to lie, claiming he had no knowledge of any such plans. As it happened, Ralegh, the veteran adversary of Spain, had no complicity in the plot itself—but he had known about it and refrained from informing on his friend's half-baked idea. Some believe he kept his silence only in order to wait for more evidence to emerge and then present it to the king as proof of his own loyalty. At any rate, he later admitted he had suspicions about Cobham's dealing with the Continental ambassador.[8] Desperate times led to desperate acts, as friend betrayed friend in the race for power and self-preservation in the new regime. Indeed, in initiating an enquiry into Cobham's activities, Henry Howard and Cecil had, it seems, deliberately set out to entrap him, hoping thereby to snare Ralegh, too.[9] And so they did. On his third interrogation, Cobham was shown Ralegh's confession of his "suspicions," which so infuriated Cobham that he accused Ralegh of masterminding the whole thing. Later, he retracted his statement, but Ralegh's fate was sealed.

His trial, which opened on November 17, 1603, is still famous as a travesty of justice if not of law.[10] It is true that he had failed to reveal what he knew of Cobham's plans. And according to Cobham's earlier testimony, Ralegh had also, in the heat of anger at some move or other by James or his allies, told Cobham he might as well urge his Spanish connections to "advise the king of Spain to send an army against England to Milford Haven." Words are not deeds, of course— as Ralegh pointed out. At the time, though, words alone, no matter how fanciful or frustrated, were acceptable evidence in treason trials.[11] Aside from this, there was no indication that Ralegh was personally involved in any treasonable activity, and the prosecutor—Attorney General Sir Edward Coke—knew it. So, presumably, did Cecil and Howard, two members of the special commission before which Ralegh was tried. Coke relied on bullying tactics instead of evidence—and on the fact that although the English had long had a form of habeas corpus and trial by jury, there was no presumption of innocence in treason trials. There was therefore no legal requirement for the prosecution to establish guilt. Consequently, Coke repeatedly refused Ralegh's requests that he produce witnesses against him. After all, he knew that Cobham, Ralegh's only serious accuser, had retracted his accusations.

Today, the Confrontation Clause of the Sixth Amendment to the US Constitution says that an accused has the right to confront witnesses. Early English common law assumed that at least two witnesses were needed for a conviction, but this was not written down until Edward VI's 1547 treason statute. Then, in the reign of Edward's sister Mary Tudor, Sir Nicholas Throckmorton—Lady Ralegh's father—had faced a charge of treason when he was implicated in the Wyatt Rebellion. Throckmorton had had as little involvement in the rebellion as Ralegh in the Main Plot and had successfully avoided conviction, in part by pointing out that at least two witnesses were required to testify against him and that the Crown had produced only one. Ralegh now tried to argue for the same right, but treason statutes had since been amended, and the two-witness rule was common precedent rather than the letter of the law.[12]

Ralegh thought he had another ace up his sleeve: an appeal to the Bible. He had been certain that there were scriptural precedents for the calling of witnesses, and Harriot on his behalf had trawled through the Old and New Testaments to find examples for him.[13] This defense only fueled Coke's righteous hatred: "O damnable atheist! He has learned some text of scripture to serve his own purpose."[14] It is clear from such comments, from both Coke and Chief

Justice John Popham, that Ralegh was condemned not for his actions but for being arrogant and atheist in their eyes. The jury took just fifteen minutes to reach their guilty verdict. In the final summing up, Popham relished his role in presiding over the fall of a man who had climbed "too high," as he put it. He adduced Ralegh's past, his "eager ambition and corrupt covetousness," as his only proof against him. Such a greedy man *must* have been taking a cut from Cobham and his Spaniards, as Cobham had alleged—a greedy man who, moreover, had been adjudged "by the world" as a defender of "most heathenish and blasphemous opinions." Think on this before you die, warned Popham. "Let not Harriot nor any Doctor persuade you to think there is no eternity in heaven; if you think thus, you shall find eternity in hell-fire."[15]

So Harriot was in danger, too. Mud sticks, as they say, and it stuck particularly hard to the scholarly Harriot. It says something about the fear with which many of his contemporaries viewed intellectuals in the arcane disciplines of mathematics and science. Finally, Popham pronounced, in graphic detail, the inevitable sentence for a convicted traitor: drawing and quartering. Ralegh's execution was scheduled for December 13, four weeks away.

THE TWO FOOLHARDY CATHOLIC PRIESTS who had planned to capture the king and force him to retract the anti-Catholic Elizabethan edicts had died hideous, eviscerating traitor's deaths, and their ally Brooke had been beheaded. Then, on December 10, first Markham, then Grey, and finally Cobham were in turn led to the scaffold and told to pray in their last moments. After a tormented half hour or so, each was led back to prison. An hour or so later, one by one they were again escorted to the scaffold. This time, after being forced to acknowledge their guilt before the crowd that had gathered to watch them die, they were told that in his mercy the king had spared their lives. Everyone was astonished and jubilant at such royal compassion. Markham was lucky enough to be exiled; the other two were to be imprisoned.[16]

Watching the charade from his window in the Tower, Ralegh felt a flicker of hope. He wrote a letter praising the king's action and begging such clemency for himself. Before the trial, Lady Ralegh had worked indefatigably on his behalf; she had even written to Cecil, utterly unaware of his betrayal, pleading with him to speak to James in support of her husband.[17] Despite it all, Ralegh was left alone to face his last days on earth. His poignant farewell letter to his "dear Bess" remains among his most famous works.[18]

> You shall now receive (my dear wife) my last words in these my last lines. My love I send you, that you may keep it when I am dead…And seeing it is not the will of God that I shall see you any more in this life, bear it patiently and with a heart like thyself.
>
> First I send you all the thanks which my heart can conceive or my words can express for your many travails and care taken for me, which though they have not taken effect as you wished, yet my debt to you is not the less, but pay it I never shall in this world.

Then he advised her on matters of his estate—including the fact that Sherborne had already been conveyed to Wat, and his hope that this would now prevent the property from being confiscated by the state. Finally, after other personal advice, he reminded her that it was he who "chose you and loved you in his happiest times."

But a last-minute reprieve was granted to Ralegh, too—without his having to play out the scaffold spectacle endured by the others.[19] Cecil had managed a qualm of conscience during the trial, admitting that in speaking against Ralegh, he had experienced "a great conflict within myself." Perhaps he helped change James's mind and so spared Ralegh's life. He certainly appeared to want to help Bess ensure that her own possessions remained untouched by the state as they carved up what they could of the estate of the "traitor."[20]

The price of Ralegh's life was permanent imprisonment in the Tower. He had been in deep despair before his trial. Now that it was all over, he held to the belief that life meant hope, even within these gloomy walls. In fact, Ralegh's reputation had already been rejuvenated. He had performed so calmly, eloquently, and reasonably at his blatantly unjust trial that many were forced to reexamine the prevailing prejudice against him. One Scottish spectator admitted he had so taken on "the common hatred" of Ralegh that he "would have gone a hundred miles to see him hanged," but by the end of the trial, he "would have gone a thousand to save his life." Even one of the judges admitted that the trial had "injured and degraded the justice of England," while Coke was reputedly shocked after being told the verdict, claiming, somewhat sophistically, to have accused Ralegh not of treason but of "misprision [concealment] of treason."[21] Life and death hung thus, on the whims of human duplicity and misunderstanding.

RALEGH ACCEPTED THAT HE HAD no option but to settle in as best he could to his new life in the Tower. Prisoners paid for their detention, and he was still wealthy enough to "rent" a pair of second-story rooms in the Bloody Tower. (The complex of buildings and towers making up the

Tower of London had had many different—and often simultaneous—uses over the centuries, including a fortress, a royal castle, and a prison. The Bloody Tower was originally called the Garden Tower; it received its sinister nickname because of the legend of the young princes supposedly murdered there in the fifteenth century by Richard III.) One of these rooms made a pleasant enough study under the circumstances, with its fireplace and large window—and it is to study that Ralegh would eventually turn. His *History of the World* is his most famous work from his years in prison. Meantime, in those early weeks and months, his letters show him wrestling with alternating moods of hope and bitterness.

His misery was mitigated somewhat when Bess and Wat were allowed to live with him in the Tower, which they did for much of the time (especially when the plague wasn't rampant). He was also allowed several servants, as well as a tutor for Wat. Harriot visited whenever he could, as did other loyal friends, such as the mathematician turned Guiana explorer Lawrence Keymis and several of Ralegh's Guianan companions. Life went on—so much so that in the summer of 1604 Bess conceived another son, Carew.

Also in the summer of that year, James made a welcome peace with Spain. Ralegh had long been a thorn in the Spaniards' side, with his successful military raids and parliamentary advocacy of anti-Spanish foreign policy, his privateering, and his encroachment in "their" New World. This may well have been why James and his key politicians had decided to get rid of Ralegh at this time. Yet he had lost his liberty for his supposed part in Cobham's *pro*-Spanish plot—and in the wake of the 1604 peace treaty negotiations, Cecil, Henry Howard, and two others who had tried Ralegh were among those who accepted secret Spanish pensions designed to shore up pro-Spanish support in England![22]

The treachery and drama of it all, the sordid trial and betrayals, the fall of a once-powerful man: this was surely too good a story for someone like Ben Jonson or Will Shakespeare to pass up.[23] In his dark comedy *Measure for Measure*, Shakespeare adapts ancient and contemporary ideas on betrayal, sex, corruption, justice, and mercy; he includes the device of a condemned man who is told to kneel and prepare for his execution the following day, while the all-powerful but disguised duke intends to save the man's life. One wonders what James made of it when it was performed for him, most likely in December 1604—just a year after the theatrical reprieves of Cobham, Grey, Markham, and Ralegh.

Jonson's *Sejanus, His Fall*, performed before the king sometime in 1603, seems to play on all the classic elements of Ralegh's story:

his rise and fall within a corrupt, autocratic system; the eager rivals ready to pounce. At the very time that Ralegh was caught in a web of plots against the king, the Catholic convert Jonson had incorporated a plot to undermine the emperor of his play. Not surprisingly, he was called before the Privy Council to explain himself.

The king was not impressed by Jonson's next effort, either: the comedy *Eastward Ho!* Coauthored by Jonson, John Marston, and Harriot's friend George Chapman, the play is a satire on the mercantile and social ambitions of London's lower middle class. But the wary new king took offense at an aside about Scots, and Jonson and Chapman were sent to the Tower for a spell. Back in 1598, part of Chapman's translation of Homer had been published as *Achilles' Shield*; it included a heartfelt poetic preface to Harriot, whom Chapman addressed as "my admired and soul-loved friend, master of all essential and true knowledge":[24]

> To you whose depth of soul measures the height,
> And all dimensions of all works of weight,
> …
> My love to you, in my desire to learn:
> Skill and the love of skill do ever kiss.
> No band of love so strong as knowledge is:
> Which who is he who may not learn of you,
> Whom learning doth with his light's throne endow?

Chapman went on to speak of Harriot's desire for truth, "far from plodding gain or thirst of glory"—a judgment borne out by Harriot's diffidence with regard to publishing his work. Now, in *Eastward Ho!*, Chapman and his cowriters included another detail related to Harriot: a scene in which an unscrupulous nobleman takes his ill-gotten gains and heads for Virginia. Fortunately, a timely shipwreck ensures his arrest.

Shakespeare's *Tempest* of 1611, often celebrated as his final work, would draw on a real-life shipwreck en route to Virginia. Ralegh's colonies might have failed, but there were many adventurers still seeking their fortunes across the sea. Although James had doomed Ralegh to a life in the Tower, he had no scruples about appropriating his prisoner's discoveries, and in April 1606 he chartered the Virginia Company of London, under whose auspices a colony of about a hundred volunteers would be founded the following year. Despite illness, hardship, and the failure to discover gold or other valuable commodities, it was the first permanent English settlement in the United States: Jamestown, near Chesapeake Bay—the very territory the Algonquians had told Harriot about and John White had intended to settle.

CHAPTER 15

Algebra, Rainbows, and an Infamous Plot

A FTER THE ORDEAL OF RALEGH'S trial, Harriot had finally returned to serious study in July 1604, when he made experimental measurements of various substances' specific weight (or specific gravity, a measure of density defined in the next endnote). Following Archimedes, he made his measurements by weighing the substance in air, and then weighing it again in a medium such as water—taking account of the weight of water before immersing the object—and then finding the ratio. The details do not matter, except to note that Harriot later used these measurements for his experiments on falling motion when a body is falling through a resisting medium such as water.[1] A significant aspect of his work on this topic was the way he applied algebra to his calculations, just as he had done with the law of refraction. He appears to have been the first to work with a purely symbolic form of algebra.[2] Before Harriot, algebra did not consist of symbolic formulae whose operations can be read and understood at a glance; rather, traditional "rhetorical" algebra was algorithmic, with each operational step being described verbally.

To take an example, today's common index notation n^2 represents the algorithm "take the unknown number and multiply it by itself." It is shorthand for $n \times n$, but in 1604, no one would have written $n \times n$, let alone n^2: the multiplication sign $\times$ would not come into use until after William Oughtred published it in 1631, while hardly anyone used letters such as n in an algebraic equation or algorithm. Harriot's older contemporary François Viète had made a significant innovation by using uppercase vowels to denote unknown quantities and consonants to denote parameters, but otherwise he generally wrote his equations in words. In order to dispense with words entirely, Harriot invented his own symbols. To denote multiplication, he sometimes used a dot, which is still used in some algebraic contexts, or else he simply wrote the multiplied terms side by side as we do today when writing an expression such as $2x$. To write algebraic equations, sometimes he used n to represent the "unknown number," especially if he was using whole numbers; more generally he used lowercase letters from the beginning of the alphabet (adapting Viète's innovation). We owe the index form for powers (as well as the use of x for the unknown) to Descartes's *La géométrie* of 1637.[3] Nevertheless, Harriot's notation thirty years earlier was almost as good. He wrote n^2 as nn, n^3 as nnn, and so on, for as many repetitions as you like. In fact, some years later, in using Harriot's work Nathaniel Torporley would anticipate Descartes by replacing Harriot's nn by n^{II}, nnn by n^{III}, $nnnn$ by n^{IV}, and so on.

Aside from Viète's there had been earlier attempts at using shorthand abbreviations for powers, unknowns, and other operations, notably by the third-century Alexandrian mathematician Diophantus and the sixteenth-century Continental mathematicians Rafael Bombelli, Michael Stifel, and Simon Stevin, but most of these were abbreviations of words rather than true symbols. Harriot was familiar with most of these earlier works, and probably drew his method of writing powers from Stifel, who had actually written expressions such as AAA on occasion. Like his peers, though, Stifel did not take his notation very far. It is only with Harriot that we have recognizable symbolic algebraic equations for the first time.[4]

For example, he worked with such equations as $4n^2 + 4n + 1 = 0$ (which he wrote as $4nn + 4n + 1 = 0$). This had arisen in his work related to the mathematics of gambling and combinations, but like many of his forerunners and successors, Harriot was also interested in exploring the properties of equations just for the intellectual satisfaction they bring. An equation doesn't have to mean anything other than what it is, a mathematical sentence that poses the challenge of finding the unknown quantity and thereby "solving" the equation. The surprising thing is that Harriot seems to have been the first to write an expression such as $4nn + 4n + 1 = 0$ at all. This is not only because of his unique symbolism but because he was the first to realize that it is easier to solve a polynomial equation if all the terms are gathered on one side of the equation, leaving zero on the other side. In other words, finding solutions by factorizing is more readily apparent for $4n^2 + 4n + 1 = 0$ than it is for the equivalent form $4n^2 = -4n - 1$. Of course, the strategy of collecting terms on the left-hand side of an algebraic equation, leaving zero on the right-hand side, is now routinely taught in junior high school algebra classes, although it requires what in Harriot's time was a sophisticated conception of zero. By using it this way, Harriot was treating zero itself as an algebraic quantity, not just a number or placeholder.[5]

Harriot's technique is even more important for solving equations of "cubes" and higher powers. (A "power" is the index value on the unknown variable n; the highest power in a "cubic" equation is 3, so that a cubic equation contains terms n^3, together with lower powers. If the highest power is 2, the equation is called "quadratic," after the Latin word for "square.") As Harriot realized, if the solutions of a cubic equation are l, m, p, then the equation can be factorized as $(n-l)(n-m)(n-p) = 0$. And so on for higher powers. Today this "factor theorem" is taught in senior high school mathematics classes. It seems self-evident in hindsight, now that we are so used to thinking symbolically. Yet even Viète had failed to find a general form representing all quadratic equations, let alone the cubics, which Harriot was the first to solve algebraically.[6]

On the topic of solving equations, Harriot was also in the vanguard in the use of what we call complex numbers and he called noetic ones (from the Greek *noetikos*, meaning "intellectual"). These involve what we call imaginary numbers—that is, square roots of negative numbers, which were generally considered "impossible" at that time, for there is no ordinary ("real") number whose square gives a negative number. Nevertheless, they arise in the solution of algebraic equations, so Harriot's term "noetic" is a good one for such purely intellectual concepts. Judging from his papers, his handling of negative numbers and complex solutions was not always consistent, but it is quite possible that the oversimplifications were to do with his teaching work, because elsewhere his papers show that he certainly understood the mathematics correctly. (For instance, he was the first to find all four solutions, including negative and imaginary ones, of a biquadratic equation. In 1579, Bombelli had published the first correct treatment of certain complex roots, but most of his and Harriot's contemporaries ignored solutions containing imaginary numbers.)[7]

Harriot's technique of writing polynomial equations with zero on the right-hand side is an innovation of his that did actually make it into print—in a posthumously published work of algebra known as *Artis analyticae praxis*—six years before Descartes published the same discovery. Not surprisingly, several later mathematicians, notably John Wallis, implied that Descartes had plagiarized Harriot—just as Huygens had implicitly accused Descartes of plagiarizing Snell's law of refraction. Wallis also suggested that Descartes was indebted to Harriot for a number of other innovations, including his symbolic notation. Descartes had visited England in 1631, the very year that Harriot's *Praxis* was published, and what, if anything, he knew of Harriot's work remains a topic of debate. Independent codiscovery is not uncommon, and Descartes had a different focus than Harriot did. Still, Descartes was notoriously vague about his sources, and even his countryman, Viète's editor Jean Beaugrand, noted similarities between Descartes's work and Harriot's.[8]

In addition to the factor theorem and the use of lowercase letters for algebraic variables, Harriot's *Praxis* popularized something Robert Recorde had suggested in 1557: denoting equality by two horizontal lines of the same length. Said Recorde, "No two things can be more equal." Recorde's lines were longer than the denotations we use today—as were Harriot's. In his own manuscripts, Harriot added two small vertical lines across the horizontal ones, as he did also in the symbols he invented for "less than" and "greater than" (which were printed, for the first time, in *Praxis*, in essentially the modern form < and >).[9] In hindsight, it seems extraordinary that in Harriot's day there simply was no formal symbolic structure for algebra, forcing him to invent his own symbols and techniques as he went along.

The significance of his symbolic innovations is readily apparent when looking more closely at the work of Viète, his immediate predecessor. Viète, who was twenty years older than Harriot, had been first a lawyer and then Henri IV's master code-breaker of enemy (Spanish) intelligence, but before he died in 1603 he was the most influential algebraist of the time. Harriot studied his work carefully—perhaps thanks to his friend Torporley, who had been a personal assistant to Viète in Paris. A letter to Harriot survives, addressed to Durham House and written on the eve of Torporley's first meeting with Viète, probably in 1586.[10] Torporley's travels in France had been frenetic, and he wrote, "I am gathering up my ruined wits" in order to prepare to meet Viète, the "renowned analyst"—a meeting dependent on "his courtesy or my boldness… What follows in [his company] I hope shortly to relate." This letter shows not only Viète's "renown" but also Torporley's playful ease. He and Harriot must have been good friends. He signed off by hoping for Harriot's "continued goodwill" and taking his leave "as yours ever in fidelity." As for Viète, he had made major advances in methods of solving equations, although his progress was limited by his lack of a fully symbolic algebra. For instance, he wrote powers in words. Instead of n^2 or nn he would have written A *quad* ("A squared"). He would have written $4n^2 + 4n + 1$ as the Latin equivalent of "four times A squared + four times A in the plane + one point." And he wrote "equals" rather than using a symbol. He also sometimes wrote "plus" instead of "+" (the symbol + was first printed in 1489 in Germany, although it did not quite mean the operation of addition; the first to use it as such seems to have been Michael Stifel in 1544, but it took decades to catch on).[11]

Viète's expression highlights the ancient geometric analogy between squares, lines, and points. For instance, the expression "A squared" arises because the area of a square of side length A is $A \times A$. Similarly "A cubed" is by analogy with the volume of a cube of side A, namely $A \times A \times A$. In writing algebraic equations like this, Viète took an important step on the route connecting Greek geometry with algebra. This connection ultimately led to the invention of analytic (or algebraic) geometry, as mentioned earlier. But much of the vitality of modern algebra lies in its symbolic flexibility, not its ancient link with concrete geometry. It is virtually impossible to visualize geometric objects in more than three dimensions, but Harriot's notation nnn ...was completely general, unconstrained by geometric visualization and conceptualization. It is because of this kind of symbolic freedom that today we are able to speak routinely of such concepts as four-dimensional space-time and ten-dimensional strings.

It was this freedom to think in any dimension that most entranced Wallis about Harriot's fully symbolic notation, and he was right in wanting to accord Harriot his rightful due as the inventor of it.[12] Its innovation was not simply its convenient shorthand. As the dimensional freedom illustrates, it also enabled an economical way of thinking. Harriot was aware of this when he used his new symbolism to solve algebraic problems in a more general and straightforward way even than Viète. A letter from one of his students mentions an algebraic problem in Viète's book for which Harriot had given a better solution, one that was, "[as] you say...more universal and more easily demonstrated."[13]

There are many other examples of Harriot's algebraic advances, including his discovery of what we now call the binomial theorem, usually credited to Newton (who developed it more rigorously). The term "binomial" refers to the fact that there are two key terms, a and b, in an expression such as $(a+b)^2 = a^2 + 2ab + b^2$. Euclid had known this result, but he used a geometric construction of a square—he did not use anything like this modern symbolic notation. Then Brahmagupta discovered, in 628 CE, that (in modern notation) $(a+b)^3 = a^3 + 3a^2b + 3ab^2 + 1$. The binomial theorem shows how to multiply out *any* number n of repeated factors in the general expression $(a+b)^n$, by recognizing that the coefficients of each successive term in the expansion follow the pattern 1, n, $\dfrac{(n)(n-1)}{2}$, $\dfrac{(n)(n-1)(n-2)}{3\times2\times1}$, $\dfrac{(n)(n-1)(n-2)(n-3)}{4\times3\times2\times1}$, and so on—to use modern

notation (which is often abbreviated further, especially to denote combinations). Harriot's notation was similar, as you can see in plate 5, which shows his manuscript page linking the "binomial coefficients" to the so-called Pascal triangle (of which more in the endnote[14]). In the middle section of the page, to uncover the numerical pattern hidden in the triangle Harriot rewrites all the entries as products of numbers—his 12 means 1×2, and so on—and finally, he takes a new step by finding the abstract form of the coefficients for $(a+b)^n$, writing the factors in his numerator one on top of the other instead of side by side in parentheses as we do today. Earlier mathematicians had known this general rule in algorithmic form, but Harriot's version is another example of his ability to take ideas beyond concrete examples by using an arbitrary n^{th} term. It enabled him to generalize binomial coefficients to include fractional and negative values of n, and to apply them decades before Newton.[15]

Another example of Harriot's mathematical innovation is his algebraic form of trigonometry, which he'd used in writing his sine law of refraction, and in his mathematical work on topics such as spherical triangles. Harriot devised his own symbols for the trigonometric quantities sine, secant, and tangent. As mentioned earlier, these quantities are treated today as mathematical functions rather than geometrically

(Image Opposite) **Plate 5**: *Harriot's algebraic symbolism for binomial coefficients*: On this page Harriot proceeds with great clarity from Pascal's triangle, through numerical examples of the pattern of the binomial coefficients that make up successive rows of the triangle, to the general form. His notation differs from ours only in that he writes his algebraic factors in a column rather than side by side in parentheses, and his numerical factorial product in the denominator is written without any multiplication sign between the numbers, as he shows explicitly via examples given on a separate page.

1.   = 1.

1. 1.   = 2.

1. 2. 1.   = 4.

1. 3. 3. 1.   = 8.

1. 4. 6. 4. 1.   = 16.

1. 5. 10. 10. 5. 1. = 32.

1. 6. 15. 20. 15. 6. 1. = 64.   &c.

1.

1. $\dfrac{1}{1}$

1. $\dfrac{2}{1}$. $\dfrac{12}{12}$.

1. $\dfrac{3}{1}$. $\dfrac{23}{12}$. $\dfrac{123}{123}$

1. $\dfrac{4}{1}$. $\dfrac{34}{12}$. $\dfrac{231}{123}$. $\dfrac{1234}{1234}$

1. $\dfrac{5}{1}$. $\dfrac{45}{12}$. $\dfrac{345}{123}$. $\dfrac{2345}{1234}$. $\dfrac{12345}{12345}$

1. $\dfrac{6}{1}$. $\dfrac{56}{12}$. $\dfrac{456}{123}$. $\dfrac{3456}{1234}$. $\dfrac{23456}{12345}$. $\dfrac{123456}{123456}$.

---

1. $\dfrac{n}{1}$. $\dfrac{n-1 \; n}{12}$. $\dfrac{n-2 \; n-1 \; n}{123}$. $\dfrac{n-3 \; n-2 \; n-1 \; n}{1234}$. $\dfrac{n-4 \; n-3 \; n-2 \; n-1 \; n}{12345}$. $\dfrac{n-5 \; n-4 \; n-3 \; n-2 \; n-1 \; n}{123456}$.   &c.

(as in figure 15a). Although the definition of a "function" was not formalized until the eighteenth century, Harriot's use of his symbols suggests that he understood the idea: he delineated the angular arguments of his sines, secants, and tangents using commas, but otherwise he wrote trigonometric equations just as we would today.[16]

A different example of Harriot's ability to think economically appears among his calculations of specific weights in July 1604. He was always looking for ways to simplify unwieldy calculations, and now he came up with the arithmetic of binary numbers. In decimal arithmetic, numbers are represented in terms of powers of 10: a decimal number such as 125 is just $100 + 20 + 5$, or $(1 \times 10 \times 10) + (2 \times 10)$ + (5×1), which can be written as $125 = (1 \times 10^2) + (2 \times 10^1) + (5 \times 10^0)$. (The explicit definition that any number raised to the power of 0 equals 1 is due to Harriot's contemporaries Briggs and Napier, in work published in 1617, although it was already used implicitly in the decimal number system and Harriot's binary system.[17]) Because decimal numbers are broken into sums of powers of 10, the ten digits 0, 1, . . . , 9 are needed. In binary arithmetic, the same number is represented as powers of 2, and so only two digits are needed, 0 and 1. The first few powers of 2 are 1, 2, 4, 8, 16, 32, 64, 128, so to represent the decimal number 125 the largest binary power needed is $2^6 = 64$. All told, we have $125 = 64 + 32 + 16 + 8 + 4 + 1$, or $(1 \times 2^6) + (1 \times 2^5) + (1 \times 2^4) + (1 \times 2^3) + (1 \times 2^2) + (1 \times 2^0)$. If any power is missing, a zero is inserted (as in the decimal number 105). So the binary form of the decimal number 125 is 1111101. Such strings of 1s and 0s are familiar today in the coding used in computers, where the 1 and 0 are represented by the on and off switch in an electric circuit.

A century after Harriot used his binary decompositions, Leibniz designed a rudimentary calculating machine, and he is also generally considered the first to have studied binary numbers mathematically. Harriot's manuscripts nonetheless show him writing out the first sixteen decimal numbers in binary form and then demonstrating the operations of addition, subtraction, and multiplication. In other words, he set out the theory of binary arithmetic. It was a significant discovery in its own right, and, as happens so often in the history of mathematics, he had come upon it in an attempt to simplify his calculations in a different topic entirely.[18] This is another example of the way mathematics developed in an ad hoc way across the centuries. Without a body of knowledge to draw upon, Harriot and his contemporaries had to make up methods and symbols as they went. The casual way Harriot did this suggests another reason that he did not publish: he couldn't judge his own originality or the value of his discoveries. It is only in hindsight—and with greater context—that many of them seem so important. At the time, he surely had no idea that posterity would see him as the discoverer of such diverse results as the arithmetic of binary numbers, the equiangular spiral and its properties, the sine law of refraction, the binomial theorem, exact algebraic interpolation, the area of a spherical triangle, and more, not to mention being seen as the first to produce a fully symbolic algebra.

HE WAS ALSO THE FIRST to offer a compelling explanation of the rainbow. People had been searching for the secret of rainbows for thousands of years, and Harriot was no exception. In the grueling year after Ralegh's arrest, he may have welcomed the chance to contemplate such uplifting natural beauty. Northumberland was interested in the topic, too, and perhaps he helped Harriot take new measurements of refraction with his astrolabe, in November 1604. By early 1605, however, Harriot was so committed to his experimental research that he had employed his own laboratory assistant, Christopher Tooke, who would be Harriot's lens grinder and glass worker for the rest of his life. In April 1605, Tooke helped with Harriot's systematic exploration of the colors produced when light is refracted by a prism. They are the same colors produced by raindrops in a rainbow, although refraction in single raindrops was not a straightforward notion.[19]

For a start, Aristotle, one of the first to write on the subject, had thought that rainbows were created by *reflections* in *whole* clouds. More than a thousand years later, even al-Haytham had made no further progress on the topic. Then, around the turn of the fourteenth century, several scholars in Persia and Europe experimented with light passing through glass spheres containing water and suggested the idea that in a rainbow light was being *refracted* through each drop of water. Their work was lost in Harriot's time, and he and his contemporaries had to rediscover this for themselves. In fact, Harriot, and a little later Kepler, were most likely the first theorists of their era to gain a clear idea that rainbows were caused by sunlight being refracted and reflected within individual drops of rain, as we'll see, and as shown in figure 18a.

As to exactly how the colors were produced, that was another matter entirely—although clearly it was somehow associated with water droplets. Everyone has admired the iridescent colors of a sunlit dewdrop on a blade of grass or a raindrop hanging from a tree twig. Prisms, too, produce colored fringes, although today's image of an entire rainbow spectrum emanating from a prism is post-Newtonian. Meantime, no one knew how these colors were produced by the prism or dewdrop, and many scholars had wondered whether they were even real. Perhaps they were just an illusion produced by the eye or by impurities in the water or the glass.

Kepler, who sometimes tended toward the mystical, had initially invoked the Pythagorean idea that the colors of the rainbow were related to the different notes in a musical octave. Newton, in his way, would be just as mystical. He, too, embraced the musical analogy when he concluded that there were seven colors rather than the six we now accept: he added indigo (though he realized there were varying shades of each color).[20] However, while Kepler had no idea that color resides in "white" light itself, Newton discovered the phenomenon of "dispersion," the splitting of "white" sunlight into the spectrum of colors. He presented his results to the Royal Society in 1672.

Almost seventy years earlier, Harriot was quietly making his own experiments on what we now call dispersion. Newton's discovery was, in fact, another example of rediscovery. Witelo had come close to the idea, by trying to relate color to angles of refraction, something Kepler (and Descartes) would also do later, since the color emanating from a dewdrop changes as you shift your line of sight. But neither Witelo nor Kepler nor Descartes had imagined that on refraction through a prism or raindrop, a *single* ray of sunlight was, in effect, refracted into six rays, each of a different color.[21] Harriot's manuscripts, by contrast, show diagrams of a single ray entering a prism and leaving as several different rays, each of a different color.

Harriot did not entertain mystical ideas such as an "octave of color," but unlike Newton he did not analyze the entire spectrum. Rather, in the diagrams and calculations in his working manuscripts, he focused on two or three colors (often just red and yellow or orange, but sometimes green, too). Still, he did set about measuring the refractive indices of these colors.[22] If his results here were less polished than his general tables of refraction, this was nonetheless an astounding conceptual leap, one in which Harriot showed quantitatively that each color had a different refractive index, so that dispersion happened because each component color in an incoming light ray was refracted by a slightly different amount. It was an impressive insight—one that Newton elaborated more fully and carefully a century later.

As for putting it all together with a mathematical theory of the formation and shape of the rainbow, that would have to wait, as external events once again intervened. On November 4, 1605, Harriot attended a dinner party that would turn out to have disastrous consequences. A small band of Catholic radicals planned to blow up the Houses of Parliament on the morning of November 5, as the king was opening the week's session. One of their leaders, Thomas Percy, had hired the vault under the House of Lords in which the plotters had stored the gunpowder for what has become known as the Gunpowder Plot—and he was a distant cousin of Henry Percy, the Earl of Northumberland. Thomas, who was also employed by Northumberland to oversee his northern estates, had been present at the Syon dinner with Harriot and other friends and family—including the MP for Carmarthenshire, Sir William Lower. Lower had recently married the earl's stepdaughter and would soon become Harriot's most loyal friend and disciple.

During the meal, Guy Fawkes had arrived with a message for Thomas Percy, and the two of them had returned to London.[23] But one of the conspirators warned a friend to stay away from Parliament, and the plot was uncovered at the last minute. Fawkes was arrested, and Percy and three others were later killed as they attempted to flee. In the fallout, many people were interrogated—including Northumberland, Lower (whom King James had knighted in May 1603), the Reverend Torporley (Harriot's mathematical friend), and Harriot himself. Cecil, whose role as secretary of state had continued under James, had never gotten along with Northumberland and viewed his links with Ralegh with such suspicion that the latter was also questioned on the matter. Cecil even placed a spy at Sherborne. He apparently hoped—fruitlessly as it turned out—to implicate Lady Ralegh, because two of the Gunpowder Plotters had family connections to the Catholic Throckmortons, Bess's kin. In fact, Cecil was so zealous in his investigation that some suggested he had orchestrated and then "discovered" the whole thing in order to reinforce his own political and religious agenda. This possibility is still being debated today. Most historians reject it, while acknowledging there are a few gaps and ambiguities in the record.[24]

None of the Syon circle had any knowledge of the plot, although it is possible that at the dinner Thomas Percy had given Northumberland and Lower an oblique warning not to attend Parliament the next morning. As Ralegh's trial had shown, however, innocence offered no protection from the law, and soon Northumberland found himself imprisoned in the Tower. Harriot was locked in the Gatehouse Prison, but Lower and Torporley were left free. Once again, the fear of mathematical "conjurors" seemed to work against Harriot, because the superstitious James wanted to know whether Harriot had cast his horoscope for Northumberland or the plotters. He also wanted to know whether Harriot had ever heard Northumberland expressing discontent with the state. The new king must have been insecure to evince such suspicion of Northumberland, who barely eighteen months earlier had been his much-valued advisor during his transition to the English throne. Or perhaps he had never really trusted the earl, given that Henry Howard had maliciously told him that Northumberland and Ralegh were both opposed to his accession.

Harriot's rooms at Syon were searched, though nothing incriminating was found. A report of the search, sent to Cecil and dated November 25, 1605, gives a rare peek into Harriot's private world. The reporter commented on the huge number of Harriot's papers—so many that their contents would have required "many days" to survey. The investigator felt this was unnecessary: a brief glance showed that they were mathematical and scientific, not treasonous. There were also "books of all sorts of learning, and many [of them, and also books] of all sorts and professions of religion"—clear evidence of Harriot's eclectic interests.[25]

On December 16, after several weeks in prison, Harriot wrote a pleading letter to the Privy Council. It is hard to imagine at least some of the lords remaining unmoved by his ingenuous sincerity, or by his touching evocation of the toll taken by his arduous labor in the name of science:[26]

> The present misery I feel, being truly innocent in heart and thought, presses me to be a humble suitor to your lordships for a favorable respect. All that know me can witness that I was always of honest conversation and life. I was never any busy meddler in matters of state. I was never ambitious for preferments. But contented with a private life for the love of learning that I might study freely. Wherein my labors and endeavors, if I may speak it without presumption, have been painful and great. And I hoped, and do yet hope by the grace of God and your Lordships' favor, that the effects [of this study] will show themselves shortly, to the good liking and allowance of the state and common weal.

So it would seem Harriot did intend to publish his results, if only he could bring all his work to fruition, in freedom. He was not in robust condition, though. His letter goes on to describe—in some detail—how "this misery of close imprisonment" had worsened the health problems he had been suffering for some weeks before, namely "great wind in my stomach and fumings in my head rising from my spleen, besides other infirmities, as my doctor knows and some effects my keeper can witness." Were he to remain in prison much longer, it would bring about his "utter undoing." He went on to speak again of the "innocency of my heart" and to say that he was anxious to continue his studies, which was why he was asking his lordships for his "liberty."

The lords were not entirely heartless, and the evidence was flimsy, and Harriot was eventually released. Northumberland was not so lucky. The plotters had been executed, but several supposed accomplices were still languishing in the Tower.

CHAPTER 16

Solving the Rainbow

ARLY IN 1606, BETWEEN VISITS to Northumberland and Ralegh, Harriot was making the most of his "liberty," taking measurements of refraction and thinking again about the beguiling mysteries of the rainbow. He'd been wondering how the rainbow came to have its uniform size and shape. He knew that rainbows are seen when the sun is behind the observer. He also knew that there was a fixed angle between an imaginary line from the sun to the tip of the observer's shadow on the ground, and his or her line of sight to the top of the rainbow (as in figure 18c)—and that three centuries earlier Roger Bacon had measured this angle to be about 42°. Fifteen hundred years before that, Aristotle had realized that the rainbow is circular because the rays from the rainbow to our eyes form a circular cone (as in figure 18d). Bacon had given this cone an angular radius, but Harriot wanted to understand why the rainbow formed in this way.

He began by pondering the fact that when a beam of light falls onto the surface of a body of water, part of the beam will be reflected off the surface and part of it will be refracted into the water (as in figure 16a). This must be so, because were there no reflected light from the surface, we would not be able to see it—and were there no refracted light, immersed objects would not look distorted or displaced. So Harriot realized that when a ray of sunlight traveling through the air meets a raindrop, part of it will be reflected away from the surface, and part of it will be refracted into the raindrop. This refracted ray, which has now slightly changed its direction because of the refraction, will pass through the water until it meets the back of the raindrop. There some of it will be refracted out into the air away from the observer, and some of it will be reflected within the raindrop. For some angles of incidence, the reflected ray will then be refracted out of the drop (as in figure 18a); otherwise, it will be reflected inside the drop again and again until it escapes (as in figure 18b, for example). Each time the ray changes direction via reflection, however, some of its light is also transmitted out of the drop and scattered away. In other words, light rays that illuminate a raindrop are eventually scattered out in all directions.

Yet this scattering cannot be uniformly random, because we see some of this light in a particularly concentrated arc of color, the rainbow itself. This means that there must be a concentration of rays that pass into the raindrop and are refracted back out again in the direction of the observer. This is why Harriot settled on the path shown in figure 18a: it has only two refractions—one in which the ray enters the raindrop and one in which it leaves—and one reflection. This single reflection is necessary to turn the ray around so that it comes back out of the raindrop and toward the observer.

Harriot discovered that the law of reflection and his new law of refraction showed just why such a concentration of rays occurs. If you draw many different rays, coming into a spherical raindrop at different angles of incidence, and if you trace each ray's path using the laws of reflection and refraction, the geometry shows that there is one particular set of angles of incidence—those around 60°—for which something unexpected happens. All these rays have a very similar path through the raindrop, so that when they emerge they all return to the observer in a concentrated bunch. By contrast, for groups of rays around other angles of incidence, the emerging rays are more spread out. This bunching was surprising, because even Kepler believed there ought to be a smooth transition from each angle of incidence to the next.[1] What is more, these concentrated emerging rays make an angle with the original incident ray of around 42° (as in figure 18a)—just as Roger Bacon had measured.

The diagrams in figure 18 show an angle of not 42° but 41°46′; this was Harriot's calculation for the maximum possible angle of the rainbow rays, and it corresponded to an initial angle of incidence of 59°17′. His method is equivalent to noting that a ray with this angle of incidence is the one that has changed its direction by the least amount during its journey through the raindrop and back to the observer. Today the idea of a "least value," or minimum, is handled by calculus. Newton would be the first to apply calculus to the shape of the rainbow—but it was Harriot who first understood the underlying geometry and physics.[2]

Harriot's angle of $41°46'$ was calculated for a single, nondispersed ray, so it is his value for the *average* size of the rainbow.[3] The colors of the rainbow are of course due to dispersion, which Harriot had studied in 1605. The top of a rainbow is red, because on being refracted into the raindrops, the red rays are bent the least, and so they emerge in the widest arc. They make a cone of not $41°46'$ but a little more than $42°$. In modern terminology, red has the lowest index of refraction for water with respect to air, and violet, which is bent the most sharply, has the highest. In a rainbow, the angle of the cone of violet rays is just over $40°$, so we see a band of color from violet at around $40°$ to red at around $42°$. (As Newton would make clear, we see each color of the rainbow emerging from a different raindrop—one that is positioned at just the right angle in our cone of vision.) Harriot did not spell out the nature of this band of color, but he certainly understood the principle because he knew that different colors have different indices of refractions, even if most of his experimental and mathematical work on dispersion focused on the difficult enough task of quantitatively differentiating between just two or three colors.[4]

Descartes would become the first to publish a correct theory of the rainbow's formation and shape, which he did in the appendices to his 1637 *Discourse on Method,* thirty years after Harriot had made his discoveries. It was yet another apparent example of independent rediscovery—and another example of Harriot's contribution being unknown to history for not being published.[5] Descartes's explanation was very clear, however, whereas Harriot's working manuscripts rarely show him bringing his ideas together in a coherent fashion. The constant interruptions played a significant role in this, but his surviving manuscripts also suggest that he was much better with symbolic mathematics and experimental records than with written explanations. On the other hand, Descartes's published versions were polished for publication, whereas Harriot's were working manuscripts.

Descartes also explained the phenomenon of the paler secondary rainbow that is often seen above the intense primary rainbow. While the rays in the most concentrated bundle have been reflected once within the raindrop, the ones in the next most concentrated have been reflected twice within the drop before being refracted out to the observer (as in figure 18b). Harriot did not discuss the secondary rainbow explicitly, although there are diagrams in his papers that seem to suggest that he understood it.[6] Descartes understood the geometry, but unlike Harriot he did not know about dispersion. Instead, he developed a complex, convoluted "theory" whereby color was due to the circular motion of the particles of "ether" that he believed filled the universe. Like most of his pre-Newtonian peers, he was given to wild but woolly speculation on theoretical matters.

When it comes to appreciating the novelty of Harriot's discovery of dispersion, we learn more by looking seventy years into the future and seeing how the young Newton's initial announcement of this phenomenon fared. The idea that sunlight was "pure" and color was an artifact was still so entrenched that after he first declared his discovery to the Royal Society in 1672, his findings were disputed by some of the best scientific thinkers—including Robert Hooke, also known for his sharp tongue, and Huygens. It didn't help that Newton favored a particle model of light whereas Hooke and Huygens preferred a wave theory, but their criticisms also stemmed from their complaints that Newton had not always explained his prism experiments adequately. This made it difficult to replicate his results, and initially few were convinced by his conclusions. Notoriously touchy and anxious, Newton alternately retreated into his shell and launched heated defenses. But he took on board the most perceptive criticisms and refined his theory and his experiments over the years, finally publishing his famous *Opticks* in 1704.[7]

HARRIOT'S INVESTIGATION OF THE RAINBOW in 1606 was an early example of a physical theory based solely on careful experiment and mathematical theory, not on untested speculation. This would be a hallmark of Newton's approach, but one reason Harriot did not put all his own work together in a polished whole was the tragedy engulfing his friend and patron Northumberland. In happier times, the earl had been eager to understand "the colors of the rainbow and the cause of his arkedness," as he had put it in a letter to his son.[8] Now, in the early summer of 1606, he had been languishing in the Tower for nearly eight months without a trial. Finally, on June 27, he was brought to the Star Chamber—with Popham as chief justice and Coke as attorney general—to face charges of complicity in the Gunpowder Plot. These charges were based entirely on suppositions and on the testimony of terrified prisoners such as Guy Fawkes. Fawkes was most likely tortured, despite Cecil's attestations to the contrary; even so, his most damning evidence against Northumberland claimed only that Thomas Percy had planned to warn his cousin to stay away from Parliament on November 5.[9]

Northumberland's case was not helped by his friendship with the convicted "traitor" Ralegh. Nor was it helped by the fact that in his secret correspondence with James before he became king, the earl had suggested that James's transition to power in England might be eased if he rescinded some of Elizabeth's anti-Catholic edicts. It was noted by a witness, however, that Northumberland was "troubled not much himself" about religion—a comment that contradicted the state's imputation of rebellious Catholic tendencies but left the earl open to dangerous accusations of atheism instead.

During the trial, it was also claimed that Torporley had cast the king's horoscope earlier in 1605 and had predicted that he would have a "troublesome reign." Harriot—"a man seen in that art [astrology] and accounted very cunning"—was said to have confessed that he had seen the horoscope and told Northumberland about it. What such a vague prognostication had to do with specific foreknowledge of the Gunpowder Plot was surely as much beyond Harriot's ken as it is ours. Coke said as much, sneering that had the stars been able to warn Northumberland he would not now be in the Star Chamber. Yet instead of dismissing the horoscope charge because of its patent absurdity, Coke used it to smear Northumberland for his "vanity" in supposedly embracing the "unlearned learning," the "deceivable doctrine," that was astrology. Northumberland hadn't made things easier by denying he had seen Torporley's horoscope, admitting instead that he had heard of a later one that had promised good things for the king.

To cut a long, drawn-out litany of accusations short—accusations based solely on hearsay and supposition because of Northumberland's relationship with his traitorous cousin—the earl was fined £30,000, a colossal sum, and stripped of his offices. Most of the lords sitting in judgment—including Cecil and Henry Howard, Ralegh's treacherous erstwhile "friends"—also believed he should be imprisoned for life. Just as Ralegh was condemned simply for who he was, now, in the final summing up, Northumberland was condemned simply as a warning against future plots like "that most execrable powder treason."

Such was Jacobean justice. Both of Harriot's patrons were condemned to life in the Tower.

CHAPTER 17

Conversations with Kepler

D ESPITE THE LOSS OF HIS liberty and offices, Northumberland still had sufficient money to ensure as comfortable a life as possible in the Tower—although this was mainly because he had negotiated a delayed payment of his huge fine. He even kept subsidizing Harriot. In fact, such was Northumberland's generosity that in 1608 he would issue, from his prison, directions so that Harriot would have his own house in the grounds of Syon. (Northumberland was not always so generous to his family. Although a degree of affection grew between them he had initially married his wife, Dorothy Devereux, Essex's sister and Sir Thomas Perrott's widow, for purely practical, dynastic reasons. He wasn't much interested in his children, either, although in 1608 he arranged for his young son and heir, Algernon, to live with him in the Tower so that he could direct his education until he went to university.[1])

In between bouts of illness and despair, Ralegh, too, was making the best of it. He and Northumberland soon set up a still house in the Tower's gardens, where they performed chemical/alchemical experiments and brewed medicines and liquors. Bess's specialty was distilled water—a necessity for safe drinking in London—while Ralegh's Guianan friends apparently helped with South American recipes, using indigenous plants such as *chincona*, or quinine.[2] Ralegh's medicinal "Cordial" became so famous that in 1612 it would be administered to Prince Henry, the king's popular young heir, who was then direly ill. To no avail: he died at the age of eighteen, of typhoid picked up during a swim in the Thames. His death was a loss to Ralegh, who had begun writing his *History of the World* "for the service of that inestimable Prince Henry." Harriot had some sort of teaching engagement with the youth, too. Henry's name is written on two of his manuscript pages: one referring to the solution of a cubic equation, and the other to a problem in combinations theory.[3]

In late 1606, however, Harriot began corresponding with someone whose learning rivaled his own: Johannes Kepler. The thirty-five-year-old Kepler was eleven years younger than Harriot and was already famous as the author of several books, notably his 1604 commentary on Witelo—*Ad vitellionem paralipomena,* or *Optics,* as Kepler referred to it—and his 1596 work of cosmology, *Mysterium cosmographica.*[4] Kepler had spent a year working as the great astronomer Tycho Brahe's assistant in Prague. (After a falling-out with his patron, the Danish king, Tycho had left his island retreat and set up an observatory near Prague in 1600.[5]) After Tycho's death in 1601, Kepler became imperial mathematician at the Prague court of Rudolf II, where Dee had been some years earlier.

Both Kepler and Harriot had been independently thinking about the rainbow, and two years earlier Kepler had published his initial ideas in *Optics.* He heard about Harriot's work on the same subject through Jan Erikson, a traveling merchant, who apparently had business dealings with Northumberland or Ralegh. Erikson told Kepler that he knew of an Englishman "most skilled in all the secrets of nature," especially optics. What is more, said Erikson, Harriot's work would show the limitations of Kepler's 1604 *Optics.* Since Kepler enjoyed feedback on his work, he was not affronted by Erikson's claim. Rather, in early October 1606, he wrote enthusiastically to this Englishman, seeking his advice.[6] (Kepler spoke German and Harriot English, but they corresponded in Latin, the common language of European scholars at that time.)

"I hear you had troubles because of your astrology," Kepler wrote Harriot, presumably referring to Harriot's being interrogated as to whether he had cast the king's horoscope. Erickson must have told Kepler of the drama of the Gunpowder Plot. "Do you think it is worth it?" He himself rejected most of traditional astrology, he explained, adding, "The only part I kept are the aspects, and I link astrology to the doctrine of [cosmic, Pythagorean] harmonies." He thoughtfully suggested that Harriot might prefer to reply in an unsealed envelope "if a sealed letter might invite suspicion."[7]

Kepler's letter outlined his latest—unpublished—thoughts about the rainbow. He had finally concluded that it depended not on the whole cloud, as he had originally thought, but on the individual raindrops within. He also admitted to Harriot that his optical theory was based on theological rather than physical principles. (As mentioned earlier, Kepler had trained in theology, originally intending to become a Lutheran minister.) For instance, he saw the sun as a visible image of God, and this not only predisposed him to accept Copernicus's sun-centered cosmology but also affected his concept of the transmission of light.[8] Giordano Bruno had accepted Copernicanism for similarly mystical reasons. He believed that the whole cosmos was literally alive, and so the planets had the capacity to move of their own accord around the sun. Harriot was not given to turning religious beliefs into scientific theories, but on a couple of manuscript pages he copied out some Latin verses celebrating the sun's divinity. There was some interest in such "pagan" views of the sun among pro-Copernicans, although copying out these verses seems to indicate the extent of Harriot's invocation (if that is what it was).[9]

Kepler wrote that he had heard of Harriot's experimental skill and now asked for help with the physical part of the analysis of the rainbow. He was clear about what he was seeking: "give me the measures of all refractions in your experiments, then everything else will be easy." He also requested that Harriot tell him how refraction causes color and asked him to share his experimental measurements "freely and frankly."[10] He reaffirmed his own view of color, expressed in his *Optics*: namely, the Aristotelian view that colors were intrusions of "darkness" due to impurities in the glass of prisms (or the water of raindrops). Nonetheless he was intrigued to learn—from Erickson— that Harriot had a theory of color and that it apparently derived from his studies of alchemy: "I desire to know the origin of colors and their essential differences, from you who have studied alchemy... If you are teaching from alchemy that all the colors inhere in the body of water, glass, crystal and so forth, then the way in which they are drawn forth and displayed on a piece of paper will be, it seems, thoroughly explained."

There is no record of Erikson's conversations about Harriot with Kepler, so no one knows exactly what he meant by telling Kepler that Harriot had "studied alchemy." But the worldview Harriot inherited from his predecessors was very different from our own, with its emphasis on studying natural phenomena by means of quantitative experiment and mathematical laws. Harriot was instinctively moving toward such a view, as most of his work shows; nevertheless, like Kepler, and even Newton a century later, he was still influenced by this ancient past, with its holistic conception of the universe—its astrological and alchemical sympathetic correspondences between sky and earth, and between the inner processes of life and the outer natural world. At that time, some alchemists still believed they might one day actually turn lead into gold as the ancients had dreamed; others saw such concepts as metaphoric. And some had no interest in lead and gold at all, other than the way they combined with other substances. Northumberland, the "Wizard Earl," has passed down an outline of this last kind of alchemy.[11] It owes more to Aristotelian concepts of material processes such as generation and decay, and to properties of matter such as "heaviness," than it does to magic or metaphor. "The doctrine of Generation and corruption unfolds to our understanding the method general of all atomical combinations possible in homogeneal substances, together with the ways possible of generating of the same substance as by semination, vegetation, putrefaction ..., with all the accidents and qualities rising from these generated substances [such as] hardness, softness, heaviness, lightness..., color, taste, smell etc."

Harriot's alchemical work was of this latter kind, although back in 1590, he and Nathaniel Torporley had written out two traditional alchemical "recipes" from a Dr. Turner—presumably the same Dr. Turner who along with Harriot and one or two others had recently received special permission to visit Ralegh in the Tower whenever he liked.[12] In traditional "hermetic" alchemy, secret recipes often functioned as spells or codes rather than as experiments designed to learn about the transformations of matter. Harriot and Torporley had used Harriot's Algonquian-inspired phonetic alphabet to write Turner's recipes, although it is likely they did this merely for amusement, not to maintain secrecy. Throughout his manuscripts, Harriot sometimes wrote innocuous scientific headings in his unique script, and he used it to sign his name elaborately on the title page of one of his treatises (shown in plate 1). He also used it to doodle his own name and those of friends such as Walter Warner, and in the same vein he occasionally made witty anagrams from his name. Evidently even someone as sober and intensely focused as Harriot needed some light relief.[13]

A decade later, in 1599, in the early years of his residence at Northumberland's Syon, Harriot had performed some detailed alchemical experiments that contrasted markedly with Dr. Turner's recipe for turning clay into pewter. These experiments show that Harriot believed in the fundamental nature of Aristotle's four elements—fire, water, earth, and air—because his goal was to explore the effects of these "elements" on a variety of different substances. For instance, in one experiment he placed containers of mercury and sulfur in heated water, earth, and hot ("fiery") ashes, then observed what happened over time. His notes read like a modern record of a chemical experiment, although his experiments were designed to investigate alchemical reactions that bear very little resemblance to those of modern chemistry. His notes also show that he was simply observing and recording; he offered no interpretation, physical or metaphorical, of his meticulously recorded transmutations.[14] Nevertheless, he was trying to document the way different kinds of matter interact physically, and in that sense, he was doing proto-chemistry.

Regardless of what Kepler had heard about Harriot's alchemy, it had nothing to do with his theory of color—that is, his discovery of dispersion, which he had deduced from his work with prisms. He was, however, working on a theory of refraction itself, using the ancient notion of atomism, which some associated with the kind of alchemy described by Northumberland in his reference to "atomical combinations." Perhaps Erikson had told Kepler something about this theory. At any rate, Kepler closed his first letter to Harriot by rephrasing his request for Harriot's refraction data and theory of color as a playful, mystical challenge: "Now you, O excellent initiate of the mysteries of nature, reveal the causes."[15]

One wonders what Harriot made of his younger correspondent's half-mocking challenge, and of his confident request for "a complete description" of the experimental labors that had cost Harriot so much. In his prison letter to the Privy Council, he had already mentioned the toll research had taken on his health. Newton would experience the same struggle, while Nunes had abandoned his study of the loxodrome altogether, as mentioned, because in doing difficult mathematics he had "irretrievably lost my health," as he'd put it in a manuscript defending his work on rhumb lines.[16] So it seems likely that Harriot was somewhat taken aback that Kepler—who was upfront about his own lack of experimentation, and as assertive as Harriot was reticent—expected so much from him at this early stage of their acquaintance. No doubt he was also wary of sharing his results with a stranger. He had told the Privy Council that he did want to publish, and Edward Wright's experience had shown how readily some colleagues could plagiarize one's work.

In his reply to Kepler, written in early December 1606, Harriot included not his complete tables of refraction, but rather a limited set of data for fifteen transparent substances: their specific gravities (or densities) and the angle of refraction corresponding to a 30° angle of incidence. He did this to show that Kepler's belief that the angle of refraction was proportional to the refracting medium's density alone was not borne out by experiment. This was an important result. It is not density per se but refractive index that expresses the medium's effect on the angle of refraction. Harriot knew the value of his painstaking measurements, and he was confident of his own worth. "This table tells you more than could be said in your whole long letter to me," he wrote. His dismissive tone can perhaps be forgiven under the circumstances: he had begun his letter by saying he had no time to reply fully to Kepler's questions at present, and his poor health meant that "it is difficult for me to write or think or argue clearly about anything at the moment."

This was surely true, after the grueling year following the Gunpowder Plot. Reading between the lines, though, it also seems he felt that if Kepler could toss off challenges so lightly, he would do likewise. Referring to his table of data, he said, "I have disclosed these things to you so that you can either share in my conclusion concerning certain optical matters [the role of density], or it can be an incentive for you to make further observations of your own."[17] In other words, if they were to be collaborators, Harriot expected Kepler to do some careful experimental work, too.

In his letter, Kepler had described his geometrical conception of the way light rays refracted through spherical drops to produce the rainbow. It was not yet correct, as Kepler himself acknowledged, and as Harriot realized immediately (it had too many reflections and refractions, in contrast with the two refractions and one reflection in figure 18a). "I would say this of the rainbow just now," Harriot wrote, "the cause is to be demonstrated in a droplet through reflection on a concave surface and refraction on a convex. However, I have said nothing in consideration of the mysteries that are concealed." Presumably these "mysteries" meant his discovery of the way the rainbow colors are produced by dispersion. For reasons of ill health—the "fumings" in his head that he had described to the Privy Council—and because he had had to put aside his rainbow research during Northumberland's trial five months earlier, Harriot had not had time to gather his results together (they are spread over many, often nonsequential manuscript pages). So he held back on revealing his discoveries, assuring Kepler that he would explain the rainbow's cause fully when he wrote up his work properly. Until then, he asked for Kepler's patience.[18]

Nevertheless, he did try to answer Kepler's challenge to "reveal the cause" of refraction. The key question was why, when a light ray shines on the surface of a transparent medium such as water or glass, the light simultaneously reflects off the surface and refracts through the medium. The surface of the medium must provide two different aspects: one that resists (and reflects) the ray, and one that allows the ray to pass through it. "Therefore," Harriot concluded, "a dense and transparent body which to the senses appears to be continuous in all its parts, is not so in reality." It must be made of tiny discrete particles, or "atoms." Harriot's theory was that when light from the sun hit a water surface, say, part of it struck a water particle and bounced off. This was the reflected part. The refracted part slipped into the water between the surface particles and then bounced between the water atoms so that "refraction is nothing else than internal reflection," as he put it. Although the refracted ray appears straight, he continued, it is really composed of "a great number" of tiny zigzagging segments.[19]

Harriot did not explain further what he meant by this, saying, "But let us make a stop here...I have led you to the doors of nature's mansion, where her secrets are hidden. If you cannot enter on account of their narrowness, abstract yourself mathematically, and contract yourself into an atom, and you will enter easily. And after you have come out, you will tell me what wonders you have seen." It was a playful metaphor for an idea in progress, which, it seems, Harriot was trying out in his letter, evidently hoping for a response from Kepler in order to help him sharpen his ideas ("tell me what wonders you have seen").[20]

This idea in progress shows Harriot speculating on a physical cause. Most of his scientific work is concerned with the *what* rather than the *why* of nature. In this case, though, he had already found the law of refraction and the refractive indices of colors in white light, and so he had begun to ask deeper questions. First he had found out why the rainbow forms, and now he wanted to know why refraction and reflection occur in the first place.

CHAPTER 18

Atomic Speculations

H ARRIOT'S THEORY OF REFRACTION INVOLVED a theory about matter itself, which he had tried out in his letter to Kepler. He was doing the same with his friend Torporley: it was during this period that Torporley wrote his notes of their discussions on the controversial subject of atomism.[1]

Torporley had the highest respect for Harriot as a mathematician and an experimentalist. In his 1602 book *Diclides coelometricae*, published not long after Harriot had discovered the law of refraction and while he was working on the trigonometry of the spherical triangle, Torporley had publicly praised him for dissipating, "by the splendor of undoubted truth, the philosophical clouds in which the world had been enveloped for many centuries."[2] But *Diclides coelometricae* is a rather odd book. It is on the topic of spherical trigonometry and combines obscure rhetoric, original theorems, and bizarre mnemonics. Unlike Harriot, Torporley did not shrink from publishing his work, no matter how slight. He seemed to have in mind astrologers rather than mathematicians as his readers. Astrology requires calculating angles between planets on the celestial sphere, and Torporley used astrological and other metaphors as mnemonic verses for remembering the required relationships between angles and distances in a spherical triangle. In view of Harriot's work for Ralegh, perhaps Torporley had sailors in mind, too. Spherical triangles are needed to work out distances traveled along great circles, as Nunes's work had also shown.

Torporley's admiration of Harriot dimmed somewhat over the years—not because of Harriot's mathematics, which Toporley would continue to praise highly, but because of his freethinking. In particular, Harriot's use of "atoms" to explain refraction was "anathema" to Torporley's religious sensibility, and he was piqued that his friend was "reviving" such a "pseudo-philosophy," as he put it.[3] As we saw in connection with Harriot's embroilment in the Cerne Abbas investigation back in 1594, the ancient idea that everything was created from pre-existing atoms suggested the heretical converse concept *ex nihilo nihil fit*: from nothing, nothing comes. For Torporley, such a concept was a denial of the biblical account in which God created the world from nothing—and he felt he needed to look no further than Aristotle for support. The ancient Greek philosopher might have believed in the eternity of the world, but he was acceptable to the Christian mainstream not only on account of his geocentric crystalline heavenly spheres but also because he had argued against the existence of discrete, indivisible atoms as proposed by Leucippus and his student Democritus in the fifth century BCE. (The Greek *atomos* means "indivisible.")[4]

Aristotle argued on such grounds as continuity. Time and space are surely decomposed into smaller and smaller parts, ad infinitum, he said; otherwise there would be gaps in space and time. These gaps would lead to paradoxes, such as Zeno's "Arrow." (Zeno's famous paradoxes have survived largely through Aristotle's discussion of them.) If space and time were made of tiny discrete units—atoms of space, moments of time, which cannot be further subdivided—then consider a moving arrow. If you take a snapshot at each separate moment, the arrow is at rest. Without the continuity enabled by infinitely subdividing moments of time, how can the arrow ever move?

On the other hand, if space and time are infinitely divisible as Aristotle believed, the famous "Achilles and the tortoise" paradox arises. If the tortoise is given a head start in a race with Achilles, then by the time Achilles reaches the tortoise's starting point, the tortoise will have moved on to a new position. By the time Achilles reaches that position, however, the tortoise will have moved on a tiny bit more. Each time Achilles catches up to the tortoise's previous position, the distance between him and the tortoise is shortened, but the argument can go on indefinitely. Achilles would have to make an infinite number of "catch-ups" to reach the tortoise—and an infinite number of steps, no matter how small, would surely take an infinite time.

Zeno had another version of this paradox: in order to run from A to B, you first have to run halfway to B, but before you reach halfway, you have to run a quarter of the way, and so on, indefinitely. Which means you can never really get started. Of course, everyone knows that Achilles will overtake the tortoise and that you can run from A to B. What Zeno seemed to be saying here is that space and time cannot be infinitely divisible, or there would be a paradox—not a physical paradox but a logical one. On the other hand, if discrete, indivisible atoms of time and space exist, there is a logical paradox when considering the motion of the flying arrow.

It took mathematicians almost two and a half thousand years to sort out what they regard as a satisfactory resolution of Zeno's paradoxes. The modern mathematical understanding of continuity, and of the analysis of motion via differential calculus, was not made rigorous until the nineteenth century, when infinitesimal and infinite "limits" were precisely defined. In a sense, the logical difficulty still remains: an infinite number of infinitesimal moves would indeed require an infinite time. So what mathematicians sought to do was to build a conception of space, time, and motion that allowed them to handle infinite and infinitesimal quantities in a way that sits most meaningfully between the intuitive and the logical.

The only way Aristotle could reconcile Zeno's paradoxes was to deny not only the existence of finite, indivisible "atoms" but also the actual possibility of infinite subdivision. In other words, to avoid all of Zeno's logical paradoxes, Aristotle suggested the idea of "potential" infinite subdivisions (and potential infinite sums). Still, a potential was not an actual, and most subsequent Greek mathematicians did not consider actual infinite sums the way Harriot did, when he found the finite length of his spirals by finding intuitive "limits" of the infinite geometric series derived from the spirals' tiny segments. In fact, Harriot had noted the Zeno conundrum in connection with his rhumb lines and the seeming absurdity of the fact that these lines keep spiraling forever even though he had shown that they had a finite length (from any given point).[5] These sorts of apparent paradoxes were deeply puzzling without the tools of modern mathematics. Harriot took a new step along the way when he gave not only the lengths of the equiangular spiral and loxodrome but also what was probably the first algebraic analysis of the Achilles paradox.

Aristotle had made a start in this direction by trying to articulate the concepts underlying the paradox. Harriot nonetheless felt that these verbal arguments were "but briefly and obscurely set down with an answer uncertain."[6] So, in good mathematical style, he made the problem concrete. To keep the calculations simple, he supposed that Achilles ran ten times faster than the tortoise and that the tortoise had a head start of one unit of distance. (If this head start were 100 meters, say, then Harriot could simply scale the other numbers in the calculation appropriately.) Using simple algebra, Harriot showed that when Achilles had run 10/9 units, he would have caught up to the tortoise (as shown in the endnote[7]). Some years later, Gregory of St. Vincent also found the exact place—10/9 units of distance—where Achilles would overtake the tortoise, and he has generally been considered the first to do this.

The point for Harriot (and St. Vincent) was not to prove what everyone knew, namely, that Achilles would overtake the tortoise. Rather, by showing this mathematically, Harriot was trying to clarify his notion of time, space, continuity, and atoms. His interest was not only mathematical but physical, however, and this led to a problem. In calculating his infinite sums of increasingly tiny spiral segments, he had implicitly assumed that space was infinitely divisible.[8] When it came to pondering the nature of matter, on the other hand, he assumed the existence of indivisible atoms with a definite size and shape. Today, the situation is similar. Mathematicians speak of infinitesimal limits, which supersede intuitive notions of infinite divisibility, while physicists speak of atoms. (Of course, we now know that the atom itself can be split, although today's fundamental particles—such as electrons—are apparently indivisible. Their underlying nature is still mysterious, and physicists think of them in terms of their energy, mass, charge, and so on, rather than as the tiny, "hard," billiard-ball-like atoms of antiquity.) In Harriot's day there was not such a clear distinction between pure mathematics and physics, and in thinking about infinitesimal atoms and infinite sums he and his contemporaries wrestled with the apparent contradictions in these two positions. One of his manuscript pages seems to acknowledge the dilemma with humor. He listed the challenges, such as whether there was a maximum and a minimum for finite quantities, and a maximum and minimum for infinite quantities—and indeed were there different "amounts" of infinity at all. (They are perceptive questions: while limit theory addresses the notion of smallest and largest finite quantities, mathematicians today do distinguish between "countably infinite" and larger "uncountably infinite" sets.[9]) Then, in the bottom right-hand corner of this page of queries, Harriot pasted a piece of paper with three lines of verse written:

> Much ado about nothing.
> Great wars and no blows.
> Who is the fool now.

The page and the pasted slip of paper are undated, so it is not clear whether Harriot wrote his verse after seeing Shakespeare's *Much Ado About Nothing* (thought to have been written in 1598–99 but not performed until 1612), which has plenty of "warring" in the verbal sparring between Beatrice and Benedick, in addition to betrayal and threats of duels. But no physical blows are landed, and the couple end up declaring their love and settling for marriage, which they had spent most of the time disparaging as an end for fools. Perhaps Harriot, still a bachelor, initially wrote this snippet about them. Whatever the origin or purpose of his slight piece of paper, he thought it fit to paste it onto his page of questions about the infinitely large and the infinitely small. In this context, presumably his "fool" refers to those who contested but did not solve ("great wars but no blows") the issue of "nothing"—of quantities so infinitesimally small that it hardly mattered whether they had an indivisible minimum. Much ado about nothing—or more correctly, about "almost nothing."[10]

A couple of particular possible "fools" spring to mind in this context. In *Ash Wednesday Supper* (Giordano Bruno's dialogues on his mystical post-Copernican cosmology, published anonymously in 1584), Bruno had written of Aristotle, "It takes a fool to believe that in a physical division of a finite body one may go on to infinity." Perhaps Harriot was referring to Bruno's lack of appreciation of the difficulty of both critiquing and replacing Aristotle's position. On the other hand, another Italian contemporary, Torquato Tasso, had written in a similarly dismissive tone in support of Aristotle and against atomism as espoused by Democritus's later disciple Epicurus: "The fool! He did not know the art and ways whereby the world was fashioned in the mold that God, its architect, had in himself." Tasso was railing against *ex nihilo nihil fit*—so perhaps he was the fool Harriot had in mind.[11]

All we can really deduce, though, is that Harriot knew a fool when he saw one, one who underestimated what mathematical and scientific inquiry involved. Indeed, at one stage in his own attempt to clarify Aristotle and resolve the issue of atoms, Harriot noted that the statements "everything comes from nothing" and "from nothing, nothing comes" were not, in fact, contradictory. He was thinking about the concept of continuity, which Aristotle had used as evidence against indivisible atoms—and he suggested that if tiny indivisible atoms were aligned side by side, touching each other, you could still have continuity. Again, he seemed to be implying that with all the scholars either contesting or supporting Aristotle, there was too much ado: indivisibles could be so small as to be almost nothing, so that everything came from "nothing"; at the same time, nothing came without these tiny atoms, so that nothing came from nothing.[12] Such were the quandaries that exercised the best minds of the ages, until experimental science developed sufficiently to produce tangible evidence of today's atoms, and theoreticians focused on mathematical rather than philosophical analyses. Meantime, the difficulty of resolving these issues meant that Harriot ended up doing what many other creative theorists did: he gave up trying to resolve logically the issue of whether space, time, and matter are infinitely divisible and whether atoms exist. Instead, he used whichever model he needed to solve the problem at hand.

TORPORLEY BLAMED HARRIOT FOR REVIVING physical atomism, but Harriot was not alone, as is evident in Northumberland's description of alchemy quoted earlier. The earl's account appears in a letter to his son, in which he also noted that while the practice of alchemy did much to further the study of material processes and qualities, it would be a "mere mechanical broiling trade" without a theory of "atomical combinations" for "explaining" its processes. His librarian Warner, too, was an atomist. Like Harriot, Warner felt that Aristotle's position on the nature of matter was "confused and irresolute." In particular, he criticized Aristotle's attempt to have it both ways—to say that matter was infinitely divisible "potentially" but not "actually." Since the universe is composed of "actual" things, wrote Warner in an unpublished manuscript, the basic stuff of matter should also be "actual," such as atoms, not some vague potentiality.[13] Warner also believed, as did Harriot, that the different natures of different materials—such as heaviness, softness, brittleness, and so on—were explained by different arrangements of atoms. In his 1638 *Dialogues and Mathematical Demonstrations relating to Two New Sciences,* Galileo would also consider an atomistic explanation of matter, while Bruno had espoused a living cosmology based on his own version of atomism— to take just two more of Harriot's "atomistic" contemporaries.[14]

Harriot invoked the idea of physical atoms as a working hypothesis to explain such things as rarefaction, as when water becomes steam, as well as the puzzling phenomenon he had mentioned to Kepler: the fact that a single light ray appears to be simultaneously reflected and refracted. Harriot sketched out his explanation of this latter process on one of his manuscript pages. He evidently had in mind the ancient mechanical conception of atoms as tiny solid balls, so that when a light ray hit the ball it was reflected away like one billiard ball bouncing off another. His diagram, represented in figure 19, also clarifies the verbal explanation of his theory on the cause of refraction that he had given Kepler. It shows the spaces between the solid atoms in transparent media, such as glass or water, through which some of the incoming light could supposedly slip, bouncing from one atom to another in such a narrow "channel" that the zigzagging refracted ray appears straight to our eyes.

Harriot's model no doubt owed something to his familiarity with the packing of spherical balls. Many years earlier Ralegh had asked him to analyze the most efficient way of stacking cannonballs or bullets (which in those days were lead balls) on board his ships. The seventh-century Indian mathematician Bhaskara had made a start on a mathematical analysis of packing in "solid heaps," though Harriot independently came up with the first modern mathematical analysis of the problem.[15] Like Bhaskara, he used "geometric" or "figurate" numbers, which had their origin with those number-worshipping Greeks the Pythagoreans. These ancient "figurate" numbers were built from symmetric arrays of dots or stones (or bullets or cannonballs). For instance, "square numbers" are made from arrays such as this:

```
*        * *       * * *       * * * *  . . .
         * *       * * *       * * * *
                   * * *       * * * *
                               * * * *
```

Adding the dots in each array, the square numbers are 1, 4, 9, 16, ... To generalize this pattern, modern mathematicians (following Harriot and Descartes) speak of an arbitrary number n, and they say that the nth square number is n^2 (or Harriot's nn).

The next figurate numbers were the "triangular" numbers:

```
*          *            *              *      . . .
         * *          * *            * *
                    * * *          * * *
                                 * * * *
```

and so on, with each new row containing an extra dot added to the bottom of the pile to generate the next number. Counting the number of dots in each pile or array, the triangular numbers are

1, 3, 6, 10, ...There is a simple pattern here: the sequence of triangular numbers is 1, 1 + 2, 1 + 2 + 3, 1 + 2 + 3 + 4, ... So, the nth triangular number is $1 + 2 + 3 + 4 . . . + n$. Moreover, this sum has a
formula, $n(n+1)/2$. While the ancients knew this formula in rhetorical, algorithmic terms, Harriot was the first to write it in essentially
this modern algebraic form.[16]

Other polygonal numbers can be made in the plane, but Harriot
was particularly interested in the three-dimensional group of figurate numbers known as the "tetrahedral" or "triangular pyramidal"
numbers. They are built from a three-dimensional array in which
triangular numbers are stacked on top of one another (so that each
stacked layer sticks out of the plane of the page). In the list of triangular groups above, imagine that the single dot on the left is placed
on top of the triangle following it. Then stack these two onto the
next one, and so on. Alternatively, imagine that the last array of dots
shown above—the one corresponding to the triangular number
10—is the bottom layer of a pile of round bullets or cannonballs.
One stacks the next smallest array, corresponding to the triangular
number 6, on top of this, and so on to build a triangular pyramid
with 10 balls on the bottom, then 6, then 3, and finally 1 on top of
the pile. As Harriot pointed out, this is the most stable way of stacking spherical objects—and stability was vital on board a rolling ship.
The sequence of these tetrahedral numbers is found by counting
down from the single ball on top of the pile and adding as you go:
1, 4, 10, 20, 35, 56, . . . , each entry being made from successive sums
of the triangular numbers: 1, 1 + 3, 1 + 3 + 6, 1 + 3 + 6 + 10, and so
on. There is a formula for the nth such number here, too, as both
Harriot and Bhaksara knew (although the latter wrote it in words,
not symbols): $n(n+1)(n+2)/6$. So if you had room for, say, a pile
twenty layers high, you could easily work out how many balls you
need in each layer—and indeed Harriot had written out for Ralegh
a chart of numbers for each layer not only of the tetrahedral pyramid
but also of square- or oblong-based pyramids, which he also considered stable and economical ways of stacking.

Today, the mathematics of packing spheres has many uses, from
efficient packing of products to error correction in data transmission
to the analysis of arrays of biological cells, or of atoms in chemical
lattices. Which leads back to Harriot's "atomic" model of matter. In
his diagram and his letter to Kepler, he had emphasized the spaces
between atoms, so that light could slip between the atoms of a refracting medium. He must have forgotten about his idea that the atoms
needed to be touching each other if matter was to be continuous. Or

perhaps he had changed his mind and believed that some kind of energy filled the spaces and enabled continuity. Following the ninth-century Arab scholar al-Kindi, whose work on "radiation" was in Northumberland's library, both Harriot and Warner spoke of "material or immaterial energy" and "radiative virtues." Harriot did not elaborate on the nature of these radiative forces, although he may well have conjectured that they held atoms in place so that they and their neighboring vacuums formed some kind of lattice, somewhat analogous to the way gravity holds a pile of bullets together.[17] At any rate, he did not pursue the logical problems of his atomic model. Clearly he was simply speculating, and the incomplete nature of his work here highlights the fact that his thinking on theoretical matters, like that of his contemporaries, is a fascinating mix of perceptive insight and untested assumption. For its time, though, Harriot's model represents a significant attempt to explain an intriguing natural phenomenon—the simultaneous reflection and refraction of light. Newton would wrestle, creatively but unsuccessfully, with the same problem, but few of his and Harriot's peers appeared to give it much thought.

Torporley, however, was not impressed. Aside from his religious objections to atoms, he wrote out a list of logical criticisms of Harriot's theory.[18] In particular, if light were an infinitesimally thin ray, the angle of refraction would depend on chance—on the random position at which the light ray zigzagged off the first material atom it encountered. Torporley was assuming that the atoms of the transparent medium were large compared with the light ray (as indeed they appear in Harriot's diagram), so that the ray could hit the atom at various angles of incidence, depending on which part of the atom's large circumference it struck. But it all depends on how closely packed the atoms are and how thin the light ray is. Today, quantum theory says that a light ray is not infinitely thin; it also posits that at the atomic level, light itself comes in "particles," or bundles of energy. So from a modern point of view, the problem with Harriot's model was not so much Torporley's objection as Harriot's mechanical, billiard-ball notion of the interaction between the light and the atoms. Instead, quantum theory asserts that refraction is caused by photons interacting with the atoms of the water, glass, or other transparent medium; specifically, they interact with the electrons in the atoms and exchange energy. Two colliding billiard balls also exchange energy, but the quantum view is of course much more complex than Harriot's simple mechanical bouncing from one atom to the next. Nonetheless, just as the light ray in Harriot's model proceeded through the medium atom by atom, so each photon interacts atom by atom. In the process, it slows down and changes direction so that the refracted ray looks bent—to put it simplistically. The truth is that no one knows exactly what is going on at this level: quantum models, and physicists' understanding of them, are still evolving.

Kepler was as unimpressed as Torporley with Harriot's atomistic model. When he replied to Harriot's letter eight months later, in early August 1607, he offered niceties such as that he eagerly anticipated Harriot's explanation of the rainbow and its colors, and that he and Harriot were now in agreement about the refractions and reflection needed inside raindrops to cause the rainbow. He also praised Harriot's data about density, although he said he would have liked more general refraction data, too. But, he added, he did not like Harriot's "mocking" allusion to contracting oneself to an atom, "in the manner of the alchemists." This is surprising, because in his first letter, he had specifically asked for information about Harriot's (supposedly) alchemical theory of color. Perhaps he was miffed that Harriot did not offer a full explanation of what he meant. He may even have felt that Harriot was concealing his knowledge as hermetic alchemists did, although Harriot *had* asked for patience until he had more time and better health. As for reflection and refraction, instead of Harriot's model of atoms and the empty spaces between them, Kepler still preferred the concept he had outlined in his *Optics*— namely, that refracting materials possess two opposite material qualities in one, transparency and opacity. "Even if it seems absurd to *you* that the same point, at the same instant, should both transmit and reflect a ray, it is not absurd to me," he wrote.[19]

Now it was Harriot's turn to be unimpressed. Kepler's notion of dual Aristotelian "qualities" was no explanation at all; rather, it was just another way of saying that refracting materials both reflect and transmit light. Harriot dutifully reread Kepler's "explanation" in his *Optics*, but when he replied to Kepler in July 1608, he noted bluntly, "I am amazed if the assumptions and arguments given there satisfy you; they do not satisfy me." He wryly filled the "vacuum" at the end of his page with an example: solid gold is completely opaque, but gold leaf is translucent. Surely this difference is explained by the relative sizes of the spaces between the atoms in the two samples, not by some bizarre composite "quality" of gold itself?[20]

Harriot was half right. Liquid gold has a little more space between its atoms than does solid gold, but gold leaf is very thin solid gold. In 1911, however, Ernest Rutherford discovered the basic structure of atoms themselves. On firing positively charged alpha particles at a piece of gold foil, Rutherford's team found that some of these particles seemed to bounce off, some were slightly deflected, but most passed straight through. Rutherford deduced that each gold atom had a heavy positively charged nucleus that repelled the incoming alpha particles; the rest of the atom contained light negatively charged electrons and empty space, which was why most of the alpha particles passed straight through the gold foil. Some light, too, is transmitted through the foil, which is why it is translucent.

Kepler's adamant rejection of atoms and interatomic spaces followed not only from theology but also from the Aristotelian belief that nature eschews voids. If you break up a lump of ordinary matter into smaller lumps, there appear to be no extraneous spaces inside; in order to "explain" this, Aristotelians assumed that matter was "compelled" to stick together, pushing aside any empty spaces. We now know that atoms of matter bind together because of the electromagnetic force between electrons and protons, while protons are held together in the nucleus by the "strong" nuclear force. Without such knowledge, though, it is not surprising that simplistic but intuitive ideas such as Aristotelian "compulsion" (and vague ideas such as "radiative virtue") remained compelling for so long. The Aristotelians also eschewed vacuums in space, which is why they believed "the heavens" were filled with "ether."

Harriot accepted some of Aristotle's physics—and, incidentally, his notion of the good life in *Nicomachean Ethics*; Harriot possessed, or cultivated, favorite Aristotelian ethical qualities such as "courage, temperance, gentleness, and affability."[21] But he questioned Aristotle's stand not only on atoms (and heliocentrism) but also on the celestial ether. So he referred Kepler to a much more "modern" thinker, William Gilbert. Like Harriot, Kepler had admired Gilbert's book on magnetism, but Harriot had recently read a manuscript version of Gilbert's cosmological work *De mundo*, in which he supported the idea that a vacuum exists in the space around stars and planets. Harriot was so impressed with Gilbert's idea that he hoped Kepler might change his mind about interatomic vacuums, too, when he read it for himself.[22]

BY THE TIME HARRIOT WROTE this letter in July 1608, nearly a year had gone by since Kepler had last written, and once again Harriot pleaded lack of time. In addition to his own research, he was regularly visiting Northumberland and Ralegh in the Tower and helping out as a conduit with their affairs in the outside world.

He had surely been helping Lady Ralegh, too. It turned out that the solicitor handling the conveying of Sherborne to the Raleghs' older son, Wat (with Harriot as one of the managers), had made a technical error, rendering the transfer invalid. This meant that the property was still in Ralegh's name and, like all the property of convicted traitors, should have automatically reverted to the Crown. Cecil was no help, despite his earlier promises to look out for Bess's interests—and despite the fact that James had actually consented to her request to overlook the technicality. Cecil simply never got around to drawing up the correct document James asked for. Then, in 1607, when the king was looking for a choice property with which to reward his current favorite, Cecil had the heartlessness to suggest Sherborne.[23]

Lady Ralegh literally threw herself at the king's feet and pleaded for justice for herself and her "poor children," but this time he just walked on by—although not, it seems, without some guilt. The case was allowed to go to court, and at the time of Harriot's letter to Kepler, the Raleghs were preparing their case for the lawyers who would try to prove the legality of Sherborne's conveyance—a task in which Harriot probably assisted, as did Lawrence Keymis. In many people's minds, the Crown's attempted theft of Sherborne underscored the shocking injustice done to Ralegh. Everyone knew he was no traitor, and a year earlier, even the king's own brother-in-law the king of Denmark had tried to have Ralegh freed so that he could serve as Denmark's admiral.[24]

Northumberland, too, was innocent, as Harriot well knew. Yet Harriot's liberty also hung on the whims of powerful people, who had at their fingertips the ever-simmering rumors of his supposed atheism. Torporley's theological objections to his atomism cannot have been comforting in such a climate, and when Harriot outlined his theory of simultaneous reflection and refraction in his first letter to Kepler, he mentioned atoms just once, in a metaphorical context ("mathematically contract yourself to an atom"). But in his latest letter, in this summer of 1608, he'd let fall a rare and enigmatic personal aside. After explaining that his theory of matter was founded on the "doctrine of the vacuum," he added, "But things are in such a state here that it is not lawful for me to philosophize freely; we are still stuck in the mud. I hope that Almighty God will soon put an end to this state of affairs."[25]

Some years earlier, Galileo had expressed a similar sentiment in a letter to Kepler. Speaking of Copernicus's theory, he wrote, "I would surely have the courage to make my thinking public if there were more people like you. But since there are not, I shall avoid such involvement."[26] That was in 1597. In 1600, Bruno had been burned at the stake in Rome, although not for Copernicanism (which was not yet deemed heretical) but for his animistic cosmology and seven other "heresies." Three decades later, of course, Galileo *would* find the courage to speak his mind about the heliocentric theory, and for his boldness he would be charged with "suspicion of heresy." Harriot's situation was different and, in a way, more invidious. The Catholic Church tried to make it clear which beliefs were heretical or prohibited, although it took until the mid-seventeenth century for its revisors to compile a definitive list of heretical "philosophical" beliefs. (In addition to thinking that heliocentrism was a fact rather than a hypothesis, such prohibited ideas specifically included the concept of atoms interspersed with tiny vacuums. The Jesuits wanted no one to think that God did not create the world *ex nihilo.*[27]) The Anglican Church was not particularly concerned with such matters. Although Torporley's criticisms show—as do those aired around the time of the Cerne Abbas heresy enquiry—that some Anglican clerics did rail against scientific issues such as atomism and mathematical "conjuring," Harriot's liberty was also subject to the unpredictable caprices of the king and of ruthless politicians such as Cecil. Perhaps this was another reason he never got around to publishing his work. He did not want to become more conspicuous than he already was, as the dangerous intellectual supposedly urging the Ralegh-Northumberland circle to their "atheistic" beliefs.

In January 1609, Sherborne's conveyance to Wat was formally declared invalid and Lady Ralegh's home was taken from her. Forty years later, her son Carew would remember bitterly his mother's anguish at the injustice done to her and her family.[28] Meantime, with her indomitable spirit, and with the help of her brother, her husband, and her friends, Lady Ralegh fought hard for compensation. She eventually managed to get James to offer her £8,000 for Sherborne, plus a pension of £400 a year. Not that the Crown ever paid this on time.

Although Harriot had revealed his fear of philosophizing freely while the English establishment remained "stuck in the mud," Kepler took over a year to respond. He had been busy with his own research, but it had also become clear that the two men were not philosophical kindred souls. In his letter of September 1609, Kepler repeated his arguments against the existence of vacuums in nature. He did express belated interest in Harriot's model of refraction, but the earlier pleasantries were absent.[29] It appears that Harriot did not reply, and that he had laid aside his atomistic research.

CHAPTER 19

Searching the Skies

I N HIS LAST LETTER TO HARRIOT in September 1609, Kepler mentioned that he had just published his *Astronomia nova* (*The New Astronomy*), in which he presented his law of elliptical planetary motion. Some months later, Harriot read the book and wrote to tell his friend and disciple Sir William Lower. A decade younger than Harriot, Lower—the son of a wealthy Cornish gentleman—had been rather a wild young man. Expelled from his legal studies for "riotous behavior" in 1591, he apparently studied languages on the Continent, but disappeared from the record until being knighted in 1603 and taking his seat in the House of Commons in 1604. His successful parliamentary career and his 1605 marriage to Northumberland's stepdaughter suggest he may have had earlier links with Essex's circle (Lower's wife, Penelope Perrott, was Essex's niece). Although he was based at his wife's estate in Wales, Lower had soon become close to Harriot and his scientific pursuits.[1]

Harriot's letters to him have been lost, but Lower wrote back on February 6, 1610, with his own opinion of Kepler's book. He had read it "diligently," although the challenging nature of the work made him realize, he told Harriot, "what it is to be so far from you [here in Wales]." Although much of their interaction was by letter, Lower had found in Harriot the perfect teacher: "In all ways of teaching, for richness and fullness, for stuff and form, your [lessons] are incomparably [the] most satisfactory," he wrote. He also sent perceptive questions about some of Kepler's arguments, adding humorously, "I am much delighted with his book but he is so tough in many places that I cannot bite him." Nevertheless, he felt he had understood enough to agree with Harriot that Kepler had established his theory of elliptical orbits "soundly."[2]

Lower was less impressed with Kepler's attempt to find a physical cause for planetary motion: "I cannot fancy those magnetical natures," he wrote Harriot. Kepler had been inspired by Gilbert's idea of planetary "magnetic souls," and one of the most radical assertions of *The New Astronomy* was Kepler's conviction that the light and energy of the sun, which he saw as the embodiment of God's power, literally caused the planets to move.[3] Most mainstream astronomers at the time believed that astronomical models were simply mathematical devices for keeping track of planetary movements. Even Tycho had believed this (although he had accepted neither the Ptolemaic nor the Copernican model; he advocated for an earth-centered hybrid in which the inner planets moved around the sun while they and the sun also moved around the earth—just as the moon circles the earth, and the earth-moon system circles the sun in the Copernican model. Ninth-century Carolingian astronomical diagrams contain a similar hybrid model.[4]) The Church had no problem with this mathematical approach. It was only when people such as Kepler and Galileo ascribed physical reality to their models that objections arose. In an attempt to forestall such criticism, Kepler took care to explain in his new book that biblical stories about the sun's apparent motion could be reinterpreted from a Copernican point of view without denying God's power. "With all due respect to the Doctors of the Church," he added, scientific proofs were more important in such matters than theological tradition.[5]

Like the idea of literal planetary motion, the notion of elliptical orbits was also radical. Astronomers and theologians alike generally favored the ancient idea that circles were requisite for "heavenly" motion: circles were "the most perfect form" (to quote Acosta's 1590 opinion), and their symmetry also suggested physical stability. Ellipses, by contrast, seemed unbalanced. Kepler had made his breakthrough about elliptical orbits by using Tycho's data to try to find support for his hypothesis that the planets, including the earth, were pushed about the sun by some sort of solar magnetic "force." Never before had so large and so accurate a body of data been available for planetary motion—although Tycho's heirs were keen to keep their own hands on this valuable information. After the great astronomer's death, Kepler had become embroiled in a battle that culminated in Tycho's son-in-law having the right to censor the more physical aspects of Kepler's book. (Tycho himself had required

Kepler to sign a confidentiality agreement; it was not just business-men such as Ralegh who wanted to protect their own investments, whether of money or hard work.)[6]

It is typical of the state of science in the early seventeenth century that Kepler's mystical, theological imperative—to place his solar manifestation of God at the causal center of the heavens—gave him the drive to follow through to a modern conclusion. He began with Mars, whose orbit was inaccurately described in both the Ptolemaic and Copernican systems. He spent years developing his intuitive hypotheses and then adapting them with reference to Tycho's empirical data, until finally, in 1609, he was able to present what are now known as his first two laws of planetary motion: first, that the orbits are elliptical, with the sun at one focus of the ellipse, and second, that as each planet moves along its orbit, the radius from the planet to the sun sweeps out equal areas in equal times. The focus of an ellipse is off-center, which means that during its orbit, the planet does not keep the same distance from the sun. It turns out that a consequence of the second law is that when the planet is closest to the sun, it is moving fastest, which is what Kepler expected: thinking of the way light radiates, he argued that the closer to the source of energy, the more power there should be.[7] He would find his third law, relating orbital distances and periods of revolution, some years later.

In an example of the unexpected ways scientific breakthroughs can be made, the second law initially arose not as an empirical deduction but as a computational device to avoid hundreds of tedious calculations.[8] It was similar to the way Harriot had dealt with his equiangular spiral. He had begun by breaking up his spiral into tiny segments, and Kepler used the same approach for a planetary orbit. And just as Harriot used the tiny triangles formed by his spiral segments and their radii to find the length and area of his spiral (figure 17c), so Kepler derived formulae for the perimeter and area of an ellipse. Although his perimeter formula was only an approximation, it enabled him to find the approximate length of a planetary orbit without having to add up the observed lengths of segments by hand; as a bonus he realized that his triangular areas were related to time, and hence the second law.

Kepler's theology rather than physical principles may have been at the heart of the quest that led to his laws of planetary motion, but he had worked hard to establish them. It is true that he used untested assumptions and some bizarre speculations, so that he did not have a proper theory of planetary motion. That is, he did not successfully explain *why* the planets move. This would lead Newton to observe that Kepler had merely guessed his ellipses—as indeed he had, on the basis of observational anomalies in the circular model. Nevertheless, using his considerable mathematical skill he had eventually squared his guess with the available empirical data, thereby successfully deducing the *what* of planetary orbital motion, expressed in his three quantitative laws.[9]

As noted earlier, Kepler was the first to conceive of planetary orbits at all—the literal paths through space taken by the moving planets. Earlier astronomers had recorded angular positions on the celestial sphere of celestial bodies at different times, while some went so far as to imagine each planet on its own celestial "orb" or sphere. But no one thought of planetary trajectories. This is largely because this notion required accepting that a force was moving the planets, just as the trajectory of a cannonball was traced out after the force of an explosion ejected the ball. Earlier astronomers, by contrast, had simply assumed that the orbs themselves rotated serenely in some divinely ordained way.[10]

In hindsight, then, the publication of Kepler's *New Astronomy* was a pivotal moment in the paradigm shift from astronomy as computation to astronomy as physics. Newton would take the final step eighty years later, when he used Kepler's laws to prove his inverse square law of gravity.[11] What is more, in contrast to Kepler, Newton showed unprecedented rigor in his use of physical principles, in his 1687 *Principia*. With gravity as the force moving all the celestial bodies, there was no need for mystical sources of motive power, no need for theology in science—and no need for inspired but untestable guesses. Kepler's model was a tour de force nonetheless. It was much simpler than Copernicus's, which had required some technical tweaking to fit circular orbits to the observed planetary motions, and Harriot and Lower were among the first of Kepler's contemporaries to appreciate his achievement.[12] Moreover, when Lower wrote to Harriot about the soundness of Kepler's elliptical law, he added, "I remember long since you told me as much, that the motions of the planets were not perfect circles."[13]

This is an astonishing claim, but Harriot did, in fact, do much research on the mathematics of ellipses.[14] Like Kepler, he studied the ancient Greek mathematician Apollonius of Perga, who had provided an analysis of ellipses in around 200 BCE. Harriot studied Apollonius's book *Conics*, a rediscovered treasure that had been published in Latin only four decades earlier. The "conics" are ellipses, circles, hyperbolas, and parabolas, which are the cross-sectional curves produced when a cone is sliced at different angles. Harriot would put his study of parabolic curves to good use when he and Galileo independently discovered the mathematical shape of the trajectory of a projectile; meantime, his manuscripts on the subject of ellipses are compelling for two reasons. First, he developed the classical Greek geometric ideas in his unique symbolic algebra. Second, his main focus was how to find an ellipse that fitted several data points presumed to lie on it, which is precisely the kind of analysis needed to find the path represented by observations of a planet at different points of its orbit. There seems to be no extant evidence of his fitting actual planetary data to his ellipses, and he did not have access to Tycho's data, so his idea of non-circular orbits was most likely an unconfirmed conjecture, albeit one presumably based on some sort of astronomical observation.[15] Either way, his work on ellipses helps explain why he was one of the first to react positively to Kepler's discovery. In fact, Kepler's elliptical orbits, and his second law of planetary motion, would not be generally accepted as empirical laws until Newton used them so fruitfully in *Principia*.[16]

Harriot parted from Kepler's theological view of the structure of the universe, however. In contrast to Kepler, he followed Gilbert's and Digges's perceptive idea that not only is there an interplanetary vacuum, but the universe may be infinite, with the stars scattered at varying distances from earth rather than being fixed to the "celestial sphere."[17]

Lower was so convinced that Harriot had anticipated Kepler on the matter of elliptical orbits that he chided his teacher for never publishing his discoveries: "Do you not here startle, to see every day some of your inventions taken from you?"[18] Furthermore, he continued, Kepler was not alone in going to print with ideas Lower believed Harriot had developed first—Viète (on certain algebraic ideas), Ghetaldi (on specific gravities), and "many others" had done likewise. Yet "too great reservedness has robbed you of these glories." Lower was obviously a close friend, because he went on to give some more forthright advice. Remember, he admonished Harriot, "that it is possible by too much procrastination to be prevented in the honor of some of your rarest inventions and speculations. Let your country and friends enjoy the comforts they would have in the true and great honor you would purchase yourself by publishing some of your choice works."

Harriot had his reasons for ignoring such advice, good as it was. Lower sensed this, because he had followed his admonition with "But only you know best what you have to do," and he closed his letter by saying, "Above all things take care of your health." Harriot's health problems and all the other contingencies certainly made it difficult to complete and compile his discoveries, and his instinct for thoroughness had created a caution that he never abandoned. In any case, at this particular moment he was too busy with new research to prepare earlier work for publication. He had a valuable new instrument at his disposal, and he had been making the most of it.

NO ONE KNOWS WHO INVENTED the refracting telescope, although the first patent application was filed by the Dutchman Hans Lippershey in October 1608. By the end of the year, telescopes were readily available for sale on the Continent, and people were using them to playfully peek into distant windows and to scan for ships on far horizons. When better instruments appeared soon afterward, stargazers marveled at how many more stars they could see in the sky.[19] It must have been an awe-inspiring experience, using the first piece of technology to improve so dramatically upon the human senses. So far as we know, though, no one left a drawing of any of these new celestial sights until the summer of 1609, when Harriot made the first telescopic drawing of the moon. He had managed to buy (or build) a telescope with a magnification of six times what can normally be seen, and on July 26, he used it to draw the crescent moon.

It was a rough and rather uninspiring sketch, compared with the map of the full moon that he would produce within the next couple of years, when he and his assistant Christopher Tooke had made their own telescopes with magnifications up to fifty times. Nevertheless, it was the very first of its kind, and in recognition of this (and his other contributions to science), there is now a crater named after Harriot. It is on the far side of the moon—a fitting tribute to the first man to attempt to map features of the moon unseen with the naked eye. Galileo's famous drawings appear to have been made a few months after Harriot's, but of course he, Tycho, Kepler, and a host of scientific and other luminaries, also have lunar craters to their name.[20]

The first to map the moon using only the naked eye seems to have been Gilbert. At the dawn of the sixteenth century, Leonardo da Vinci had made an attempt to sketch the moon accurately, and other artists, too, had drawn the strange patches of shadow that give the man-in-the-moon appearance. But modern scholars have provided compelling evidence that both Gilbert and Harriot were actually mapping rather than merely drawing the moon.[21] Certainly their images look like two-dimensional maps of continents and seas; they both thought that the moon's darker and lighter patches were seas and islands or continents, as the labels on their maps show. Gilbert's map, like Harriot's first sketch of the crescent moon, was fairly basic,

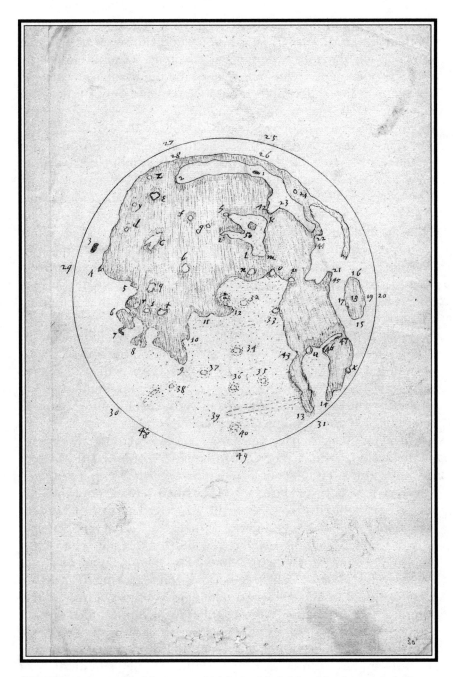

Plate 6: *Harriot's map of the full moon,* on which he has labeled forty-nine features including craters.

(Petworth House/Leconfield, HMC 241 IX f30; reproduced with the kind permission of Lord Egremont and the West Sussex Record Office.)

but it no doubt influenced Harriot, who was one of the few people who had seen it. It was in that unpublished manuscript he had told Kepler about. Harriot's experience in mapping "Virginia" would aid him in producing his telescopic map of the full moon, on which he labeled forty-nine different features, including not only the "seas," or *maria*, but craters and other topographical features. Its accuracy would not be equaled for another three decades.[22]

Harriot also recorded one of the first observations of the phenomenon of lunar libration.[23] This is a slight wobble or rocking in the moon's orbit, produced by a number of factors related mainly to the moon's simultaneous rotation and orbital motion. It means that over the lunar month we can see slightly more than half the moon. Galileo would describe it well in his famous *Dialogues* of 1632, where he showed that during the moon's monthly orbit, key markings near the edge of the moon seemed to move slightly toward and away from the edge. He did not know that first Gilbert and then Harriot had already noticed this change. Aristotle had assumed that the moon did not rotate on its axis because he believed we only ever see exactly the same half of the moon, but Gilbert argued that the monthly movements of features near the edge of the moon could only be explained if the moon was rotating and its axis was tilted. Gilbert's naked-eye observations were rudimentary, but his deduction was inspired.[24] Harriot probably made his own careful observation of lunar libration after reading Gilbert's manuscript.

Harriot may also have been spurred on to map the moon with greater accuracy after seeing Galileo's relatively polished drawings of the moon and its craters, published in his *Sidereus nuncius* (*The Starry Messenger* or *Message*) in March 1610. Harriot first read about Galileo's various telescopic discoveries when he received a copy of Kepler's *Conversation with the Starry Messenger*, published two months later to support Galileo's claims.[25] Like Harriot and Tooke, Galileo had made his own telescope, which was five times more powerful than Harriot's first 1609 instrument. Galileo had long been searching for physical proof of Copernicus's heliocentric model (like most of his contemporaries, he showed little interest in Kepler's elliptical adaptation), and he used his moon drawings as evidence that there was no physical distinction between the heavens and the earth. Like Gilbert and Harriot's apparent lunar seas and islands, Galileo's depictions of craggy lunar craters suggested that celestial bodies were not "perfect" otherworldly orbs but were made of the same stuff as earth—and this made it easier to conceive of the earth moving through the "heavens," too. Galileo also discovered that Jupiter has its own moons circling about it. This was a remarkable new key to the universe, because it showed that not all heavenly bodies revolve about the earth.

Galileo's willingness to challenge mainstream Aristotelian cosmography publicly made him instantly famous—and controversial. Indeed, Kepler had generously published his *Conversation*, at his own expense, so that "against the obstinate critics of innovation, for whom anything unfamiliar is unbelievable, for whom anything outside the traditional boundaries of Aristotelian narrow-mindedness is wicked and abominable, you [Galileo] may advance reinforced by one partisan."[26] Galileo would find to his cost, though, that it was one thing to acknowledge the relative conceptual simplicity of the Copernican model and to make supporting conjectures for its physical reality, but quite another to prove that the earth really did move around the sun. The first clear evidence of such motion would not be found for another century; it was deduced from the "aberration of starlight" not long after Newton's death in 1727, by his disciple James Bradley. (The "aberration" is the discrepancy between the observed direction of a star and its true direction, discovered in the late seventeenth century when astronomers noticed that over the course of a year, the relative positions of the stars with respect to earth change in a way that is "out of phase" with expected parallax effects. While the latter depend on the earth's changing *position*, Bradley explained the puzzling discovery of aberration in terms of the earth's orbital *motion*. It is analogous to driving in heaving rain: if the rain is actually falling vertically, then as you drive forward it will appear to fall toward you, as though the rain is falling for some place ahead of you rather than from directly above.) Meantime, craters on the moon and Jovian moons did not convince Jesuit scholars, and in 1616, the Catholic Church would make it clear that belief in a motionless sun at the center of our planetary system was both scientifically "false" and theologically "heretical." The idea of a moving earth was not specifically condemned as heretical but considered "at least to be erroneous in faith."[27]

Since Harriot already accepted Kepler's model and there was no Anglican requirement to "prove" it, it was not Galileo's speculations that excited him but the actual discovery of Jupiter's moons. He immediately wrote to his friend Lower in Wales, telling him the exciting news. Once again, we only have Lower's reply to go on. Harriot had sent him a telescope some months earlier, and Lower and his friends had become eager sky watchers. Now, in his letter dated "the longest day 1610," Lower told Harriot that he had been speculating about the possibility of unseen moons circling the other planets, just as our moon circles earth—and that just as he was explaining this idea to a friend, Harriot's letter had arrived. "When I had read [about Jupiter's moons]," wrote Lower, "Lo, said I, what I spoke probably, experience has made good; so that we both with wonder and delight fell to considering your letter. We are both so on fire with these things that I must render my request and your promise to send more of all sorts of these Cylinders [telescopes]."[28] The excitement occasioned by the new wonders being opened up by the telescope couldn't be clearer.

Lower went on to say, "My man shall deliver you money for any charges requisite [so] send me as many [telescopes] as you think needful [and] in requital I shall send you a store of observations." His letter indicates that Harriot and Tooke were becoming accomplished at making their own telescopes. They evidently made a good team, and they had been making astronomical observations together even before they had their telescopes. At Syon House in 1607, for example, they had made observations with Harriot's cross-staff of the path of the comet we now know as Halley's. Kepler and others had also tracked the comet, but Harriot's record was so extensive and accurate that two hundred years later the German mathematician and astronomer Friedrich Wilhelm Bessel would use it, along with later telescopic observations of the comet's appearances in 1682 and 1759, to make an extremely accurate estimate of the orbit of Halley's Comet.[29]

Lower asked if, in addition to more telescopes, Harriot could "send me also one of Galileo's books if any yet be come over and you can get them." This suggests that such books from the Continent were not yet widely available in England, so Harriot may well have received his copy of Kepler's *Conversation with the Starry Messenger* from Kepler himself. At any rate, Kepler's book made reference to another interesting telescopic discovery of Galileo's, one that gave many hours of distraction to Harriot and his friends, just as it had done for Kepler. It was common for early modern scientists to send each other anagrams that when solved would announce a new discovery. Galileo had sent one to Kepler, who published his own attempt at a solution in his *Conversation*. Kepler's rearrangement suggested that Galileo had observed two moons circling Mars. In fact, the solution was "I have observed the most distant planet to be in three parts." The most distant planet known at that time was Saturn, and Galileo had observed Saturn's rings, albeit not as rings but as two small bumps on either side of the planet. Harriot and his friends Torporley and Warner spent hours trying to solve the puzzle. At one stage, Torporley thought it might involve Jupiter, while one of them apparently

managed a bawdy rearrangement of Galileo's cryptic sentence—
after which the author crossed out his initials in embarrassment. It
seems, though, that none of them had any more success than Kepler
in unlocking Galileo's secret via the anagram.[30]

With his new telescope, Harriot studied not only the moon and
comets but also sunspots. His first surviving drawing is dated December
8, 1610, so his telescopic study of sunspots is contemporaneous with,
and independent of, Galileo's.[31] Sunspots had long been observed
with the naked eye, but by 1611 a number of astronomers besides
Harriot and Galileo—notably Johannes Fabricius and Christoph
Scheiner—were using telescopes to observe these strange apparitions
in detail. Harriot generally saved his eyesight by recording the changing shapes and positions of sunspots through fine mist or cloud, pref-
erably when the sun was low in the sky. He tried using "colored
glasses," too, but occasionally he looked at the clear sun—after which
"my sight was dim for an hour."[32]

To measure the positions of his sunspots at different times, presumably Harriot used the same instruments with which he observed
the stars and comets: his cross-staff, and perhaps also the beautiful
armillary sphere that Northumberland owned, which is now in the
Museum of the History of Science at Oxford. In February 1610,
Lower had asked Harriot how to make such a sphere, and later he
told his teacher that he had fashioned "a very fine" one using the
iron hoops from a barrel.[33] Harriot noticed that although sunspots
were often transient, they also appeared to move in a regular way
over several weeks—and he realized that this motion could be used
to calculate the time the sun takes to rotate once on its axis. Solar
rotation had not been part of either Ptolemy's or Copernicus's system,
but its discovery implicitly suggested that if the sun could rotate, then
so could the earth. Fabricius and Galileo, too, deduced the existence
of solar rotation from their sunspot observations, and Galileo later
went so far as to correctly conclude that the sun's axis was tilted, something Harriot did not do. Nevertheless, although he left few details
explaining his two hundred sunspot drawings and his notes describing
a total of 450 observations, Harriot made an impressive compilation
of sunspot data—and he deduced from these data the most accurate
period of solar rotation of his time, and probably the earliest.[34]

He also seems to have been the first to calculate the periods of
revolution of Jupiter's moons, which he did using Galileo's data from
The Starry Messenger, as well as taking his own observations. The patience
and technological expertise needed for such observations is highlighted by the fact that back in Wales, Lower had had little luck in

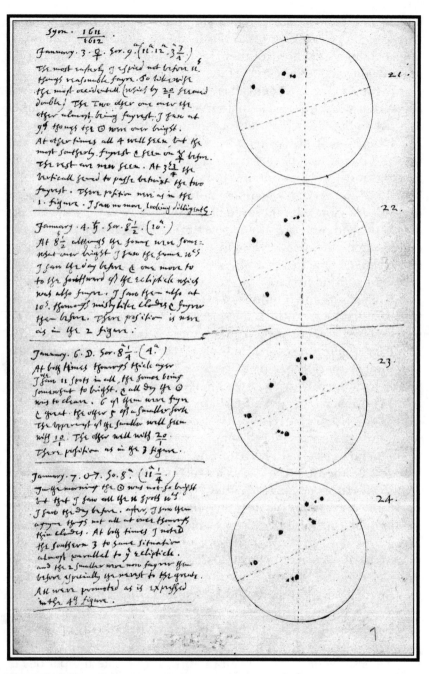

Plate 7: *One of Harriot's sunspot records:* The spots appear to be moving to the right when observed over a number of days, as if the sun is rotating.

(Petworth House/Leconfield, HMC 241 VIII; reproduced with the kind permission of Lord Egremont and the West Sussex Record Office.)

seeing the moons at all. He told Harriot that perhaps his eyesight was too poor, but he was not able to see them as he had done at Syon House with Harriot's "great glass," which apparently had a magnification of 20.[35] As with Harriot's sunspot data, modern scholars have confirmed the high degree of accuracy (given the still-rudimentary instruments of the time) of his analysis of the Jovian moons—and of his moon maps, too.[36]

While it is clear that Harriot was the equal of Galileo in taking astronomical observations and making astronomical calculations, some modern scholars have criticized him for focusing only on the descriptive and mathematical aspects of these phenomena, and not offering a philosophical, Copernican interpretation of them as Galileo did. However, maverick English astronomers had been chipping away at Aristotelian cosmology for decades. In the 1550s, Robert Recorde had published the subtly pro-Copernican dialogue discussed earlier, and Dee—whose world-class library contained two copies of Copernicus's *De revolutionibus,* as well as a copy of the first printed Greek edition of Ptolemy's *Almagest*—had urged his friend John Feild to publish a Copernican ephemeris because of the inaccuracies in the geocentric model. Then Digges and Dee had made their parallax investigations, and Digges published his two groundbreaking works of astronomy (his translation of Copernicus, and his *Alae seu scalae mathematicae*).[37] Thanks to all this work, Harriot had inherited a uniquely English Copernican mindset, albeit one that was mostly centered outside the universities. The establishment of the first chairs in mathematics and science at Oxford and Cambridge was still a decade away; meantime, Harriot was at the forefront of innovative English science, and as mentioned he had no religious need to prove heliocentrism, as Galileo did—or to disprove it, as the Jesuit Scheiner did. To add to the evidence of lunar mountains and orbiting Jovian moons, Galileo argued that sunspots were part of the sun's surface or atmosphere, whereas Scheiner claimed they were satellites of the perfect, unblemished heavenly sun. (Today we know that these darker spots are cooler areas in the sun's photosphere, caused by magnetic changes that slow down the transfer of heat. Gilbert's idea of planetary and solar "magnetic souls" had a grain of truth in it!) Like Galileo—and Fabricius—Harriot, too, accepted that they were "spots in the sun," as he put it. Yet notwithstanding his Copernican inheritance, his critics suggest that by not specifically speculating on the anti-Aristotelian implications of his sunspots and other astronomical observations he did not contribute to the Copernican debate the way Galileo, Fabricius, and Scheiner did.[38]

On the other hand, speculation isn't the only driving force in the development of ideas, and if Harriot had managed to publish his moon map and his calculations of the periods of solar rotation and Jupiter's moons, he might have contributed to the debate simply by providing accurate observations and mathematical deductions that implicitly supported the heliocentric position. His lack of publication can readily enough be explained by his perfectionism and the dramas in his life, but his personality and position had something to do with it, too. Galileo actively courted the limelight, while Harriot did not. Being generously retained first by Ralegh and then by Northumberland, Harriot had long been comfortably off, whereas Galileo had always been on the lookout for ways to impress wealthy patrons. In the spring of 1610, he had finally landed himself a prestigious and well-paid appointment as mathematician and philosopher to his former pupil Cosimo de Medici, the Grand Duke of Florence, after whom he'd named Jupiter's moons. Of course, *The Starry Messenger*'s flowery dedication to Cosimo also helped Galileo's case, as did the military advantages of his new, improved telescope, which he'd successfully demonstrated to the Venetian senate.[39] This habitual psychological and pecuniary need to impress meant that Galileo was always ready to publish his discoveries, and his newfound fame and position meant that establishing his priority was more important than ever. In 1613, he followed his *Starry Messenger* with *Letters on Sunspots*—his letters responding to Scheiner, in which he presented his data and gave his opinion that sunspots were like clouds in the sun's atmosphere. Scheiner eventually conceded that the sun was not "unblemished," although he went only so far as to accept Tycho's hybrid earth-centered system.

Sunspot Letters, and the associated dispute with Scheiner, did not make Galileo popular with the Jesuits. In 1616, he would be given a warning not to make unproven claims that contradicted the scriptures.[40] Not surprisingly, he would hold off publishing a full account of his speculations on the matter of the moving earth until 1632. *Dialogue concerning the Two Chief World Systems* outlines Galileo's case for Copernicus versus Ptolemy and Aristotle. As we've seen with his explanation of the trade winds, some of his arguments were wrong, while his telescopic discoveries were perceptive if not watertight strikes against the Aristotelian/Ptolemaic system.[41] By styling his account as a dialogue between fictitious characters, Galileo had hoped he'd kept himself safe, because the Church did allow theoretical discussion of Copernicus's ideas, but the Inquisitors felt that he had supported Copernicus rather too passionately in his *Chief World*

Systems. The fact that he had called his Aristotelian character "Simplicio" didn't help: his former ally Pope Urban VIII felt that Galileo was satirizing him personally.

Scholars are still debating the personal, legal, and theological issues at play in Galileo's trial. Suffice to say that he was charged with "a vehement suspicion of heresy" in 1633. He was not burned like Bruno, because he agreed to plead guilty to presenting Copernicanism in a way that could possibly be seen as "calculated to compel conviction," and because he denied believing in heliocentrism and agreed to formally "abjure" it.[42] Instead, he spent the rest of his life under house arrest.

DESPITE HIS FAILING EYESIGHT, Galileo made good use of his imprisonment, developing and writing up his earlier work on such topics as falling motion, in *Two New Sciences,* which he managed to have smuggled to Louis Elzevir, a publisher in Protestant Leyden (in Holland).

One of the most important aspects of *Two New Sciences* is the analysis of motion, particularly motion under the influence of gravity. Galileo's astronomy may not have proved that the earth moves about the sun, but his work on falling motion was a landmark in physics. Unbeknown to him—and to almost everyone until recently— Harriot had independently come up with essentially the same results that have gone down in history as Galileo's sole preserve, and for which he is so famous today.

CHAPTER 20

Gravity

I N HARRIOT AND GALILEO'S TIME, the term "gravity" still
had its Aristotelian meaning: the tendency of bodies to fall toward
the center of the universe—and therefore, according to Aristotle,
toward the center of the earth. Although the latter part of this definition seems intuitively obvious, it did not imply Newton's concept
of a gravitational force associated with matter itself (let alone an
Einsteinian curving of space-time). Surprisingly, perhaps, the nature
of falling motion had not been studied in any detail before Harriot's
and Galileo's systematic explorations in the opening years of the seventeenth century. A few over the centuries had challenged the belief
that heavy objects fall faster than light ones—a belief that Aristotle
himself had held. But no one had managed to adequately quantify
gravitational acceleration so that the way things fall could be properly analyzed.

To understand falling motion, it was necessary to know how to
mathematically analyze motion itself. A remarkable group of medieval scholars based at the universities in Paris and Oxford—notably
the fourteenth-century Parisian bishop Nicole Oresme—had taken
the first step by developing a numerical and graphical analysis in
terms of the distance traveled during a given time. They considered
motion in the abstract, so their approach was mathematical rather
than experimental. It was intuitively apparent, for example, that
when traveling at a constant speed the distance traveled is proportional to the time of travel. They also paid particular attention to
motion in which speed increases at a constant rate—as they assumed

happens in falling—although they tended to conflate speed changing with respect to time and speed changing with respect to distance traveled.[1] The former refers to "acceleration," although this is a later term for the rate of change of velocity with respect to time ("velocity" refers to the speed *and* the direction of motion). Nevertheless, the term "acceleration" is often used anachronistically by commentators and translators, and I will use it here, too, although both Galileo and Harriot had trouble coming to grips with the concept. Like their medieval predecessors, at first they were not sure how to measure constantly increasing motion, either.

Harriot tried out a number of graphical approaches, but none of them yielded a definitive result to his question: In falling motion, was the constant rate of increase in speed constant with respect to time or distance or both? In other words, was the speed of a falling body proportional to the time elapsed or the distance traveled? So he turned to experiment as the ultimate arbiter of reality. He measured the time it takes to fall a given distance, by dropping objects from various heights (up to 55.5 feet, or about 17 meters). Because there were no precise timepieces available, he was aware that his results could not be very accurate. He recorded, for example, that a bullet fell 55.5 feet (to use anachronistic decimal point notation) in a time that was "more than two pulses, less than three pulses." He tried this twenty times and concluded that the average time of fall was 2.5 seconds (a pulse beat was generally taken to occur every second). He repeated the experiment for a variety of objects falling from different distances, and he also did experiments designed to measure the relationship between speed and distance, which he hoped would be more accurate than trying to measure time and distance. (Galileo overcame this problem by slowing down falling motion by sliding objects down an inclined plane.) He set up an elaborate balance system with a counterweight on one pan; then he dropped bullets—lead balls—from various heights onto the other pan, to determine when the force of percussion was balanced by the counterweight. Following his ancient predecessors, he believed that this force was a measure of the object's acquired speed (rather than its acceleration).[2]

Nevertheless, Harriot's assumption was approximately satisfied under the conditions of his experiment.[3] This meant that he could use these results, together with those of his experiments on the time of fall, to amass a considerable set of data, which he was able to compare with his theoretical speed-time and speed-distance graphs. He correctly deduced that in "free fall"—idealized fall in a vacuum, which in practice meant neglecting minor air resistance and small changes in altitude—speed is proportional to time, not distance. His manuscripts contain some impressive examples of hypothesis testing, where he worked out the theory for rates constant in time and for rates constant in distance, then carefully plotted speed-time and speed-distance diagrams showing both the expected theoretical result and the actual experimental result.[4]

As so often happens in science—and as happened especially in that transitional time between the medieval and the modern periods—Harriot was able to deduce correct results even though his understanding of the physics underlying his experiments was not always correct. Galileo, too, combined correct conclusions and premodern physics. His reputation today is due partly to the work of later scientists who were inspired by his insights and able to reformulate them—in particular, Newton and his disciples, who put the science of everyday motion into its modern form. This is the way of science; scientific "revolutions" are never the work of one person.

Galileo did indeed leave a marvelous legacy for his successors to build upon. He spelled out in *Two New Sciences* that his analysis of motion was confirmed by experiment, unlike his predecessors'. (At least, he confirmed it as well as could be expected at the time.) He focused especially on falling motion, on which, he said, only "superficial" pronouncements had previously been made—pronouncements such as that the gravitational acceleration was "continuously increasing." Yet "to just what extent this acceleration occurs has not yet been announced; for so far as I know, no-one has yet pointed out that the distances traversed during equal intervals of time, by a body falling from rest, stand to one another in the same ratio as the odd numbers beginning with 1."[5] What this means is that in free fall, a body moves 1 unit of distance in the first unit of time, 3 units in the second unit of time, 5 units in the third unit of time, and so on.[6] And what this means is that if the unit of time is 1 second, after 2 seconds the distance fallen is $1 + 3 = 4$ units of distance; after 3 seconds, the distance fallen is $1 + 3 + 5 = 9$ units; after 4 seconds, the distance is $1 + 3 + 5 + 7 = 16$ units, and so on. In each case the total distance fallen is (proportional to) the square of the total time the body has been falling. This was a new law of nature, the *law of free fall*, sometimes called the times-squared law. This law shows that the acceleration of a freely falling body depends only on the time and the distance—it does not depend on the body's weight.

In terms of basic calculus, the law of free fall arises by integrating, twice, the constant gravitational acceleration. Galileo did not have calculus, but he explained that it made sense for gravitational acceleration to be constant, because a constant rate of change is the simplest possibility.[7] It was an assumption rather than an empirical deduction, but it was an assumption borne out by empirical results. (Newton would show later that the simplest assumption is not always the best—but it is good enough for everyday falling motion.[8]) An intuitive times-squared rule was implicit in some of the abstract diagrams of medieval scholars; Galileo not only spelled it out clearly but also realized the importance of experimental confirmation.

When he wrote up his results in 1638, Galileo believed he was the first to point out the 1, 3, 5, 7,...rule. He did not know that Harriot had discovered this very same law of free fall (and had done so more rigorously than Galileo himself. Harriot also attempted to quantify the fact that weight, or rather the related concept of specific gravity, does influence the rate of fall in a denser medium than air, although he was not entirely successful.[9]) Unlike Galileo, though, Harriot did not write up his results in a conclusive form; later scholars have had to piece together his scattered writings on the topic.[10] It seems that there had been no correspondence between the two men, so this is another example of the way scientists often make independent codiscoveries, as though there is something "in the air" that makes it the right time for such a leap forward. Like Galileo, Harriot discovered the law of free fall from a mix of mathematical argument using medieval-style velocity-time graphs, and experiment. By today's standards, neither man had a full proof of his discovery, and each made some faulty assumptions. For their time, though, both showed uncommon skill and rigor, and they both came up with fundamental new results.

Yet another example of this involves their independent study of the effect of gravity on projectile motion. It is a fact of history that the needs of war sometimes provide the spur for scientific discovery. Harriot and Galileo lived in an era of widespread religious and nationalistic violence, one in which the weapons of warfare were constantly being improved upon. Cannons and musketry were becoming increasingly common, and for accuracy and strategy gunners needed to know the range of their bullets and cannonballs (as well as arrows from a crossbow or a sling, as Galileo mentioned).[11] In England, Digges and Bourne had published the leading works on the subject, building upon the analysis of the earlier Italian mathematician Tartaglia. Gunners' manuals contained measurements of angles of elevation and ranges of shots gleaned from experience, but neither the gunners nor the theoreticians knew the actual shape of a projectile's trajectory. If you knew this shape, as well as the angle of the cannon and the speed of projection, you could predict the range of the shot more accurately. You could also work out the angle of elevation that would ensure the greatest range.

Northumberland, who had been sent by the queen on an investigative military tour of duty in the Low Countries in 1600, likely encouraged Harriot to try to solve this problem. He knew that if anyone could successfully quantify a physical problem it was Harriot. For instance, sometime around 1602, Northumberland had had some repairs done on the plumbing at Syon House, and Harriot's early research on gravity was put to good use in an analysis of water flow through the pipes. On the basis of his analysis, he worked out a siphon system so that a newly installed cistern would not overflow and damage nearby infrastructure, and he furnished his patron with the required diameter for the pipes and the costs of the project.[12] The earl himself had read up on what Archimedes had had to say about water pressure and had engaged in experiments on water flow and written his own analysis on the "weight of water."[13] As for ballistics, Northumberland was fascinated by military strategy and before his imprisonment had purchased some four thousand toy soldiers made of lead. Now, in the Tower, he spent much of his time reading classic writers, such as Cicero and Plutarch, on how to find peace of mind in difficult circumstances; key to this was focusing on the life of the mind, and for Northumberland this included playing a military game akin to chess. It probably helped him hone ideas for the unpublished manuscript he was writing on the art of war. The game was played on an inlaid table, and the pieces included 140 soldiers made of brass, complete with wire pikes, and another 320 lead soldiers carrying tiny muskets.[14] No doubt Harriot joined in the play on occasion. He, too, had some interest in military matters—or perhaps his half-dozen manuscript pages on military organization and fortification were the result of work done on behalf of one or other of his patrons.[15] Either way, his manuscripts show that his main interest in such things was mathematizing ballistic trajectories.

Galileo had, in fact, begun his own gravitational research by exploring ballistic curves, from which he deduced his times-squared law. Harriot had studied free fall first and then applied his law to ballistics. Once again, from a modern perspective there were limitations to each man's work, but both again used experiment, as well as knowledge of Archimedes's rediscovered work on mechanics, and their own mathematical analyses, to deduce that the shape of a projectile's trajectory is a parabola.[16]

In the course of this work, Harriot (like Galileo and, later, Descartes) also produced a fledgling version of what would become in Newton's hands the law of inertia (or first law of motion). In Harriot's words, when a bullet is fired at a given angle, then "because of the bullet's gravity the crooked line is made" (that is, the bullet does not continue on in a straight line but curves downward). "If the gravity be abstracted, the motion would [continue on in a straight line], and if the resistance of the air or medium be also abstracted, his motion would be infinitely onward."[17] Harriot used his intuitive law of inertia in correctly deriving the parabola for point-blank (horizontal) projectiles, but he did not use it for oblique angles of ejection. Rather, he used the Aristotelian concept that there are two types of motion: "natural" (that is, free fall) and "violent" (or forced). Newton's first and second laws of motion would show that this is an artificial distinction, because all changes in motion are produced by forces. This meant that Harriot's solution for oblique angles of projection was not quite correct; he had assumed *decelerated* "violent" motion rather than *constant* inertial motion in the oblique direction, and as a result he derived a tilted rather than an upright parabola. Galileo achieved the correct shape but with incorrect reasoning: not only the flawed Aristotelian appeal to natural motion and violent motion but also a faulty analogy between a projectile's path and a hanging chain. (The latter's shape is not a parabola but a hyperbolic cosine.)[18]

WITHIN THE LIMITS OF HIS PARTIALLY incorrect physics, and with the help of his algebraic symbolism, Harriot was able to mathematically deduce a projectile's time in the air, maximum range, and parabolic trajectory.[19] Such a rigorous blend of mathematics and experiment was his hallmark. Galileo's strength was his big-picture thinking, his attempts to find connections between apparently different phenomena, as exemplified in his *Chief World Systems* and *Two New Sciences*. These two works reveal Galileo's ambitious attempt to reform astronomy and physics by overturning the ancient Aristotelian conception of reality that still held sway. In hindsight, some of his "reformulation" was erroneous, and much of it was incomplete, but as noted it was an influential bridge between medieval and modern science because it inspired later researchers who developed Galileo's ideas more fully.

Harriot did not appear to attempt such a grand project at all, so some scholars have consigned him to a lesser intellectual rank than Galileo, and also than Kepler and Descartes. It is part of a debate that has indulged in controversial claims on both sides. (I touched on this in the previous chapter and have given a little more detail in the next endnote.[20]) In any case, retrofitting Harriot into a celebrity star system is fraught and misguided. It is hard to compare published and unpublished work, or that which was accomplished in the process of specific experiments about the *what* of nature and that which was designed for articulating universal principles. Harriot's work concerns mostly the *what* of science and mathematics, but he was not blind to the need to articulate the underlying principles of natural philosophy. He had shown this in his explanation of the rainbow and in his atomistic hypothesis about the cause of simultaneous reflection and refraction—something very few had attempted to explain. He had also determined the parabolic shape of a projectile's trajectory with the help of "first principles" (fundamental experimental and mathematical facts and concepts), notably the law of free fall and an implicit concept of inertia. Later he would also spell out his interest in natural philosophy with respect to his analysis of the way colliding bodies ricochet off each other. In a note to Northumberland, he would write that he had intended to include the underlying first principles of motion with his analysis but would do so "in a time of better ability."[21]

Harriot had mentioned feeling unwell several times since his letter to the Privy Council in 1605. No one knows if he had any specific ailment at that time, although it has been conjectured that he might have picked up some sort of mysterious disease in the New World.[22] As he would tell Northumberland, and as he had told Kepler in 1606, this intermittent illness prevented him from going more deeply into the foundations of physics. It takes time and physical stamina to gather together all the results of cutting-edge research— to sort out which ones are valid, to work out how they challenge existing assumptions, and then to tie them all together in some sort of theoretical framework. Mathematical analysis itself, however, came relatively easily to Harriot, and despite his bouts of ill health, his paper on collisions would be yet another of his greatest works.

CHAPTER 21

Mathematics, Jamestown, Guiana

ARRIOT'S WORK ON COLLISIONS AND falling motion developed over many years (as, of course, did Galileo's). So did his research on mathematical combinations, which was related to his earlier work on gambling. Now he began exploring the relationship between "triangular" and other "figurate" numbers on the one hand and "binomial coefficients" on the other. As described earlier, the triangular numbers are made from triangular arrays of dots, and they form the sequence 1, 3, 6, 10, 15,..., which Harriot had applied to the stacking of spherical balls. The triangular numbers also arise in the mathematics of combinations, be they combinations of cards or dice or anything else. The ancients had known that the number of unordered pairs that can be made from a set of five things, say—for instance, the number of ways you can choose two cards from a set of five—is the fourth triangular number. Like a handful of his predecessors, Harriot set out to explore this pattern more generally.

To begin with, he applied a systematic approach using the letters a, b, c, d, and e to represent the five "things," so that all the possible pairs can be written as ab, ac, ad, ae, bc, bd, be, cd, ce, de. This list shows that there are ten such pairs from five things. Ten is the fourth triangular number, and through trying many other combinations, Harriot confirmed the general pattern: the number of pairs of n things is the $(n-1)^{\text{th}}$ triangular number, and it has the form $\dfrac{n(n-1)}{1 \times 2}$ (except that Harriot wrote 12 instead of 1×2, and he wrote the factors n and $n-1$ one on top of the other instead of using parentheses, as in plate 5).

As indicated earlier in connection with Harriot's work on the binomial theorem, $\dfrac{n(n-1)}{1 \times 2}$ is also a "binomial coefficient." It can be generalized to the number of ways of choosing three things from a group of n things, then four things, five things, and so on, and Harriot used his new symbolic notation to express all these quantities. Although a few of his ancient and early modern predecessors in both the East and West had known the essence of these combinatorial rules, no one had developed a concise notation in which to show them in such crystal-clear generality.[1]

That was not all. Harriot went on to make a completely new use of these formulae when he discovered that the binomial coefficients turn up in a way far removed from probability and combination theory. This is the key work in which he applied binomial coefficients with fractional values of n, unconstrained by concrete applications requiring whole numbers of cards or dice throws, or the simple binomial expansions from Pascal's triangle of whole numbers. Half a century later, Newton and his Scottish contemporary James Gregory would rediscover (and further develop) the same result—and unlike Harriot, they would publish. Today Harriot's discovery is therefore called the "Newton-Gregory forward difference formula" or the "Newton-Gregory interpolation formula." The idea of "interpolation" is to find additional approximate numerical values of mathematical functions, and, as discussed earlier, simple forms of it had been known since ancient times. But Harriot's discovery went beyond simple approximate interpolated values to theoretically exact ones. When rediscovered by Gregory and Newton, this formula would lead to the development of what is now known as Taylor's series, a vital mathematical tool for representing complicated functions. Nowadays, both the interpolation formula itself and Taylor's series are also extremely useful for programming computers to handle functions.[2]

Although he had not yet written it up in its final polished form, Harriot had developed his interpolation formulae by 1611, when Lower wrote to him about his "doctrine of differences... or triangular numbers."[3]

The year 1611 brought not only mathematical successes but also personal difficulties for Harriot and his patrons. First, there was a court case. Ralegh had his executors sue his former friend and business partner William Sanderson over outstanding funds owing from the Guiana venture. Sanderson had fallen on hard times in recent years, thanks to some unsound investments, and he was still paying interest on his Guiana loans. Not surprisingly, he was in no mood to negotiate over accounts he believed had been settled years earlier, and he immediately countersued Ralegh and his two executors, as well as Harriot, who still held the relevant accounts. It was a sorry affair that dragged through the courts for a couple of years and culminated in Sanderson accusing Ralegh or his agent of having forged his signature on a release document that made Sanderson liable for any debts on Ralegh's return from Guiana. There is insufficient evidence to draw any conclusions about the supposed forgery or the other disputed matters, and there is no known outcome of the case. Because of other debts, Sanderson ended up spending seven years in a debtors' prison, and Ralegh was already incarcerated, so it was a sad and perhaps telling slide into ignominy for both these once-hopeful New World entrepreneurs. The bitterness between them was such that their sons would later carry on the feud (via polemics rather than litigation).[4]

It cannot have been an easy time for Harriot, being caught in the middle of such an acrimonious dispute and having to give evidence in court, but he was also distressed on account of Northumberland's deteriorating situation. On the basis of some spurious new "evidence" against him regarding the Gunpowder Plot, the earl was interrogated in the Tower, and the councilors reported that he was in "weakened spirits" and expressed "great fear."[5] He had been trying to make the best of his unjust imprisonment, focusing on scholarship and science to keep his mind alive, and remotely overseeing the affairs of his pensioners and estates, but his anger and resentment redoubled in the wake of this new interrogation. James now demanded that Northumberland pay off more of the £30,000 fine imposed for his supposed role in the plot, but as the earl pointed out, in a letter to the chancellor of the exchequer, it was difficult to raise money as a prisoner whom no one trusted. James's response was merciless: he took over several of the earl's leases, which ensured that Northumberland would be unable to raise money without selling his estates, which he regarded as his own and his son's patrimony. He and his wife, Dorothy—who had remained loyal to him, despite their difficult marriage and their separate lives—began a desperate battle of negotiation with James, similar to the one Bess had fought over compensation for Sherborne.

Harriot did what he could to help his patron. Presumably he also helped reinforce the psychological or spiritual qualities the earl needed to survive in the Tower, as he had done for Lower on the death of Lower's beloved son the previous year. On that occasion

Lower had written Harriot that, "among other things, I have learnt from you how to settle and submit my desires to the will of God."[6] Like Lower, Northumberland took comfort in study during trying times, so Harriot soon returned to work. Algebraic interpolation was not the end of his research on binomial coefficients, and he began exploring yet another application of them: compound interest.

Earlier peoples, as far back as the Mesopotamians, had studied compound interest, but early modern Europeans knew nothing of such work. In fact, for hundreds of years, the Church had forbidden usury. Aristotle had thought it immoral, too, and charging interest on a loan had not been legal in Europe until Henry VIII allowed it in England in 1534 (the maximum allowable rate in Harriot's time was 10 percent). Nonetheless, a century or so earlier the needs of a flourishing mercantile class in Renaissance Italy had led to its tacit acceptance in many districts, and this, in turn, had given rise to studies on the mathematics of interest—in particular, "interest upon interest," or compound interest.[7] Harriot was no doubt drawn to this topic because he was still doing Ralegh's accounts, as the Sanderson case also illustrates. Chasing creditors was an occupational hazard in an era without a banking sector, as noted.

If an amount of money, L, is borrowed (or invested) at an interest rate of, say, 7 percent compounding annually, then the amount owing after t years is $L(1+0.07)^t$. This is an application of the binomial theorem, and the formula for compound interest had long been known in essence, although not in its symbolic, fully general form. In pioneering the algebraic form of the binomial theorem, Harriot was the first to show how to symbolically multiply out the product of the brackets in an arbitrary binomial expression, as explained earlier. This in turn enabled him to discover the formula for when interest compounds *continuously*. This was another example of his ability to handle infinitesimal and infinite quantities, because continuously compounding interest is added every instant, which means that first Harriot adapted the binomial formula for interest compounding annually to account for when interest is paid n times a year, and then he showed what happens as n becomes larger and larger, eventually approaching infinity.

Harriot's formula for continuously compounding interest turns out to be a special case of the Taylor series expansion of the function now known as e^x. The modern formula for the amount owing on an investment or loan of L after t years at 7 percent compounding continuously is $L\,e^{0.07t}$, which Harriot wrote in its general, algebraic series form for any arbitrary interest rate. (Astonishingly, Indian mathematicians in the fifteenth century had verbal algorithms for the equivalent of Taylor series for sine, cosine, and arctan, although they remained isolated discoveries that were rediscovered later in the seventeenth century.) In deriving his general formula for Taylor series a century later, Taylor would apply the very same method Harriot had used when he found his formula for continuously compounding interest and his analogous forward difference interpolation formula. The key to both is the use of the binomial theorem in exploring an increasing number of ever-decreasing increments—be they "differences of differences" or "interest upon interest" or the later notion of "differentials" and derivatives—and then in effect taking limits as the increments become infinitesimally small and the number of them approaches infinity. In Harriot's work on interest, his intuitive handling of limits is evident both in his symbolic working and his numerical examples of the amount owed on a given value of $L invested at a particular continuously compounding rate for a specified time. He calculated the latter correctly to within 1/500th of a penny; he knew that he could get more accurate answers by including more terms in the series, but he figured he'd shown well enough that although the interest compounds an infinite number of times, the amount accruing is definitely finite. This was another new result.[8]

There was nothing, it seems, that Harriot did not seek to mathematize, from the Syon plumbing to compound interest, from rhumb lines to Zeno's paradox, from magnetic variation to the shape of the rainbow to falling motion and more, and he made original contributions in every field he touched. He even tried his hand at mathematical population analysis, two centuries before Thomas Malthus. Malthus argued that population increases exponentially—that is, at an increasing rate—while food can only increase at a constant ("arithmetic") rate. It was a pioneering wake-up call, although today there are competing theories, depending on the chosen assumptions about human behavior and the role of new technologies in food production.

In general terms, the need to balance population growth and resources was not a new idea in Harriot's day. The sixteenth-century Italians Niccolò Machiavelli and Giovanni Botero, for example, had discussed it in connection with matters of state and the planning of cities.[9] Unlike his predecessors, however, Harriot used mathematics to provide a quantitative analysis of the problem. In showing that unchecked population growth is exponential, he used a sophisticated version of the now-famous Fibonacci problem about a rabbit population. Start with a pair of baby rabbits, and assume they can reproduce at the age of two months, after which they (and subsequently their mature offspring) produce a pair of babies each month, one male and the other female. After the first month, there is still just the original pair of rabbits, who will produce a new pair of babies at the end of the second month, so there will then be two pairs. In the third month, the original pair produce another set, and the total is now 1 + 2 = 3 pairs. By the fourth month, the first litter of babies can reproduce, too, and the original pair also has produced another pair, so there will now be five pairs, and so on. The rabbit population each month forms the "Fibonacci sequence" 1, 1, 2, 3, 5, 8, 13, 21, … , where each term is the sum of the previous two. This is widely known today for its appearance in nature (although it also has other uses): for instance, in the spiral arrangements of seeds in sunflowers and pinecones, the number of clockwise spirals and the number of anti-clockwise spirals are successive numbers in the Fibonacci sequence. Although Fibonacci (the nickname of Leonardo of Pisa) published the problem in 1202, the mathematical significance of the Fibonacci sequence was not generally recognized until the eighteenth century.[10] It is not clear whether Harriot knew of the rabbit problem, although quite possibly he did, and Kepler certainly knew of the Fibonacci sequence. (Kepler showed that the ratio of successive terms in the sequence tends toward the so-called golden number, $\left(1+\sqrt{5}\right)/2$, which is approximately 1.618.)

In terms of population growth, the assumptions in Fibonacci's problem are clearly unrealistic because they assume that all the rabbits remain alive and fertile, and they ignore the perils of interbreeding. Harriot made similarly simplistic assumptions, although his calculations are necessarily more complex, and interesting in their own right. They include an application of the figurate numbers (triangular, tetrahedral, and so on) to give the number of children produced each generation.[11] He concludes that between the first and second decades, the number of births in his idealized population has doubled; after three decades it is sixfold, after four decades fifteen-fold, and so on at an increasing rate.

Harriot also made an estimate of the maximum population that the earth could support. Given the limited information available at the time, of course, his assumptions were very rough—but his calculations showed, for the first time, an attempt to quantify the essential "Malthusian" features of population versus food production. He assumed that England represented the average fertility of the earth. Though his own trials had suggested it was not as fertile as Virginia, it was clearly more fertile than a desert. Then he estimated England's surface area, from "Saxton's great map," as he put it, and its population, which he took to be 5 million (although at that time it was closer to 4 million[12]). This gave him an average of about 5.5 acres per person. Extrapolating this to the size of the earth—using a known estimate of the earth's radius and the well-known formula for the surface area of a sphere, and assuming half the earth's surface was land (an overestimate, too, as we now know that oceans cover 70 percent of the earth)—he calculated that the maximum sustainable world population was a little over 7 billion. This is the figure we reached four hundred years later, in 2012. Some modern analysts predict the population will not climb to much more than 10 billion. In another calculation, Harriot also assumed that the population would eventually remain steady; he thought it would stay at around 7 billion, the maximum number he thought the earth could support.[13]

No doubt Harriot discussed his ideas with Ralegh, who had been toiling away in the Tower on his monumental *History of the World*, in which he included reference to the problems that arise when populations outgrow their territories' resources.[14] Ralegh's first (and, in the end, only) volume of his *History*, published in 1614, was an eight-hundred-page contribution to the development of historiography, politics, and the English language. It covered ancient secular history, though Ralegh also included biblical events; he was taking no chances of being branded an atheist again. For instance, he emphasized God's role in Creation, and he was firmly against *ex nihilo nihil fit*, the atomistic idea that Harriot had explored. Harriot had continued to help with research, comparing different biblical texts to ascertain the best translations and compiling a list of key biblical and classical events and dates. He also used astronomical data to analyze the English (Julian) calendar for Ralegh. In particular, he calculated the earliest and latest possible dates between which Easter would fall and calculated the date of Easter at various times in the past, beginning in 46 BCE when the Julian calendar was instituted. He even made up his own rule—a mnemonic combined with finger counting—to work out such things as the day of the week in any year (following an old calendric rule used by the Romans and then by the Catholic Church).[15]

In his *History*, Ralegh also discussed the importance of human reason and science. He used Harriot's calendric-astronomical chronology, and his and others' knowledge of geography and the New World, to reinterpret traditional stories such as the Flood and to square them with secular history. He also analyzed the nature of miracles, distinguishing between what he saw as Christ's humanly impossible miracles and those accounted miraculous by people who lacked an adequate understanding of experimental science and natural philosophy or "natural magic."

He discussed ethics and politics, too. A significant theme in the *History*, perhaps originally included with young Prince Henry in mind, was the idea of kingship expressed in Queen Elizabeth's "Golden Speech" to Parliament in November 1601: "And though God has raised me high, this I count the Glory of my Crown, that I have reigned with your loves." This had nothing to do with democracy, of course; but given the terrible religious divisions that had followed the Reformation and Counter-Reformation, Elizabeth was notable for keeping her country relatively united for nearly half a century. Ralegh's telling of history highlighted the folly of despots and observed that stable government needed the support of the people—a notion that prefigured the constitutional monarchies to come. Prince Henry's father, James, on the other hand, would have preferred to be an absolute monarch, and felt that Ralegh's version of history was "too saucy in censuring princes." Ralegh indeed highlighted the brutality rather than the glory of such classical heroes as Alexander the Great, and of modern kings such as Henry VIII, too. James attempted to suppress the book, but this only added to its appeal: it ran through five editions in three years. By the end of the century it had been reprinted twice as many times as the collected works of Shakespeare, and ironically was second in sales only to the King James Bible (which had been published in 1611). It also proved influential on some of the radical Parliamentarians who would bring down the Stuart dynasty, including Oliver Cromwell.[16]

POSTERITY HAS TENDED TO PRAISE Ralegh's scholarship, prose, and methodology while naturally criticizing his theological philosophy of history.[17] Nevertheless, his book was the first world history in English, and there is much that is remarkable about it—including the witty asides. This one gives the flavor: "It is not truth, but opinion, that can travel the world without a passport." As for those inclined to the kind of judgmental piety that had led to the Cerne Abbas enquiry, he had little left to lose now—"I am on the ground already, and therefore have not far to fall"—and brazenly castigated the way zealots condemned as atheists anyone who was "not transported with the like intemperate ignorance."[18]

While Ralegh turned to satire to express his anger at the enemies who had brought him so low, his fellow prisoner Northumberland was aiming for personal stoicism in the face of James's arbitrary justice.[19] By the end of 1614, the year Ralegh saw his *History* into print,

Northumberland finally paid out his fine. He'd had to borrow money to do it, but he'd had a measure of success in his negotiations with the king. James had reduced the fine to £20,000 and then had finally agreed to a lump sum of £13,200 in lieu of interest on a later payment by installments. Northumberland was still furious at what he saw as the king's depredations on his estate; he equated it with undermining the rights of nobles, which had been enshrined in the Magna Carta four hundred years earlier. The young lawyer, playwright, and pamphleteer John Ford was one of those who had been doing his bit to raise the earl's plight publicly, and in 1614 he published an expanded version of *The Golden Mean*, his 1613 tract on "Nobleness of perfect Virtue in Extremes." The new edition contained an explicit dedication to Northumberland, outlining the worthiness of such a "great peer" despite his loss of prestige and power—a loss that had happened through no fault of his own. Ford also spoke of the importance of friendship independent of fortune and attacked the "inconstancy" of kings and their sycophantic favorites. Having received some kind of patronage from Northumberland for his literary work, he thanked the earl accordingly.

This might have been the time for Harriot to publish something of his own with a supportive dedication, but he was not the kind of person to enter the political fray the way Ford did with his barely concealed criticisms of James. Besides, had Harriot done so his words would not have carried much weight: he was under a cloud himself, having been interrogated after the Gunpowder Plot. Instead, 1614 was the year he made the final breakthrough in his analysis of the loxodrome. Perhaps this was reward enough for Northumberland, who was trying to keep up his spirits through intellectual satisfaction and stimulation.

As discussed earlier, Harriot had long ago discovered Mercator's secret "stretch factor," which showed how to place parallels of latitude at increasing distances from the equator in order that spiraling rhumb lines appeared as straight lines on a map. In using this stretch factor ($\sec\varphi$) to calculate the first comprehensive tables needed to make such a map, Harriot, like Wright, had calculated by hand the distance on the map from the equator to any given parallel of latitude φ—a distance, called the meridional part as mentioned earlier, which is equal to the sum of the secants of incrementally increasing values of φ. Today, such a sum is worked out using integral calculus, and the required integral is $\int_0^\varphi \sec x \, dx$.

It turns out that this integral is equal to $\ln\left|\tan\left(\dfrac{\varphi}{2}+\dfrac{\pi}{4}\right)\right|$, a forbidding enough expression that took accomplished mathematicians decades to work out. The first published clue to it had come about completely by chance, years before Newton and Leibniz worked out the algorithms for calculating integrals in the 1660s and 1670s. A navigation teacher, Henry Bond, had been studying Wright's published tables, and for some reason he decided to compare them with Napier's tables of logarithms, which Napier had published in 1614. As noted earlier, the logarithm "to the base 10" of 100 is 2, because a logarithm is a power: 100 is 10^2, so its logarithm is 2. The expression *ln* in the integral formula above stands for the "natural logarithm" or the "logarithm to the base *e*," where *e* is the mysterious transcendental number that implicitly appeared in Harriot's work on compound interest and the equiangular spiral. In 1645, Bond published his conjecture that Wright's tables of meridional parts could be calculated from the formula $\ln\left|\tan\left(\dfrac{\varphi}{2}+\dfrac{\pi}{4}\right)\right|$, rather than by all that tedious adding up by hand. It was a conjecture that tantalized English mathematicians for the next twenty-five years, before it was proved by Gregory and then by Isaac Barrow.[20] No one knew that half a century earlier—and thirty years before Bond made his accidental discovery—Harriot had already proved it. He had, of course, long been seeking a way to use mathematics rather than brute-force arithmetic in calculating meridional parts. Now, in an incredibly sophisticated step-by-step tour de force of logical argument and proof, he managed to obtain the equivalent of the formula $\ln\left|\tan\left(\dfrac{\varphi}{2}+\dfrac{\pi}{4}\right)\right|$ for the sum of the secants in a meridional part.[21]

With the algorithms of calculus, together with the definition of a logarithm as a function, and a clever trigonometric substitution, the integral of $\sec\varphi$ can be obtained in a dozen lines. Harriot's precalculus proof took many years and many dozens of pages; needless to say, it was neither elegant nor polished, and not every step was proved as fully as would be expected today. Nevertheless, for its time, it is a masterpiece of rigorous conception. Together with the many new results that Harriot discovered in order to solve the problem— the equiangular spiral, infinite geometric series, binary arithmetic, trigonometric breakthroughs, the binomial theorem, interpolation, and the conformality of stereographic projection—it remains a testament to his genius.[22] It is a testament to Northumberland, too, for recognizing and generously supporting such genius.

AFTER HIS LOXODROME BREAKTHROUGH, which he used to produce highly accurate tables of meridional parts for navigators, Harriot continued working on algebra.[23] One of his goals was developing algebraic methods for solving complicated polynomial equations, in which he continued the work of his early modern predecessors, notably Cardano and Viète, using his own method of factors to build up a systematic structure of polynomials and their solutions. Understanding polynomials is important mathematically, but there is also a unique thrill in developing a logical mathematical proof. It was what drove Harriot, Viète, and others to produce so many pages of work that today appears arcane, if not tedious, to most of us.[24]

Harriot also continued to observe the skies with his telescope and to think about the physics of motion. But by 1615, his health had taken a turn for the worse. A nasty sore in his nostril had been growing and causing increasing discomfort, and Harriot's doctor sought a second opinion from the king's chief physician, Theodore de Mayerne. Harriot, now fifty-five years old, first saw Mayerne in May 1615. Mayerne's report describes Harriot as "melancholy." This would not be surprising, given the constant pain he was in, but "melancholy" was also the supposed effect of an excess of black bile, one of the "humors" popularized by Galen 1,400 years earlier. Mayerne was a more "modern" doctor, though. He still conformed in his report to the idea of humors, and he revered its creator, Hippocrates (as had Galen), but he also followed some of the teachings of the iconoclastic Swiss doctor and alchemist Paracelsus.[25] In 1527, Paracelsus had apparently burned a copy of Galen outside the University of Basel, where he lectured. Galen had been an avid anatomist, gaining knowledge from practice via dissections (and, unfortunately, vivisections), and in his time he had made some significant contributions. Paracelsus advocated a new kind of experimentation, pioneering the use of proto-chemistry in medical treatment and diagnosis (for instance, he was the first to connect goiters with minerals in the diet). He also advocated allowing the body to heal itself where possible, disdaining medieval potions that seemed more quackery than medicine.

At his first visit, Harriot discussed with Mayerne the medical benefits of tobacco, in which Paracelsus had taken an interest. Mayerne believed it was a "noble" plant but should be used with caution. Such advice was too late for Harriot, habitual smoker that he'd been for the past thirty years: Mayerne diagnosed a malignant tumor in his nostril, which had grown large enough to affect his lip, making it so hard and numb that it would soon surely impact on his ability to speak and eat. The doctor would have preferred to remove the tumor surgically, if his patient could bear the agony in those days before adequate anesthetics. Harriot's opinion on the matter is not known. Mayerne, in any case, hesitated and recommended some sort of plaster, which seemed to improve things.

With his Paracelsian interest in reforming medicine, together with his outsider status as a recent Huguenot immigrant, Mayerne was possibly a kindred soul for Harriot. Mayerne knew that Harriot, too, was an outsider, with both his patrons in prison and he himself still under suspicion, and it seems that the doctor confided to his extraordinarily intelligent and sympathetic patient some of his own troubles. Although he had the support of the king, initially the English medical establishment was wary of his Paracelsian approach. Courtiers jealous of his place in James's esteem were quick to undermine him when any of his eminent patients died. Harriot wrote him to say that he hoped the doctor would consider him as not just a patient but also a "loving friend." After updating Mayerne on the course of his health, he added, "Your situation concerns me as much as my own. My recovery will be your glory too, but through Almighty God, the author of all things."[26] So much for Harriot's atheism.

In the spring of 1616, Mayerne revised his opinion of Harriot's malady, suggesting it was an ulcerous skin disease rather than a cancerous tumor. Harriot seemed to be otherwise in good health at this time and continued working with vigor. Ralegh, too, had been working energetically in his enforced capacity as a full-time writer, pouring out poems and crafting arguments on matters of state. He had continued to develop the idea that kings should rule by gaining the respect of the people and to emphasize the role in public life of reason and the use of evidence rather than blind obedience. He also considered such practical matters as import duties on luxuries, which he supported as long as they were fairly collected, and he discussed what he saw as the reasons for war. There were "natural" wars, which he defined as arising when more territory was needed in the face of overpopulation, disease, or hunger, as well as for defense; otherwise wars were "unnatural" and unnecessary, due merely to the designs and ambitions of kings and popes or of powerful, equally ambitious individuals able to foment civil rebellions.[27]

But Ralegh was longing for freedom, not for posthumous fame as a writer. The high-minded intellectual and man of action had become a restless, captive lion. He had long been dreaming up schemes that might secure his release, including writing his *History* for young Prince Henry, who had admired Ralegh immensely and might well have secured his release had he not died in 1612. (Mayerne had been his doctor, but unlike some of his courtiers, the king did not blame

Mayerne for the death.) Eventually Ralegh came up with a plan that appealed to the cash-strapped king, whose profligacy was often criticized but who also had to contend with the Tudor legacy of rising inflation and outdated fiscal policy: he would mount another expedition in search of El Dorado.[28] He had been talking about the idea for years, but finally it caught the king's fancy, and after bribing a couple of courtiers to press his case Ralegh was released in early 1616. He was to be kept under surveillance, of course, although he tried to secure a pardon at the outset. Why, he argued, should a thousand men follow a convicted traitor to the ends of the earth? James in turn promised Ralegh his freedom—but only on his successful return. On this dream of fabled fortunes, Ralegh wagered his life.

At sixty-three years old, with no legal status and a death sentence still dangling above him, Ralegh went about London without his former swagger and style. After thirteen years of captivity, he savored seeing once again the old familiar sights, amazed by the changes, too. While he and Bess were feverishly raising money for the Guiana venture, calling upon relatives and selling what little they had left (including Bess's property in Mitcham), Northumberland was entertaining in the Tower the Powhatan princess Pocahontas.[29] The Powhatan (or Virginia Algonquian) people hailed from the East Virginia area, where Jamestown had been established in 1607. Northumberland's younger brother George Percy had been a colonist and a councilor there until 1612, having served briefly as deputy governor in 1611. He lived at Syon and before his departure to Jamestown had no doubt asked Harriot about his own experiences in America. Harriot also probably taught him some Algonquian.[30]

The chief of the Powhatans was Pocahontas's father, Wahunsenaca (or Wahunsonnacock), whom the English simply called Powhatan. Pocahontas was his favorite child; her given names were Amonute and Mataoka, but she adopted her nickname Pocahontas, which meant "playful, joyous one." Six months after his arrival in America, the Jamestown councilor John Smith had been captured by Powhatan's brother, Opechancanough, and later Smith famously claimed that Pocahontas had saved him from death. Today scholars believe it is more likely that he had unknowingly been going through a ritual initiation rather than a preparation for execution, and Smith certainly became a trusted friend of Powhatan.[31] His initial kidnapping likely arose because a few years earlier a white man had captured several of Powhatan's men. This could have been during the 1603 expedition led by Samuel Mace, who had unsuccessfully searched for the Lost Colonists in 1602 and 1603 and had received some linguistic training from Harriot. (Whoever took them, several Powhatans were seen demonstrating their canoeing skills on the Thames in 1603, much to the delight of spectators.)[32] After interrogating and then befriending Smith, at first Powhatan believed he could accept the colonists without ceding power to them. An ambitious man, he had been expanding his influence over neighboring tribes for years, and he probably assumed that with their advanced weapons the new settlers could be a useful part of his confederacy. Unfortunately, the English had not learned from Ralph Lane's mistakes at Roanoke. Within a couple of years, lack of adequate supplies had led to increased pressure on Powhatan to supply food, at the very time when a long drought was affecting indigenous crops. George Percy recalled that disease and famine had killed most of the original set-tlers, so it was a desperate time, and the English used threats, theft, and reprisals to obtain food from the local people. Not surprisingly, just as Wingina/Pemisapan had turned against the settlers, so too did Powhatan. He removed his village further inland from Jamestown and refused to trade. Smith had returned to England in 1609 because of an injury, but by 1613 hostilities between the two peoples were such that Captain Samuel Argall kidnapped the eighteen-year-old Pocahontas, intending to ransom her for Powhatan's English prison-ers and stolen goods. What happened next is not clear, but in 1614 she married John Rolfe. She did this apparently with the consent of Powhatan, perhaps in the hope of bringing peace with the English or perhaps simply to obtain her freedom. Rolfe had arrived in Jamestown in 1610; en route, he and his fellow passengers had been ship-wrecked by the storm that had inspired Shakespeare's *The Tempest*. By 1614, Rolfe had become a tobacco planter with good prospects. Now, in 1616, he and Pocahontas, and a dozen or so of her people, made a visit to England under the auspices of the Virginia Company, and meeting Northumberland—and staying with George Percy at Syon—was on their agenda.

Presumably Ralegh met the Algonquian delegation, too, and perhaps their presence in England comforted him with the belief that his efforts to build a peaceful alliance with the original inhabit-ants of Virginia had paid off. Perhaps it encouraged him to hope that his old Guianan friends would remember him kindly when he returned. But the omens for Jamestown—and especially for the Algonquian people—were not good. Pocahontas died, possibly of tuberculosis or pneumonia, just after setting off home to America.

The South Americans did, however, receive Ralegh well when he and his fleet made their way into Cayenne in November 1617, after a voyage dogged by ill winds, desertions, and disease. Ralegh, who was suffering from a tropical fever, was overjoyed to meet up with his old companion, a chief he called Harry, who had gone to England after the first English visit to Guiana, and according to Ralegh's journal of the current venture, "had lived with me in the tower for 2 years." This must have been soon after Ralegh's imprisonment in 1603, because he noted that Harry had almost forgotten his English. Ralegh also recorded in his journal that Harry and his people—"the Indians of my old acquaintance"—"fed and assisted" him "with a great deal of love and respect."[33] He wrote Bess, his "dear heart," with the touching pride of a man who suddenly finds the recognition so long denied him in his own homeland: "To tell you that I might be here King of the Indians were a vanity; but my name has still lived among them. There they feed me with fresh meat and all that country yields." Then he added that all his Guianan acquaintances "offer to obey me."[34]

This warm Guianan welcome was not enough to turn rotten luck and a foolish dream into good fortune. Ralegh was still too sick to undertake the expedition in search of gold, but the men he chose to lead in his stead proved utterly incompetent, while the rest were "scum," as he later put it.[35] His leaders included Lawrence Keymis, veteran of Ralegh's first two Guiana expeditions two decades earlier and loyal friend to the Raleghs. Then there was the twenty-three-year-old Wat Ralegh, in the role of captain of the flagship. Sir Walter adored his son, but for all Wat's Oxford education and his tour of the Continent (albeit a tour chaperoned by the irascible, promiscuous Ben Jonson), he seems to have been an immature young man with a penchant for ribaldry and swagger.[36] Ralegh's instructions were that Keymis and his men should travel up the Orinoco River until they were twenty miles from the Spanish fort at San Thomé, which had been built after Ralegh's first Guiana expedition. Here, in the jungle, they were to leave a contingent of soldiers ready to defend the English if the Spaniards attacked, while Keymis and his party were to head off in search of the legendary gold mine. On no account were the English to attack the Spanish first. King James had given the Spaniards his word on this.[37] Soon after Keymis set out with his party of several hundred men, however, he picked up from friendly locals that not everyone in San Thomé was happy with its governor and his iron rule. Keymis gambled that with such internal divisions, the frontier town would readily surrender, and, disobeying Ralegh's orders, he landed his men just five miles away and advanced toward the settlement. What he did not know was that the Spanish were ready for him, and they fired on the English. A wiser captain would have retreated, but Wat roused his terrified men and charged forward. An instant later, he fell to the ground, fatally wounded.

Keymis opted to continue on to San Thomé, where he found that most of the population had already been evacuated. The English easily took the fort, in which, to Keymis's astonishment, they discovered a copy of Ralegh's plans.[38] The Spanish had been ready for the English because their own king had betrayed them. James had developed a close friendship with the able Spanish ambassador, Count Gondomar; he also had in mind a Spanish marriage for his son Charles, which might enrich his depleted coffers with a dowry of half a million pounds.[39] When Ralegh was first released, the Venetians and Savoyards had approached the English to discuss a possible alliance against Spain's activities in the Mediterranean, with Ralegh as organizer and captain of an English fleet. In the end, James, a peacemaker at heart, had decided against such a strategy and redirected Ralegh to Guiana. When Gondomar protested that this was Spanish territory, James had reassured him by showing him a complete list of Ralegh's ships, men, and itinerary, and the written promise he'd extracted from Ralegh not to attack Spanish settlements. Consequently, Gondomar was able to send King Philip III precise details of Ralegh's destination even before he had left England. (Philip II had died in 1598.)[40]

For twenty days, Keymis and his men rowed up the Orinoco in search of the illusory mine. As his morale sank, and with it that of his men, he became brutal in interrogating his captives about gold, but El Dorado was a fantasy, and the deaths of his men, and of the Spaniards, too, had been for nothing. When what remained of the party arrived back at the mouth of the Orinoco where the fleet was anchored, Ralegh's despair knew no limits. His beloved son was dead, and buried with him in the San Thomé chapel was his own last hope, for he realized that Keymis's encampment so close to the Spanish fort, and Wat's charge against the Spanish, could be seen as an attack, in contravention of James's orders. In desperation, Ralegh turned on Keymis the full force of his fury. Keymis stammered out lies and excuses, but Ralegh was in no mood for forgiveness. Keymis went into his cabin and killed himself.

There was nothing left for Ralegh other than to write a justifying letter to the government and a heartbroken letter to his "dear Bess," telling her of their son's death. When grief gave way to anger and the need to explain more fully to his wife, he unsealed the letter and added a long postscript, outlining Keymis's excuses and his suicide and enumerating various other betrayals before and during the expedition. Another man might have deserted then and there. Ralegh told Bess that "if I live to return,…it is the care for you that has strengthened my heart."[41] In late spring 1618, he set sail for England.

CHAPTER 22

The End of an Era

W HILE RALEGH WAS AWAY, HARRIOT had been reading up on comparative religion. Doctrinal debates were raging on the Continent, concerning such matters as free will and how it fitted with the notion of eternal salvation. Harriot may have felt that he was heading ever closer to the afterlife. His loyal friend Lower had died two years earlier, and the tumor in his nose had continued to grow. On Mayerne's advice he would eventually seek a specialist surgeon's opinion; meantime, Mayerne had nothing more to offer. Harriot began making an unprecedented effort to get his intellectual affairs in order, writing up his work on triangular and binomial numbers and their use in interpolation.

Whether or not he was preoccupied by his own death, this work shows that Harriot's mental energy was undimmed, so perhaps his eclectic religious reading at this time arose not so much from a sense of his mortality but simply out of interest in keeping up with current debates. Some of this consolidating was likely done on behalf of Northumberland, too.[1] To repay his patron's support, he sometimes did research for him—and perhaps it was also time to repay him by preparing his mathematical work for publication, with the traditional flowery dedication that scholars offered their patrons.

At any rate, Harriot was clearly preparing his work on triangular numbers and interpolation for circulation if not publication, because he gave it an impressive-looking title page:

De Numeris Triangularibus
et inde
De Progressionibus Arithmeticus:
Magisteria Magna
T. H.

This translates as *On Triangular Numbers and Thence of Arithmetic Progressions: The Great Doctrine*. It was indeed "magisterial"—*the* authoritative tract on the subject—and Harriot wanted to make that clear. His pride was not misplaced: nothing like it, either in form or content, had ever been published. Its thirty-eight densely packed pages of mathematics are written entirely in symbolic form; there is barely a word of text. It is a masterpiece of economy.[2]

The work sets out Harriot's method of interpolation—the very first known algebraic method (aside from Brahmagupta's fledgling verbal version), which, when rediscovered by Gregory and Newton, would prove so fruitful in the development of mathematics. Nevertheless, in the six decades between Harriot's discovery and Gregory and Newton's, Harriot's results, though unpublished, would have considerable influence among English mathematicians. His manuscript circulated via Torporley and Warner, who also wrote manuscripts of their own based on Harriot's work. For instance, in 1651, an acquaintance of Warner's, Charles Cavendish—a hobbyist rather than a research mathematician, but one with influence—would write to the mathematician John Pell, offering to send him Harriot's doctrine and confessing that he was "so far in love with it that I copied it out; though I doubt I understand it not all, much less the many uses which I assure myself you will find for it." Others who utilized or knew of Harriot's result include Nicolaus Mercator (a different Mercator from Gerardus of mapping fame), possibly Henry Briggs—who praised Harriot's formula for the area of a spherical triangle and published early results on interpolation similar to some of Harriot's—and John Collins, the librarian of the Royal Society.

John Wallis, the best English mathematician of the generation immediately preceding Newton, had been introduced to analytic mathematics mainly through Harriot's posthumous book *Praxis*, but he had not seen the *Doctrine of Triangular Numbers*. He rediscovered Harriot's formulae for figurate numbers and used them to develop an interpolation method of his own (although it was not as good as Harriot's). He published his results in 1656, and they happened to be read by young Newton, who was inspired to develop his own approach. Newton's initial algorithm was remarkably similar to

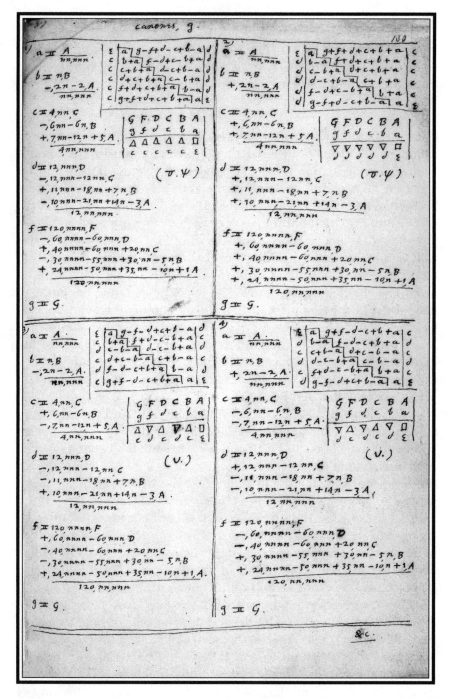

Plate 8: *Magisteria magna*: This is a sample page on Harriot's forward difference method of interpolation. Plate 5, showing his algebraic symbolism for binomial coefficients, is also from his *Magisteria magna*.

Harriot's, although it seems to have been another case of independent rediscovery. Had Newton had access to Harriot's superb little tract, though, he might have saved himself considerable effort.[3]

HARRIOT HAD TO PUT ASIDE his work before completing and publishing it, however, because Ralegh's fate was about to engulf him. In a letter written on March 22, 1618, Ralegh had asked Bess to apprise "my Lord of Northumberland" of the Guianan disaster, so presumably Northumberland or Bess filled Harriot in on what had happened. The rest of the drama unfolded over the ensuing months.

Bess traveled down to meet Ralegh at Plymouth in July. He was still ill but frantically working on an account of the disaster that might yet exonerate him. In his *Apology*, he wondered why, were he now to be charged with crimes against Spain, had James encouraged his venture in the first place. Surely the gold mine he was authorized to search for was as much—or as little—a part of Spanish territory as San Thomé. He also pointed out that he had claimed Guiana for the English Crown twenty years earlier and that aside from San Thomé, there were no Spanish settlements in the country. The Spanish, moreover, had attacked first.[4]

Lady Ralegh, under no illusions that her husband's words would fall on sympathetic ears, helped arrange for the couple to flee to Huguenot protection in France. Before leaving for Guiana, Ralegh had, in fact, almost succeeded in persuading the Huguenots to join the venture.[5] Now that it had all gone so disastrously wrong, Ralegh yielded to his wife's entreaties—but the escape failed. Indecision on his part meant that the first attempt was aborted; he still wanted to try to clear his name. The second attempt was foiled by yet another betrayal from someone he trusted: his plans were leaked, and just as he was being rowed away, the boat was intercepted and Ralegh was escorted back to London, and ultimately to the Tower.[6]

Bess was imprisoned, too—and she was proven right about Ralegh's *Apology*: despite its desperate self-justifications, its acuity helped to seal his fate. Against the advice of most of his commissioners, James was not about to give Ralegh an open trial. He had not forgotten how well Ralegh had acquitted himself in 1603. The *Apology* only reinforced the point, and the die was cast. The king of Spain wanted blood, reminding James that marriage negotiations for his son were at stake. James signed the death warrant even before the first formal hearing on October 22.[7]

It must be said that James had waited many weeks before deciding on this course. Ralegh's enemies who had spearheaded the 1603 charges, Robert Cecil and Henry Howard, were both dead; so was Chief Justice Popham, and in his place was a judge more sympathetic to Ralegh. Coke was still in the background, arguing for immediate execution on the grounds of his previous conviction; nevertheless, he and the other commissioners had urged for more transparent justice this time—including letting Ralegh confront witnesses! (Coke would have a strong influence on the development of common law. Several years earlier, he had fallen foul of James because he felt that even the king should be subject to such law.) But when it seemed that Ralegh had been negotiating secretly with the French for asylum, James believed he had little choice if he wanted to keep peace with Spain. This was despite the fact that Francis Bacon, one of James's six chosen privy councilors who were to examine Ralegh in lieu of a proper trial, reported that "in what concerns the French, Sir Walter Ralegh was rather passive than active."[8]

And so, at his final hearing on October 28, 1618, Ralegh did not stand a chance, despite his eloquent defense and his courageous composure—and despite Chief Justice Edward Montagu's compassionate summing up: "I know you have been valiant and wise, and I doubt not but that you retain both these virtues...Your faith has heretofore been questioned, but I am resolved you are a good Christian, for your book [*History of the World*], which is an admirable work, does testify as much."[9] His execution was scheduled for the following morning. Ralegh was granted his last wish, to die publicly, in the light of day, rather than in some dark corner of the Tower. He wanted the chance to speak from the scaffold in the hope of justifying himself to his compatriots.

Ralegh proceeded to his execution with such dignified courage that some argued that he would have the final victory over his sovereign. If his conduct during the trial of 1603 had turned public opinion in his favor, his forty-five-minute speech on the scaffold turned him into something of a legend. Harriot was there, taking notes to remember his old friend and patron's final words, and others, too, helped immortalize the last speech of this "last of the Elizabethans."[10] What resonated among his listeners wasn't so much what he said but the general recognition that he had been wrongfully imprisoned for all those years in the Tower and now was to die without a retrial of the 1603 accusations, for recent actions that few believed were his fault. Later generations of scholars have sought to unravel the complexities of Ralegh's character—the contrary mix of humanity, ambition, self-pity, brilliance, political naivety, piracy, literary flair, deception, candor, pride, humility. But to most of his contemporaries, he was now a hero. Even the executioner begged his forgiveness, so it was said. When the moment came, Ralegh "kneeled down upon the executioner's block...and groveling along on his arms and hands [tried] to reach his neck to the block," as an eyewitness put it, while the executioner was "busied a ripping [Ralegh's] shirt and waistcoat with a knife, that he might more conveniently bring the axe to his neck."[11]

Harriot left no such record of his friend's final moments. This was not the way to remember him. Bess wrote, "God hold me in my wits."[12]

LADY RALEGH HAD LITTLE TIME for mourning. She now had to contend with lawsuits from investors in Guiana and to fight for compensation for herself and her thirteen-year-old son, Carew. The government had impounded her husband's purpose-built ship, which was all that remained of the Raleghs' investment in Guiana. She also had her husband's reputation to restore, and she would soon become renowned for regaling listeners with tales of Ralegh's days as favorite of that most extraordinary sovereign Elizabeth. Most likely it was Bess, too, who circulated copies of his letters explaining his actions in Guiana, and even the poignant letter he had written to her back in 1603 on the night before he expected to be executed. It showed the tender, vulnerable side of Ralegh that few people had seen. Meanwhile she continued to fight for her rights. In the end, she won not only financial compensation for her own investment in the Guiana venture but also the right for her son to be "restored in blood." This was what she had fought hardest for, because it meant that Carew was freed of the stain of being a traitor's son and thus able to claim compensation for what had been taken from his father. Parliament approved the decision in 1628, ten years after Ralegh's execution.[13]

As for Harriot, in that awful autumn of 1618 he was unable to work. He did not even use his telescope—except to observe a comet in November.[14] It is clear that what had happened to Ralegh was devastating to him, though to measure his grief all we have is the gaps in the dates of his manuscripts.

He and Galileo were not the only ones to be diverted from scientific research by the religious and political dramas of their era, though. In 1620, Kepler would take over the legal defense in his mother's trial for witchcraft in southern Germany. The trial had already dragged on for years, and it would drag on for yet another harrowing year, during which time seventy-something Katharina Kepler was chained to the floor of her cell. Kepler somehow managed to bring reason to the defense, and finally she was freed.[15]

In 1619, apparently at Northumberland's request, Harriot began to prepare a new treatise. It was about his work on the nature of motion—not falling motion but motion in response to the impact of a collision, such as between two billiard balls. He must have been in considerable distress over his tumor at this time, because this is the work, addressed to Northumberland, in which Harriot said he would have included "first principles" if he were in better health.

The way colliding bodies rebound is so fundamental in the modern science of motion that Newton, who had not mentioned Kepler when he derived his law of gravity for elliptical planetary motion, had nevertheless acknowledged the work of his predecessors on the theory of material collisions. He named his immediate forerunners Christopher Wren, Wallis, and Huygens; he did not know that Harriot had already developed the theory to a remarkable degree. (Nor did he know about Harriot's work when he acknowledged Galileo's "times squared" rule for local falling motion, and his proof of the parabolic path of projectiles, in the same scholium in which he named the pioneers of collisions.)[16] Harriot himself realized the significance of his work on this topic, which in fact treats more dynamically complex collisions than Wren, Wallis, and Huygens half a century later.[17] He prefaced his tract by saying with confidence that "these results show by their own force the whole knowledge concerning the [behavior of rebounding] bodies."[18] Therefore, he concluded, his results were authoritative; like his work on triangular numbers, these results, too, were "magisterial," and they surely "must rank among our foremost guides to the inner chambers or mysteries of natural philosophy."[19]

Harriot was nearly correct in his claim that he had developed the "whole" knowledge of the topic. As a modern scholar put it, while Harriot had successfully described what happens when two balls of equal size collide from any direction, he had made a slight error when examining the more complex case of collisions between balls of unequal size, an error that "easily might have been corrected by a competent editor."[20] Nonetheless, Harriot appears to have been the first to carry out a mathematical study of mechanical collisions at all. It's possible that his and Northumberland's interest in this topic first arose in 1613, when the earl's workmen had constructed a paved bowling alley in the grounds of the Tower. Harriot's analysis was entirely mathematical, so perhaps it was informed by experimental evidence gathered when playing bowls and billiards with his patron.[21]

It is also likely that Harriot was drawn to study the "reflection" of "round bodies," as he described his rebounding balls, because of his atomistic theory of the reflection and refraction of light. He had studied earlier attempts at geometrically analyzing the law of light reflection, such as those by al-Haytham and Kepler, but mechanical collisions required a considerably more sophisticated analysis, and Harriot was justly proud of his tract. It is one of the most polished of his works, being a summary of earlier work on the subject that has not survived.[22] It anticipates all three of Newton's laws of motion and implicitly foreshadows conservation of momentum.[23] This is not to say that Harriot thought in exactly the same way about motion as we do, post-Newton or even post-Descartes. (In the 1630s, Descartes would take perhaps the earliest steps toward explicitly generalizing the laws of impact to universal laws of motion and toward articulating the conservation of what we now call momentum.[24]) Although Harriot successfully and deliberately developed a new, non-Aristotelian analysis of mechanical impacts, he and his contemporaries had inherited an Aristotelian mental framework. Their work—with its trial and error and intuitive, often flawed assumptions, as well as its "modern" mix of experiment and mathematics—highlights how difficult it is to find even the simplest laws of nature without a road map, a body of proven techniques, concepts, and definitions. Harriot's ability to express and analyze physical problems mathematically was a crucial step in this process, and in his work on collisions he takes both Aristotle and "recent philosophers" to task for presenting pronouncements without showing any working. As Newton would make clear, it is only when a quantitative analysis is provided that theories can be tested to a specific accuracy.

Harriot's paper on collisions is yet another indication of what he might have been capable of accomplishing had he enjoyed better health and more tranquility. Years earlier, he had made it clear to Kepler that he did intend to publish when he felt better, and when prevailing attitudes stopped being "stuck in the mud" so that he could undertake in peace the arduous process of writing up his results. Now, in his late fifties, he had begun this process, beginning with his papers on collisions and interpolation via triangular and binomial numbers. Galileo would only manage to achieve the necessary solitude to write up some of his best work when he was in his seventies, after he had been sentenced to house arrest. It is tempting to wonder what Harriot might have achieved had he had the same longevity and the same impetuses to publication.

Indeed, if Harriot had managed to find the time to bring together what remains scattered among his manuscripts, and explained more fully his conclusions and assumptions, he might well have had a similar impact on the development of science as Galileo.[25] He, too, critiqued the prevailing Aristotelian physics, but as indicated earlier he did this through the persuasive power of individual results rather than by attempting a reformation of physics. These results included his law of refraction and his theory of the rainbow, which overturned Aristotle's conception not only of refraction and the rainbow but also of color in general; his infinite sums of infinitesimal quantities, which showed that the human mind could, after all, calculate the length of a curve such as the loxodrome; his astronomical discoveries and calculations; and, of course, his experimental and mathematical analyses of falling motion, projectiles, and collisions, which, if published, would have helped pioneer the fundamental laws of motion and gravity.[26] Moreover, many of his mathematical methods and discoveries were in advance of his time and should have had wider influence.

But it was not to be. The cancer had continued to grow, disfiguring his face and invading his bones. His new specialist offered purges, cauterization, and surgery, but little hope. Harriot's time had run out.

CHAPTER 23

All Things Must Pass

ARRIOT'S LONDON WAS A CITY of bells, chiming away the time every hour, caroling joyously on festive or holy occasions, or tolling somberly when someone was mortally ill. When the local church bell tolled, people would drop to their knees and pray for the soul of their dying fellow parishioner. In 1624, Lady Ralegh's friend and brother-in-law John Donne would immortalize this custom when he wrote, "Perchance he for whom this bell tolls may be so ill, as that he knows not it tolls for him... [Yet] no man is an island, entire of itself...: any man's death diminishes me, because I am involved in mankind, and therefore never send to know for whom the bell tolls; it tolls for thee."[1]

At the end of June 1621, Harriot was staying with an old friend from Roanoke days, mercer Thomas Buckner of Threadneedle Street. The local church stood in the same street, and surely its bells would soon toll for Harriot as his life ebbed away, prematurely and painfully. He had suffered with dignity and courage during the cancer's inexorable progression. Five years earlier, he had written to his doctor, Mayerne, saying that he hoped for a recovery, but that he would fight for life boldly while acknowledging that everything happens "in its own time, according to God's providence." He had concluded with a Latin conjugation added to the famous religious dictum *Sic transit Gloria mundi*: "So passes away the Glory of the world, all things shall pass away, we shall pass, you will pass, they will pass away."[2] Now, in the summer of 1621, it was time for the first person singular: "I am passing away."

He had not yet made a will, and he struggled to set down his last words.[3] He was, he said, "troubled in my body with infirmities, but of perfect mind and memory Laude and praise be given to Almighty God for the same." And so he proceeded to wind up his mortal affairs. He had only two living relatives—a nephew and a cousin—and to each of them he bequeathed £50 (about $14,000 today[4]). His laboratory assistant, Christopher Tooke, was to receive £100, while his house servants, John and Joan, were to be given £5 in addition to their wages, and Joan's assistant, Jane, was to be given two years' additional wages (her yearly wage being £1). He even remembered two of his former servants, to whom he gave £5 and £2, respectively.

Most of his friends were wealthy enough not to need monetary bequests, but he gave £15 to Mrs. Buckner for the costs of tending him as he lay dying. He suggested that his executors—including Buckner—should take something for themselves after his debts were cleared and his bequests were paid, including £20 to three charitable institutions for the poor. If there was anything left after that, he wanted it to go to the new Bodleian Library at Oxford University.

The money was to come partly from the sale of his "books and other goods." He must have had an impressive collection of books—he owed his bookseller £40 just for his most recent purchases—and he asked that his friend Robert Hues oversee the pricing of them. And with that "perfect state of mind and memory" he went on to list his other debts and to ask that they be duly discharged.

He did not forget his friends, of course—not even Ralegh, although he'd been dead for three years. As his former accountant, Harriot was able to do his old friend one last favor by asking his executors to burn all Ralegh's accounts—"for all of which I have discharges or aquittances lying in some boxes or other." These included "a canvas bag" of accounts relating to Ireland, of which Harriot added cryptically, "the persons whom they concern are dead many years since." It is a comment that hints, perhaps, at some secret from Ralegh's days as a colonizer in Ireland.

As for his closest living friends, they were to receive personal gifts, and the measure of this man of science and simple tastes is that these personal treasures were entirely scientific. To his friend and generous patron Northumberland, he bequeathed "one wooden box full or near full of drawn Maps standing now at the Northeast window of that room which is called the parlor at my house in Syon." Additional gifts, to Northumberland and other friends, included telescopes and other scientific equipment.

Most precious of all, though, were Harriot's "Mathematical papers," which ultimately were to be stored in "a convenient trunk with a lock and key" and placed in Northumberland's library. But first, Harriot asked that his old friend Torporley sift through these papers, sorting out the "Chief of them from my waste papers, to the end that after he understands them, he may make use in penning such doctrine [found therein] for public uses as it shall be thought convenient by my executors and himself." If Torporley needed any help in understanding the symbolic notation or other details, then he was to ask Warner or Hues for assistance.

Three pain-wracked days later, the present tense irrevocably gave way to the past and Harriot passed away. Buckner arranged for his friend to be buried in the church in Threadneedle Street.

Two weeks later, Northumberland was released after nearly sixteen years in the Tower. Apparently, one of the first things he did in his new freedom was to sponsor a memorial plaque on Harriot's grave, inscribed with a tribute to his friend and protégé:[5]

> Stay traveler, lightly tread;
> Near this spot lies all that was mortal
> Of that most celebrated man
> THOMAS HARRIOT.
> He was that most learned Harriot
> Of Syon on the River Thames;
> By birth and education
> An Oxonian
> Who cultivated all the sciences
> And excelled in all—
> In Mathematics, Natural Philosophy, Theology.[6]
> A most studious searcher after truth,
> A most devout worshipper of the Triune God,
> At the age of sixty or thereabouts,
> He bade farewell to mortality, not to life,
> The Year of our Lord 1621, July 2.

Resurrecting Harriot

I N 1666, THE GREAT FIRE destroyed Harriot's plaque on Threadneedle Street. Although the church that had housed it was rebuilt after the fire, it was demolished in 1782 to make way for the expansion of the neighboring Bank of England, whose headquarters are still located there. It seemed that fortune was doing its best to erase any signs of Harriot's existence.

The public memories and memorials that did linger were often tainted by his supposed impiety. John Aubrey's late seventeenth-century collection *Brief Lives* perpetuates this in a trivializing anecdote about Harriot:

> The bishop of Sarum (Seth Ward) told me that one Mr. Haggar (a countryman of his), a gentleman and good mathematician, was well acquainted with Mr. Thomas Harriot, and was wont to say that he did not like (or valued not) the old story of the creation of the world. He could not believe the old position; he would say *ex nihilo nihil fit*. But, said Mr. Haggar, a *nihilum* killed him at last: for in the top of his nose came a little red speck (exceeding small), which grew bigger and bigger and at last killed him.

Aubrey adds that Harriot was a Deist and taught this "doctrine" to Ralegh, Northumberland, and others, so that "the divines of those times looked on his manner of death as a judgment upon him for valuing the Scripture at nothing."

In fairness to Aubrey, he does mention a handful of Harriot's achievements: his reputed prediction of the appearance of seven comets, his "description of Virginia," and "an alphabet that he had contrived for the American language, like Devils." He also quotes a poem that shows the high status accorded Harriot by his more open-minded contemporaries. It was written in December 1618 by the Anglican bishop and poet Richard Corbet, in a letter to Thomas Aylesbury, who had been a student of Harriot's in the last years of his life and whom Harriot had named an executor of his will.[1] (In fact, Harriot's will highlights just how sociable he was: Aylesbury and Buckner appear there, along with other little-known friends.) The lines from Corbet to Aylesbury that Aubrey quotes conclude:

> Thine own rich studies, and deep Harriot's mine,
> In which there is no dross, but all refine.

Unfortunately, his friends failed to produce a suitable exemplar of Harriot's "refined deep mine" of learning from his "Mathematical papers." Not that they didn't try. Nathaniel Torporley may have disapproved of Harriot's atomistic *ex nihilo nihil fit*, and he may have thought that as a "mortal man" Harriot "erred signally," but he immensely admired his mathematical work and enthusiastically set about planning an ambitious, five-part series to propagate it.[2] The gargantuan nature of the task of sorting through Harriot's papers is reflected in the list that Aylesbury dutifully compiled. It begins with sixteen bundles of papers on "analytics" (algebra), three bundles on centers of gravity, a "great bundle" on Jupiter's moons and sunspots, and yet another great bundle on observations of the moon—and this was just the beginning. There were dozens more topics on the list.[3] Walter Warner was involved, too, but it was a lengthy process. Four years later, in 1625, Henry Briggs told Kepler that a posthumous publication of Harriot's work was expected "any day."[4]

Somewhere along the line, Torporley gave up—or was pushed aside, since Warner apparently took over in order to speed things up.[5] Still the work dragged on: the *Artis analyticae praxis* was not published until 1631, ten years after Harriot's death. Its introduction proclaimed that it represented the first-ever fully symbolic algebra, but overall it offered a flawed version of Harriot's work; Warner was not as capable a mathematician as Torporley, who was furious when he saw it. He set about writing a scathing review, *Analytical Corrector to the Posthumous "Analytical Art" by Thomas Harriot*, but he died shortly after venting his spleen, and his *Corrector* remained unpublished.[6] In any case, in just six years' time, Descartes's *La géometrie* would eclipse the *Praxis*.

Warner and Aylesbury intended to publish more of Harriot's work, with financial support from Northumberland, so the papers remained with Aylesbury. Through him and Warner, younger mathematicians such as John Pell and John Wallis were introduced to some of Harriot's manuscripts, and Warner had also told Pell that he had learned the law of refraction from Harriot years before Willebrord Snell's discovery.[7] In 1662, under the auspices of the newly formed Royal Society, Pell and his colleagues initiated a search for Harriot's papers among Aylesbury's effects (he had died several years earlier). They found nothing, and by 1669 everyone assumed that aside from the *Praxis*, Harriot's scientific work was irrevocably lost.

And so it remained for over a hundred years, until a young Austro-Hungarian astronomer named Franz Xaver Zach made an exciting discovery. Zach was a tutor in the household of Count von Brühl, Saxon ambassador to England. Von Brühl had recently married the dowager Lady Egremont, and in 1784 he and Zach visited the country mansion of his stepson, the 3rd Earl of Egremont. As luck would have it, the earl was a descendant of Northumberland, and the mansion was Petworth House, in Sussex. On his release from the Tower in 1621, Northumberland had been banished from London and instructed to stay within 30 miles of Petworth, his country estate and ancestral home. He died there in November 1632. Nearly twenty-five years later, Aylesbury had honored Harriot's wishes and returned his papers to his mentor's estate.[8] And there they had lain hidden for more than a century. The family knew about Harriot's connection with their famous forebear, and it seems they even knew that his manuscripts were somewhere at Petworth—but no one had found them until the summer of 1784, when Zach and von Brühl began rummaging through the library. Under a pile of old stable accounts, they discovered thousands of disordered but remarkably preserved sheets of diagrams, intricate mathematical calculations, and detailed experimental observations.[9]

Zach immediately realized the scientific value of the work. Like Wallis before him, he thought that the novel algebraic symbolism and methodology had surely anticipated the work of Descartes. He was also amazed by the detailed optical and gravitational measurements, not to mention the observations of sunspots and other astronomical phenomena, which suggested to Zach that Harriot had anticipated even Galileo. The young visitor thought how glorious it would be for England to discover its forgotten genius.

Of course, the rediscovery of Harriot would bring Zach fame, too, and he eagerly approached Oxford University, Harriot's alma

mater, believing the dons would jump at the chance to publish
Harriot's papers. And so they did: in the first flush of excitement
over the Petworth discovery, Zach's efforts earned him an honorary
degree from Oxford and the title of Baron von Zach. Unfortunately,
the man charged with making a decision on the matter of publication—Abraham Robertson, Oxford's Savilian Professor of geometry
and, later, astronomy—was not interested in history. He reasoned
that England had produced Newton, the greatest scientist of them
all, and that his and others' work meant that Harriot's discoveries
were now outdated. In other words, the publication of Harriot's work
"would not contribute to the advancement of science," as Robertson
put it in his 1798 report to the university.[10]

In 1810, the Earl of Egremont gave the majority of the manuscripts found at Petworth to the British Museum. (These are the ones
I studied in the British Library, where they are now housed. Most of
the astronomical and navigational papers, and some of the algebraic
and gravitational researches, remain in the Petworth House Archives.)
Zach continued to agitate for the public resurrection of England's forgotten scientific hero, writing letters and outlining his understanding
of the importance of Harriot's contributions, but he never actually got
around to editing Harriot's manuscripts himself, or to completing his
promised *Life of Harriot*. This was partly because he had received a post
as astronomer back in Saxony and was preoccupied. Still, his lack of
effort with the manuscripts and his cocksure pronouncements about
Harriot anticipating Descartes and Galileo antagonized scholars like
Stephen Rigaud (Robertson's successor in the Savilian chairs).

In 1831—nearly fifty years after Zach's discovery of Harriot's
manuscripts—Oxford University Press appointed Rigaud to make a
further assessment. (In the wake of Zach's dedicated proselytizing,
there had been public criticism of Oxford's failure to honor Harriot's
legacy.) Whatever his feelings about Harriot's stature, Rigaud was a
meticulous scholar, and he set about trying to sort and edit Harriot's
manuscripts. Their disarray was partly due to the fact that Harriot's
trustees had picked through some of his mathematical papers; Zach
had made things worse by sending batches to Oxford without making
a note of their place and context. All this well-meaning carving up of
Harriot's body of work made it impossible for Rigaud to judge
Harriot's contributions to science. Nevertheless, he made a detailed
study of Harriot's observations of sunspots and Jupiter's moons and
acknowledged that Harriot had made most of these discoveries independently of and contemporaneously with Galileo. But Rigaud wrongly
conceded technical sophistication to Galileo.[11]

As for the folios on optics, gravity, and mathematics, Rigaud made little attempt to study them; he was not a young man and lacked the time and energy for further study. He sincerely hoped that later scholars would vindicate Harriot's priority in mathematics, and accepted William Lower's claims for Harriot's originality—his report contains full transcripts of several of Lower's letters to Harriot. Rigaud also recognized that Harriot had completely solved some biquadratic equations (although he failed to acknowledge the generality of Harriot's method), and he considered that what he had seen of his mathematical works was of "high character." But he concluded that Zach had made excessive claims for Harriot's work on astronomy and that Harriot's manuscripts were not worth publishing.[12] When Zach died in 1832, the greatest British mathematical scientist before Newton seemed doomed to remain in obscurity.[13]

WHILE HARRIOT'S CONTRIBUTIONS TO MATHEMATICS, physics, and astronomy were soon all but forgotten, memory of his role as a pioneering ethnologist and explorer in Ralegh's Roanoke colony had never quite faded, especially in America. The nineteenth-century American book lover and collector Henry Stevens, who had read Harriot's *Brief and True Report* on Virginia, traveled to England to track down more of Harriot's papers. One of his most significant finds was Harriot's will, excerpts of which were published for the first time in Stevens's book, which was posthumously and privately printed in 1900.[14] Half a century later, the American scholar John Shirley also set about searching for more information about Harriot. Beginning in 1949 he published a number of scholarly works related to Harriot's life and work, which ultimately helped initiate a new wave of Harriot scholarship.

There must have been something in the air, for just as Shirley was pursuing his investigations, so were a number of others. In the 1950s, such research included the French historian Jean Jacquot's investigation into Harriot's supposed impiety; the Irish historian David Quinn's exhaustive study of the Roanoke ventures; the English literary scholar Ethel Seaton's attempt to decipher Harriot's Algonquian-inspired alphabet; the English geographer and historian Eva Taylor's early research on Harriot's navigational discoveries; the work of Donald Sadler, superintendent of the British Nautical Almanac Office, on Harriot's analysis of the loxodrome (expanded upon in the 1960s by Jon Pepper); and the Norwegian mathematics teacher and historian of science Johannes Lohne's analysis of Harriot's optics.[15]

Lohne had chanced on Harriot's existence while studying a copy of Friedrich Risner's 1572 edition of Witelo's mediaeval *Optics* in the library at the University of Oslo. Witelo, as we've seen, had tabulated angles of refraction for various substances, though it was clear to a modern reader that his tables were inaccurate. To Lohne's amazement, however, he discovered that an earlier scholar had noticed this, too. What is more, in 1597 this mysterious scholar had produced his own very accurate table of refractions, which he had written out in ink "on the last page but one" of the copy of Witelo's *Optics* that Lohne was now studying. The table was not only dated—there was a place name, too: Syon. Lohne had never heard of such a place. He searched through the book for other annotations and at first found that "although notes appeared in several places, the annotator had left his name nowhere in the book." Then he noticed a brief note in Latin, headed by the initials T. H. Who could this remarkable T. H. be, he wondered? From Kepler's correspondence, Lohne knew that the German scientist had had a correspondence with an Englishman named Thomas Harriot, whom he looked up in the *Dictionary of National Biography,* which informed him that the Earl of Northumberland had given Harriot a home and a laboratory at Syon. Lohne realized that he had been reading Harriot's very own copy of Risner.[16] Three and a half centuries earlier, Harriot could have had no inkling that this book would provide such a crucial key to the resurrection of his fame, but it inspired Lohne to travel to London to study Harriot's manuscripts for himself. (The *Dictionary of National Biography* had mentioned Zach's discovery of the Petworth papers.) There he found Harriot's work on refraction, color, and the rainbow—all the discoveries Harriot had told Kepler about in 1606. Lohne published his analysis of Harriot's work in 1959, the eve of the four-hundredth anniversary of Harriot's birth.

Regarding the Oxford saga, Lohne concluded that although Zach may have been too eager to praise Harriot's achievements at the expense of those of Descartes and Galileo, the assessments of Rigaud and Robertson were seriously flawed. In 1963, Lohne published an analysis of their reports in which he suggested that Robertson's lack of historical context had clouded his judgment, while Rigaud had been more concerned with nitpicking Zach's claims about priority than with properly assessing Harriot's achievements. On the basis of his own analysis of the manuscripts, Lohne provided detailed refutations of their conclusions. A decade later, John North gave a similar critique of Rigaud's pronouncements on Harriot's astronomy.[17]

Nonetheless, Rigaud and Robertson were correct in one thing: most of Harriot's manuscripts were in no state to be published without a great deal more scholarly annotation and editing. Over the past fifty years, thanks to the work of Shirley, Lohne, and many others, a growing number of Harriot seminars, symposiums, and lectures, on both sides of the Atlantic, have inspired new generations of researchers.[18] As more details of his remarkable life and work have been uncovered, Harriot's story has fired the imaginations of scholars who have sifted through his scattered papers, reconstructing his treatises, following his interrupted trains of thought, and polishing those long-hidden gems of early modern science. Their efforts show that Rigaud was also right that the task of editing Harriot's manuscripts for publication was too much for one person. Instead, the manuscripts have been described, excerpted, and analyzed in numerous scholarly articles, while his *Doctrine of Triangular Numbers* and his *Treatise on Equations* were edited, annotated, and published in the opening decade of this century. All these scholars and their works are listed in my bibliography and gratefully acknowledged in my endnotes.

Zach may have been eager to see Harriot's manuscripts published just as they were, but he could not have foreseen the possibility of digitizing Harriot's entire collection. Conceived and edited by Jacqueline Stedall, Matthias Schemmel, and Robert Goulding, the digital files have been made available through European Cultural Heritage Online (ECHO), in a collaborative project of the British Library, the University of Oxford, the Max Planck Institute for the History of Science (Berlin), the University of Notre Dame (US), and the Petworth House Archives.[19] The rehabilitation of Harriot has been truly an international effort.

His hometown has not forgotten him, either, and an annual series of Harriot Lectures takes place at Oriel College, with support from the current Lord Egremont of Petworth House.[20] Oxford University Press has also played its part, publishing in 1974 a collection of papers presented at a Harriot symposium at the University of Delaware and edited by Shirley, followed by Shirley's scholarly biography of Harriot in 1983 and Harriot's *Treatise on Equations,* edited by Jacqueline Stedall, in 2003. Now OUP is publishing this book, which aims to bring Harriot to a wider audience. In 1798, Robertson may well have sincerely believed that the publication of Harriot's work "would not contribute to the advancement of science." Today, it is certainly contributing to the advancement of the history of science, as the Harriot scholars have shown. In synthesizing their research as

ACKNOWLEDGMENTS

My journey with Harriot has been a long one, and I'm immensely grateful to Tim Bent, at Oxford University Press in New York, for believing in this project from its first fledgling conception. He has also worked creatively and tirelessly during the editing process, offering numerous invaluable suggestions, and it has been a great pleasure working with him.

It's also been a pleasure working with India Cooper, to whom I'm indebted for her meticulous copyediting and polishing of the manuscript. Her care and clarity, and her constructive suggestions, are very much appreciated.

A big thank-you for their help and approachability is also due to assistant editor Mariah White and senior production editor Amy Whitmer—and thank you, too, to all the team at OUP New York. In addition, I want to acknowledge that a work such as this cannot be written without access to good libraries—in this case the marvelous British and Monash University Libraries, and their helpful librarians. I also want to thank my agents Jenny Darling and Brandi Bowles.

I'm extremely grateful to Lord Egremont of Petworth House, for his generous permission to publish Harriot's moon map, sunspot diagram, and calculations on spherical triangles.

Finally, very special thanks are owed to Morgan Blackthorne, who not only offered helpful suggestions on an early draft, along with much inspiration and support, but also turned my sketches into the diagrams printed in the appendix. And I want to thank all the friends, family, colleagues, and readers who have encouraged my writing over the years. I won't hazard a list here, but please accept my gratitude. You have all been very important to me.

APPENDIX

Figure 1: Observing the zodiac

Intuitive earth-centered view:

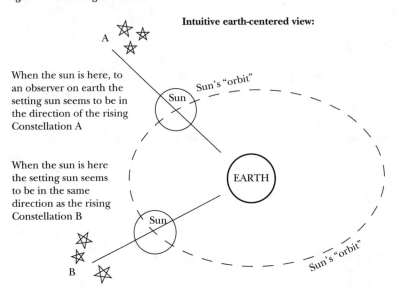

A

When the sun is here, to
an observer on earth the
setting sun seems to be in
the direction of the rising
Constellation A

Sun's "orbit"

Sun

EARTH

When the sun is here
the setting sun seems
to be in the same
direction as the rising
Constellation B

Sun

B

Sun's "orbit"

Modern sun-centered view:

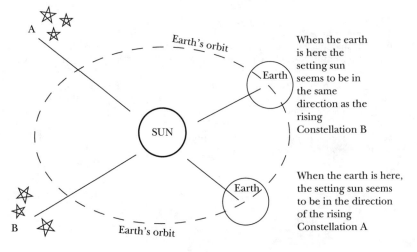

A

Earth's orbit

Earth

When the earth
is here the
setting sun
seems to be in
the same
direction as the
rising
Constellation B

SUN

Earth

When the earth is here,
the setting sun seems
to be in the direction
of the rising
Constellation A

B

Earth's orbit

Figure 2: Terrestrial latitude and longitude

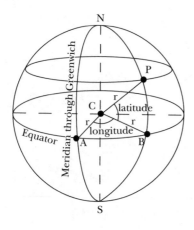

C is the center of the earth.
P's latitude is north of the equator and
its longitude is east of Greenwich.

The latitude of P can also be decribed
as the (north-south) distance PB from
the equator, and its longitude is the
(east-west) distance AB.

PB and AB are the arcs of circles with
radius r = radius of the earth (that is,
arcs of "great circles").
The ancients knew the circumference
of a circle is (in effect) $2\pi r$
and so the arc length PB is a fraction
of the circumference; this fraction is
$$\frac{\text{Latitude in degrees}}{360°}$$

or $\frac{\text{Latitude in radians}}{2\pi}$ (2π radians = 360°)

So PB = $\frac{\text{Latitude angle}}{\text{whole circle angle}}$ times the whole circumference

$\qquad = \frac{\text{Latitude in radians}}{2\pi} \times 2\pi r =$ Latitude in radians x r

If the earth's radius is known (and in 230 BCE Eratosthenes had made a good
estimate of it) then P's latitude can be measured by measuring the actual distance
PB, and the latitude angle is then PB/r. Similary for AB.

The equator or line of zero latitude is midway between the North and South Pole.
The angle from the equator to the North Pole is 90 degrees (because the line through
the Poles and the center of the earth is perpendicular to the plane of the equator). So
the latitude of the North Pole is 90 degrees N (and the South Pole is 90 degrees S).

The distance from the equator to the North Pole (or South Pole) is about 6,222
miles, so for every degree of latitude north (or south) of the equator, the distance is
$\frac{6,222}{90}$ or about 69 miles.

So if PB is known in miles, its latitude is approximately $\frac{\text{PB}}{69}$ degrees. (69 miles is
about 111 km.)

Figure 3: Celestial sphere centered on the observer

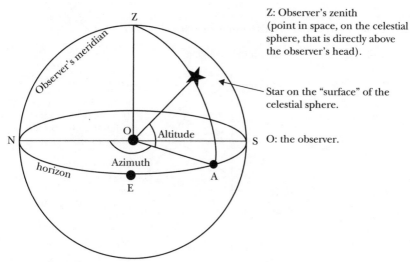

Z: Observer's zenith
(point in space, on the celestial
sphere, that is directly above
the observer's head).

Star on the "surface" of the
celestial sphere.

O: the observer.

For technical details see any textbook on
spherical astromony (eg Smart p25–34).

Figure 4: Celestial sphere centered on the center of the earth

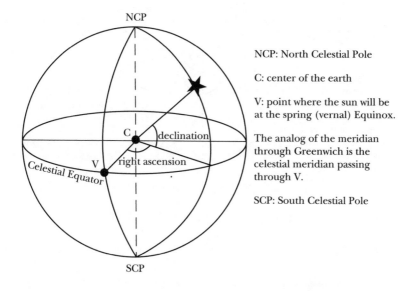

NCP: North Celestial Pole

C: center of the earth

V: point where the sun will be
at the spring (vernal) Equinox.

The analog of the meridian
through Greenwich is the
celestial meridian passing
through V.

SCP: South Celestial Pole

Figure 5: Angle between observer's horizon and celestial equator

Figure 5a: North Celestial Pole

O: Observer on the
surface of the earth
with latitude $\varphi°N$

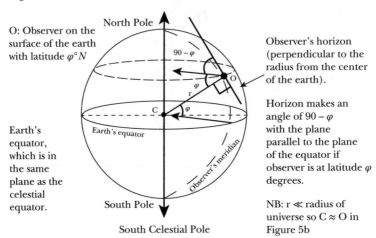

Observer's horizon
(perpendicular to the
radius from the center
of the earth).

Horizon makes an
angle of $90 - \varphi$
with the plane
parallel to the plane
of the equator if
observer is at latitude φ
degrees.

Earth's
equator,
which is in
the same
plane as the
celestial
equator.

NB: $r \ll$ radius of
universe so $C \approx O$ in
Figure 5b

Figure 5b:

Earth (small circle) is
at the center of the
celestial sphere;
observer O is at
latitudeφ; $O \approx C$

Angle DCA = BCE
= $90 - \varphi$ degrees
= angle between the
planes of the
celestial equator and
the celestial horizon
of an observer at
latitude φ degrees

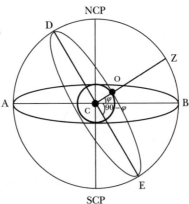

Z Observer's zenith

AB: Celestial
equator diameter

DE: Celestial
horizon diameter,
parallel to the
plane of the
observer's horizon
and therefore
perpendicular to
CZ

Figure 6: In this model, the sun always sets due west

This is a bird's-eye view, looking down from above the table, at the bowling ball sun and the golf ball earth.

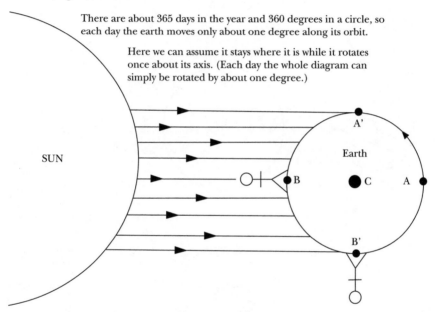

There are about 365 days in the year and 360 degrees in a circle, so each day the earth moves only about one degree along its orbit.

Here we can assume it stays where it is while it rotates once about its axis. (Each day the whole diagram can simply be rotated by about one degree.)

When the observer is at position B, the sun is diretly overhead and it is noon. When the earth has rotated a quarter of the way in its daily 24-hour rotation, the observer is at B', and it is 6pm. The sun's rays are now parallel to the observer's horizon, so it is sunset. Twelve hours later, at A', it will be sunrise, at 6am.

Because the plane of the earth's daily rotation is that of the equator, in this model all observers, no matter their latitude, see the sun rise and set at 6am and 6pm. In other words because the earth's N-S axis is vertical, the same argument applies no matter where the earth is in its yearly orbit.

Figure 7: The earth's tilted axis of rotation produces seasons (and changing sunset/sunrise directions)

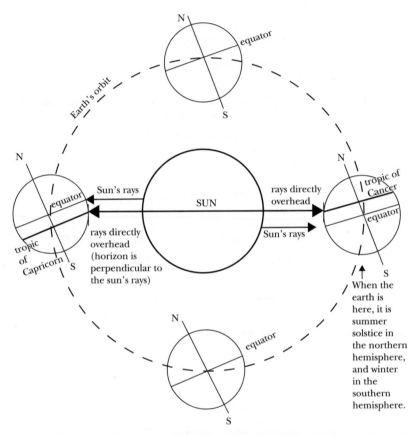

When the earth is in the position shown on the right, the northern hemisphere is tilted towards the sun. This means that the sun's light and warmth are more concentrated in the northern hemisphere than in the south, and it is summer in the north and winter in the south. The situation is reversed at the left-hand position. The other two positions shown are at equinox.

As the earth rotates each day, the sun appears to travel through the sky in the same plane as the earth's rotating equator—that is, at an angle to the table top in my analogy. This is also why the directions and times of sunset and sunrise change throughout the year: the earth's daily rotation is no longer in the same plane as its yearly orbit, as shown in Figure 10.

Figure 8a: Measuring the angle of the sun

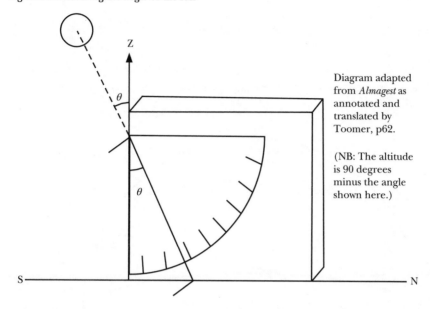

Diagram adapted from *Almagest* as annotated and translated by Toomer, p62.

(NB: The altitude is 90 degrees minus the angle shown here.)

Figure 8b: Calculating the obliquity ϵ of the ecliptic

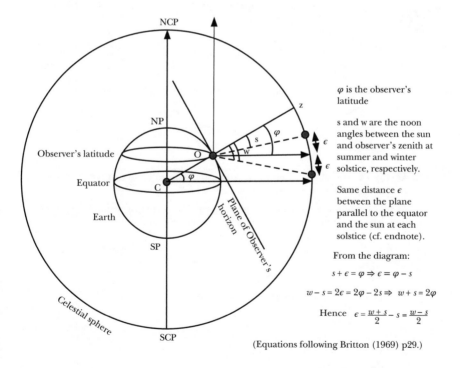

φ is the observer's latitude

s and w are the noon angles between the sun and observer's zenith at summer and winter solstice, respectively.

Same distance ϵ between the plane parallel to the equator and the sun at each solstice (cf. endnote).

From the diagram:

$$s + \epsilon = \varphi \Rightarrow \epsilon = \varphi - s$$

$$w - s = 2\epsilon = 2\varphi - 2s \Rightarrow w + s = 2\varphi$$

Hence $\epsilon = \dfrac{w+s}{2} - s = \dfrac{w-s}{2}$

(Equations following Britton (1969) p29.)

Figure 9: Celestial sphere with ecliptic and equator, showing the seasonal positions of the sun on the ecliptic

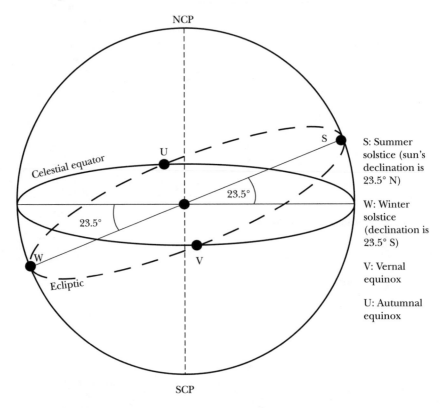

At U and V, the ecliptic meets the equator, so the sun at these points is "on" the equator. Its declination is therefore zero here.

(The seasonal positions of the solstices and equinoxes are reversed in the southern hemisphere, so that V is the autumnal equinox, and so on.)

Figure 10: Celestial sphere for an observer at latitude φ°N

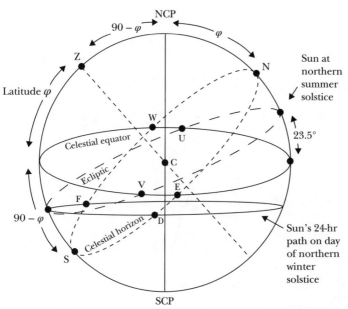

The observer's celestial horizon and the celestial equator intersect at the east (E) and west (W) points of the horizon. The observer's meridian is the circle through the zenith (Z) and the North Celestial Pole NCP; it meets the horizon at the north (N) and south (S) points on the horizon. The ecliptic and the celestial equator intersect at the equinox points (V and U).

The earth rotates about the axis through its center C and the poles, so the sun's apparent daily journey through the sky is parallel to the equator. On the equinox days, the sun appears to travel around the celestial equator itself, and it rises – that is, it crosses the horizon – due east at E, and sets due west at W. (The exact time of the equinox is not necessarily sunrise or sunset, but when the sun crosses the equator at V or W.) (Directions, solstices and equinox dates are reversed in the southern hemisphere.)

On the day of the northern winter solstice, the sun crosses the horizon at D and F, so it appears to rise to the south of due east, and set to the south of due west. The sun moves about 1 degree along the ecliptic each day, so all of its daily paths can be drawn as circles parallel to the equator; the intersections with the horizon show the positions of sunrise and sunset through the year.

Figure 11: Calibrating the cross staff

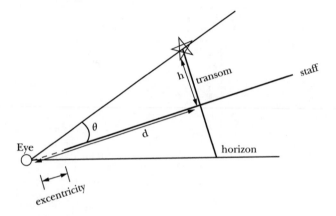

2θ is the altitude of the star

d is the distance from the eye to the transom

2h is the length of the transom

The scale is calculated from $\dfrac{h}{d} = \tan\theta$

but as Harriot pointed out, most navigators used their staffs as if d referred only to the distance to the end of the staff, on the observer's cheek.

Figure 12: Theory for Harriot's tables of solar amplitude (for determining magnetic variation)

Figure 10 gave a "3-D" view of the Celestial sphere; project this view (cf. figure 12b) onto the vertical plane through the center of the celestial sphere (leaving out the ecliptic except for the 2 solstice points, which are not drawn to scale).

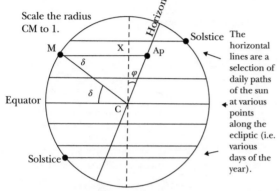

Scale the radius CM to 1.

Ap marks sunrise on a typical day. M is the position of the sun at midday on the chosen day. So δ is the declination of the sun at noon. φ is the latitude of the observer (the horizon makes an angle of $90 - \varphi$ degrees with the equator).

The horizontal lines are a selection of daily paths of the sun at various points along the ecliptic (i.e. various days of the year).

Depending on the day chosen, δ varies between 0° at the equinoxes, and ± 23.5° at the solstices. CAp represents the sine of the deviation from true east (E) of the sunrise direction on that day, or in Harriot's terms, the sine of the solar amplitude α (to see why it is the sine, see figure 12b).

Figure 12b:

Ap is the projection of A; E is due east.

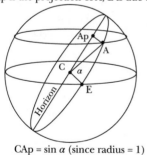

CAp = sin α (since radius = 1)

From the triangle CXA_p

$$\frac{CX}{CA_p} = \cos \varphi$$

So $CA_p \; (= \sin \alpha) = \dfrac{CX}{\cos \varphi}$

From triangle CMX, $CX = \sin \delta$

so $\sin \alpha = \sin \delta \sec \varphi$

My figure 12 is an amplification of Pepper's Figure 4.9 (Shirley (ed.) p73)

Figure 13: Deriving the stretch factor for a Mercator map

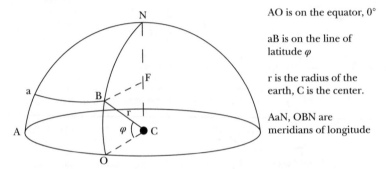

AO is on the equator, 0°

aB is on the line of latitude φ

r is the radius of the earth, C is the center.

AaN, OBN are meridians of longitude

aB is parallel to AO, BF is parallel to OC (= r), angle OCB = φ = angle CBF. By similarity AO/aB = OC/BF. But OC/BF = r/rcos φ = sec φ so AO = aB sec φ.

On a map with meridians equally spaced, aB has to be streched to a'B' = AO = aB sec φ Similarly, Aa and OB are stretched to Aa' = Aa sec φ and OB' = OB sec φ (See Fig. 14 for more explanation.)

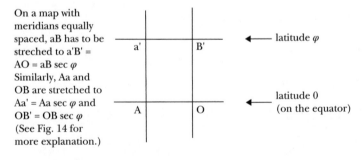

latitude φ

latitude 0 (on the equator)

Figure 14a: Rhumb line (loxodrome)

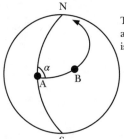

The curved path sailed from A to B and beyond spirals around the pole if α is a fixed compass bearing.

Figure 14b: Nautical triangles on the globe →plane triangles on the Mercator map

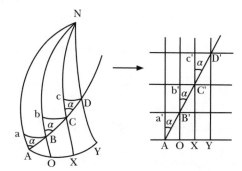

From Fig.13,
aB→a'B'
= aB sec φ.
So, Aa → Aa'
= Aa sec φ, so
that the angles
aAB and a'AB'
are the same,
and similarly
for the other
triangles.

Assume aAB, bBC,... are infinitesimal nautical triangles, so that the rhumb line ABCD..., which makes a constant angle α with each curved meridian line, is in effect broken into tiny straight line segments AB, BC, ... Only four meridians are shown here, but they are all assumed to be spaced so that AB=BC= ... (= m, say).

AOXY is on the equator.
aB, bC, cD, ... are segments of different latitude lines crossed by the rhumb line.

By the time the ship is at D, it has increased its latitude from zero at A to Aa + Bb + Cc (and so on for higher points). See Figure 15b for the final formula.

The distance AD along the rhumb is AB + BC + CD = m + m + m (and so on when more points and meridians are added).

Figure 15a: Trigonometric ratios in a right-angled triangle

In Harriot's time, secants and tangents were generally conceived as shown here:

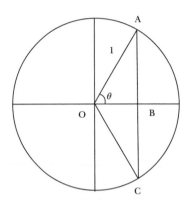

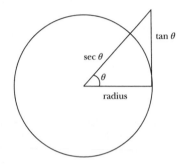

AC = chord for angle 2θ
sin θ = AB (half chord)
cos θ = OB.

Today we think of trigonometic functions and ratios rather than chords and other lines.
We also write
sec θ = 1/cos θ,
tan θ = sin θ/cos θ.

Figure 15b: Formulae for changes in longitude and latitude along a rhumb line

If, in figure 14b, A, B, C,... are so close together that m is almost zero, the "nautical triangles" ABa, BCb, ... can be treated as plane triangles. (In other words, AB, BC, ... and aB, bC, ... are treated as straight line segments.)
Then Aa = Bb = ... = m cos α, and aB = bC = ... = m sin α.

So the change in latitude from A to any point on the rhumb line is
Aa + Bb + ... = cos α times the sum of the m's, the incremental distances along the rhumb.

The change in longitude is not aB + bC + ... but AO + OX + ... where AO/aB = sec φ (from figure 13), and φ is the latitude of aB. So the change in longitude
= sin α [m sec φ + m sec 2 φ + m sec 3 φ + ... + m sec k φ + ... + m sec n φ], where n φ is the required latitude, and φ is the increment in latitude for each segment along the rhumb.

Rearranging the formula for the change in longitude gives
tan α times the sum of (each incremental change in latitude times sec k φ at each latitude k φ in the sum) = tan α times the meridional part (for latitude n φ).

Figure 16a: Reflection and refraction

i = angle of
incidence of
incoming ray
 = angle of
reflection

r = angle of
refraction

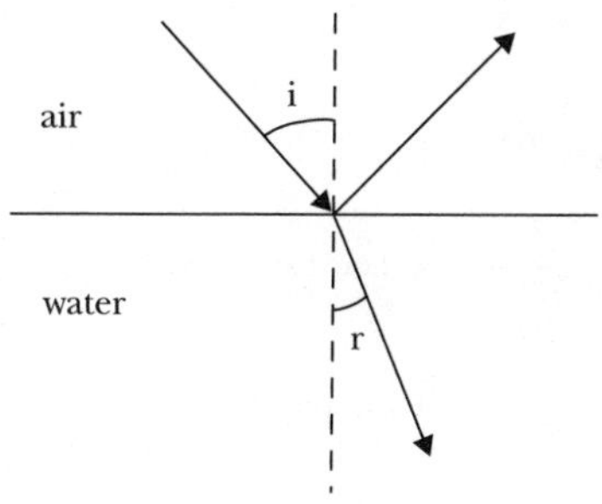

At the water surface, light is
partially reflected away, and
partially transmitted into the
water, where it is refracted.

air

water

Figure 16b: Harriot's experimental set-up

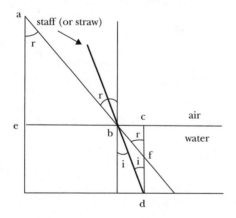

The immersed segment bf of
the staff appears bent
upwards compared with the
actual staff (bd).

The line abf is the line of
sight to the refracted image
as the ray passes from water
to air (and to the observer's
eye).

Harriot measured bc, cd, and
bf.

Note: in figure 16a, the light ray is
going from air to water (a denser
medium), and i > r.
In figure 16b, Harriot is tracing the ray
from the water to the air (from d to a)
so i < r.

sin r = bc/bf, tan i = bc/cd,
so i and r can be found from
trigonometric tables.

Figure 17a: Stereographic projection

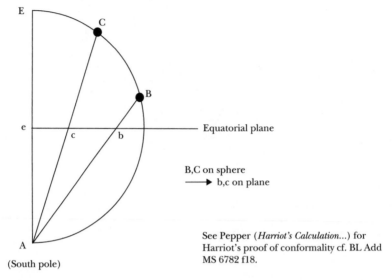

(South pole)

B,C on sphere
⟶ b,c on plane

See Pepper (*Harriot's Calculation...*) for
Harriot's proof of conformality cf. BL Add
MS 6782 f18.

Figure 17b

Stereographic projection of lines of
latitude (circles) and longitude (radial
lines) and rhumb line (equiangular spiral).

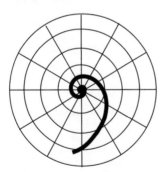

The angle between each radial line
and the tangent to the spiral is the
same. The diagram is a bird's eye
view of the equatorial plane in Fig.
17a.

Figure 17c

Harriot's approximation to the equiangular spiral (from his MS HMC 240 II
f211).

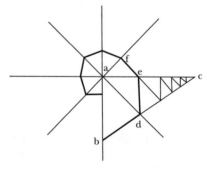

He used the geometry of similar
triangles to find the lengths of the bold
lines, whose sum is the spiral's length;
the area of the triangle abc gives the
area enclosed by the spiral.

Figure 18: Refraction and reflection in a raindrop; rainbow formation

(a) Primary rainbow

(b) Secondary rainbow

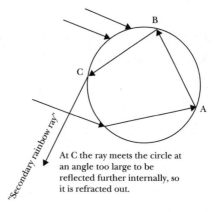

At C the ray meets the circle at an angle too large to be reflected further internally, so it is refracted out.

(c) Geometry of the rainbow

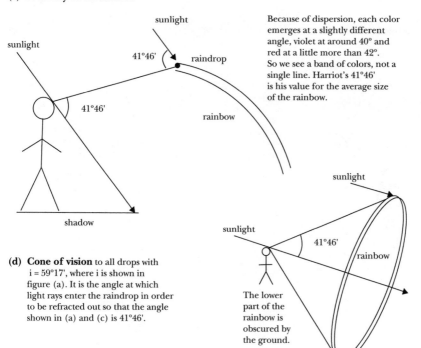

Because of dispersion, each color emerges at a slightly different angle, violet at around 40° and red at a little more than 42°. So we see a band of colors, not a single line. Harriot's 41°46' is his value for the average size of the rainbow.

(d) Cone of vision to all drops with i = 59°17', where i is shown in figure (a). It is the angle at which light rays enter the raindrop in order to be refracted out so that the angle shown in (a) and (c) is 41°46'.

The lower part of the rainbow is obscured by the ground.

Figure 19: Harriot's atomistic model of reflection and refraction cf. BL. Add MS 6789 f336r

Reflection at the surface

Light ray being reflected
from a water "atom".

Refraction

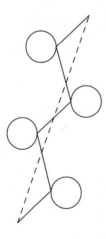

The light ray
enters the
water and is
reflected
from each
water "atom"
it encounters.

The zigzag path
of tiny segments
looks to the eye
like a straight line,
(which I have
drawn as a dotted
line not shown in
Harriot's
diagram.)

NOTES

PROLOGUE

1. *Environmental context:* The Roanoke expert D. B. Quinn (*Set Fair for Roanoke*, 158) makes this point, saying that Harriot acted as an ecologist, "convey[ing] a sense of the whole environment, man and nature, even if the concept had not yet found a term to describe it."

2. *Greatest British mathematical scientist before Newton:* In terms of depth, method, and diversity; e.g., Whiteside's review of Shirley ("Essay Review," 61); Seltman, "Harriot's Algebra," 184; Flood and Wilson, *Great Mathematicians*, 80; also Schemmel, *The English Galileo.*

3. *One of the fathers of modern algebra:* In an 1883 letter to Arthur Cayley (quoted in Seltman, "Harriot's Algebra," 154), J. J. Sylvester called Harriot "the father of current Algebra." Harriot's friends similarly appreciated the new clarity and methods of his symbolic algebra, as indicated in their letters and their introduction to his posthumous *Praxis*, as this book will show.

4. *Possible Harriot portraits* are discussed by Batho ("Possible Portraits") and Swan ("Portrait"). Swan says the portrait in the President's Room of Trinity College London was likely a self-portrait by a left-hander, but there is no evidence so far that Harriot ever spent time painting. However, P. Molarno has recently advanced a theory suggesting that the portrait *is* of Harriot, although he acknowledges this as still conjectural ("Thomas Harriot at the National Gallery?" *Astronomische Nachrichten* 339 (2018): 103–108.

1: HARRIOT'S LONDON

1. *Contemporary descriptions of Durham House* by Nordern, Aubrey, and Browne, quoted in Coote, *A Play of Passion*, 67–68. See also Quinn, *Set Fair for Roanoke*, 3.

2. *Harriot's background:* Shirley, *Biography*, 50–54. *Yeomen's educated sons:* Salter, *Elizabeth I and Her Reign*, chapter 3, document 4a. *Relative Elizabethan incomes incl. craftsmen, yeomen:* Singman, *Daily Life in Elizabethan England*, Table 2.3.

3. *A note on dates and spelling:* Dates of birth for Ralegh and Harriot are not certain; nor is it known precisely when they moved into Durham House, or when Ralegh first hired Harriot. Here, as elsewhere in the narrative, I will follow the scholarly literature, where any uncertain dates are based on scholarly approximations or deductions.

   Note, too, that Elizabethan spelling was not systematic, so at the time there were variants on the spelling of the names Harriot and Ralegh, although I will keep to the spelling used today. Also, in most of my direct quotations from Elizabethan sources, I will use modern US spelling, but in general I will keep the Elizabethan style and grammar (apart from occasional changes of punctuation and clarification, including changing "hath" to "has," etc.). Those interested in original spelling can consult my references to the original sources.

4. *Dragging offenders along the Thames:* John Harrison, *Description of England*, 1587, excerpted in Kinney, *Elizabethan and Jacobean England*, 98.

5. *"As pleasant, perhaps, as any in the world":* John Aubrey, *Brief Lives*, quoted in Coote, *A Play of Passion*, 67.

6. *Durham House dates:* Lambert, *History and Survey of London,* 476; Gatehouse Record (online); British History Online digital library.

7. *Cardinal Allen's Defense of English Catholics* (1584) republished in Kinney, *Elizabethan and Jacobean England*, 241–244. For an analysis of the political role of Elizabethan Catholic and Protestant polemic, see Peter Lake, *Bad Queen Bess?* NY: OUP, 2016.

8. *Freedom of speech*: In 1576, when House of Commons MP Peter Wentworth gamely challenged the queen's autocratic style as an erosion of the Commons' ancient right of freedom of speech, his fellow MPs were outraged at his disloyalty to the queen and had him tried in the Star Chamber, which sentenced him to imprisonment in the Tower! Original documents are reprinted in Salter, *Elizabeth I and Her Reign*, chapter 6, documents 3d and 6e. See also document 3b. As another example, when a "Puritan commoner," John Stubbs, had the audacity to write a pamphlet urging the queen not to marry the "syphilitic" French Catholic duc d'Alençon, Elizabeth had his right hand struck off (see, e.g., Coote, *A Play of Passion*, 41).

9. *Figures for religious executions under Mary and Elizabeth*: Collinson, "Post-Reformation Religion," 193. As in Mary Tudor's reign, however, many hundreds more were executed for their role in plots against Elizabeth: six hundred were hanged as a result of one rebellion alone, the 1569 Northern Rebellion, another failed plan in support of Mary Queen of Scots (Perry, *Elizabeth I*, 207).

10. *Campion*: Salter, *Elizabeth I and Her Reign*, chapter 2, document 6c. Campion was later canonized as St. Edmund Campion.

11. *On Faustus*: The first printed form of the medieval legend—*Faustbuch*, printed in Frankfurt in 1587—identified the main character with a German necromancer, Dr. Georg Faust, who was active at the turn of the sixteenth century (Ousby, *Cambridge Guide to Literature in English*, 262).

12. *Spies: Marlowe*. See, e.g., Honan, *Christopher Marlowe*. *Dee:* Katherine Birkwood, "Was Dee 'The Original 007'?, *Notes and Queries*, June 2017, 248–9. *Bruno:* John Bossy, *Giordano Bruno and the Embassy Affair*, Yale University Press, 1991.

13. *Ralegh and the Second Desmond Rebellion*: Coote, *A Play of Passion*, 49, Nicholls and Williams, *Sir Walter Raleigh*, 13–17, and Wallace, *Sir Walter Raleigh*, 16–17.

14. *Spenser* quoted in Coote, *A Play of Passion*, 49. Note that some scholars see an apologia for Grey in *The Faerie Queen* 5.12.26 (Hadfield, "Another Look at Serena and Irena," 295). For more on Spenser and Ireland: Jenkins, "Spenser and Ireland" and "Spenser with Lord Grey in Ireland."

15. *Spenser* quoted in Ashley, *Elizabethan Popular Culture*, 47.

16. *Ralegh* quoted in Coote, *A Play of Passion*, 53; the figure of thirty thousand dead in six months is also given here.

17. *Criticizing Grey*: Quinn, *Set Fair for Roanoke*, 7.

18. *Censorship re publishing*: See, e.g., Elizabeth I's *Proclamation Announcing Instructions for Religion* (1559), document 51, and *Proclamation Ordering Arrest for Circulating Seditious Books and Bulls* (1570), document 7, both in Kinney, *Elizabethan and Jacobean England*.

19. *Elizabethan "self-help" books*: See, e.g., Bushnell, "Early Education," 498.

20. *Market described by German visitors*: Quoted in Richardson, "Social Life," 298. *On "common- wealth" reforms*: Martin, *Francis Bacon, the State, and the Reform of Natural Philosophy*, 15–22.

21. *Elections*: In every shire and city, any man of the status of yeoman or above could vote for two local knights as his parliamentary representatives; see Sir Thomas Smith's *De republica anglorum* (1565), excerpted in Salter, *Elizabeth I and Her Reign*, chapter 6, document 3a.

22. *London's population, Venice*: Singman, *Daily Life in Elizabethan England*, 85, 90. *Quote from German visitor*: Excerpted in Salter, *Elizabeth I and Her Reign*, chapter 5, document 1a.

23. *Ships in the Thames*: Singman, *Daily Life in Elizabethan England*, e.g., 87. William Harrison, *The Description of England* (1587), in Kinney, *Elizabethan and Jacobean England*, 342, attests to the abundance of silver, gold, and other luxury goods in England.

24. *Cecil's policies*: Martin, *Francis Bacon, the State, and the Reform of Natural Philosophy*, 7–8; Thirsk, "Seeking National Prosperity and Personal Survival," 407–408. *Unpopular taxes/ monopolies*: Thirsk, 408; Coote, *A Play of Passion*, 7. *Jesuits*: Coote, 95.

25. *Piracy*: William Harrison, *Description of England* (1587), in Kinney, *Elizabethan and Jacobean England*, 98–99. Note that relatively few convictions for piracy were handed down.

26. *Deteriorating Anglo-English trade relations*: J. Leitch Wright Jr., "Sixteenth Century English-Spanish Rivalry in La Florida," 266; Quinn, *Set Fair for Roanoke*, 15–16, 85. See also Wright, 266–267: by the 1570s, Elizabeth gave English traders in the West Indies tacit approval, and sometimes financial support, but if they were captured by the Spanish, she would "disavow their voyages and do little to save them from the Inquisition and the galleys."

27. *Trade with Ottomans*: In this paragraph I have drawn from Woodhead, "England, the Ottomans and the Barbary Coast." Note also that when England established an ambassador in Constantinople, it was only the third European country to do so, after France and the Venetian Republic.

28. In 1600, Britain's GDP was 1.86 percent of the global total (Tharoor, *Inglorious Empire*, 216). By the 1870s the British Empire's GDP was 21 percent (Chris Matthews, Fortune, October 5, 2014). Today China has 19 percent and the US 15 percent (IMF 2018).

29. *Florida, and nominal Spanish influence*: Quinn, *Set Fair for Roanoke*, 13. *Trade with China*: See, e.g., Atwell, "International Bullion Flows and the Chinese Economy." India and China accounted for around three-quarters of the world's industrial output in 1600 (Tharoor, *Inglorious Empire*, 216).

2: SEA FEVER

1. *Sons of plebeians at Oxford*: Between 1567 and 1622, plebeian student numbers equaled those of all other classes; figures from Clark's *Register of the University of Oxford (1887)*, in Ashley, *Elizabethan Popular Culture*, 245. Burrow discusses university expenses, scholarships, and says wealthy students were the majority ("Higher Learning," 501–502). Fewer than a quarter of Harriot's classmates graduated: Shirley, *Biography*, 54.

2. *Social hierarchy and numbers of aristocrats*: Singman, *Daily Life in Elizabethan England*, 16.

3. *Hakluyt*: Quinn, "Thomas Harriot and the Problem of America," 11; Mandelbrote, "The Religion of Thomas Harriot," 272. Incidentally, Hakluyt had an older cousin also named Richard Hakluyt, who was a lawyer, and who was also a proponent of English colonialism. The elder Hakluyt supported a Spanish-style conquering of the Indians by force (Quinn, *Set Fair for Roanoke*, 50).

4. *Was Drake the first after Magellan?* Yes, if you count only single voyages.

5. *Tourists' destruction of the* Golden Hind: Singman, *Daily Life in Elizabethan England*, 91.

6. *Elizabeth's and Mendoza's responses to Drake*: For details of negotiations over restitution of Drake's plunder, and Elizabeth's eventual knighting of Drake, see Perry, *Elizabeth I*, 249–250.

7. *Private schools*: Quinn, *Set Fair for Roanoke*, 20. *Dee and better training of navigators*: See, e.g., his preface to the 1570 translation of Euclid's *Elements*. Note that despite the important message about training in Dee's *The Perfect Art of Navigation*, it is a rambling and blatantly propagandist imperialist tract, rather than a mathematical account of navigational theory such as Harriot was writing in his *Arcticon*.

8. *Dee's ambition*: He wanted to be a modern Aristotle; quoted in Pumfrey, "John Dee," 453n11. *"Ungrateful" for advice*: Woolley, *The Queen's Conjuror*, 195, 93.

9. *Dee, portents, "new" star," "blazing star"*: Woolley, *The Queen's Conjuror*, 161–162, 164, 55–56.

10. *Digges (and others, including Tycho) using "new" star of 1572 to test Copernicus's theory*: Note that Digges seemed to accept both Copernicanism and aspects of Aristotle's moral cosmology, notably that the sublunary world was doomed to decay: Pumfrey, "'Your Astronomers and Ours," 50. Note, too, that Digges's Copernican argument, and especially that of Tycho later, were more complex than I can indicate here. See also Johnson, "The Influence of Thomas Digges"; Hellman, "The Role of Measurement in the Downfall of a System"; Gingerich and Voelkel, "Tycho and Kepler."

11. *Biblical references*: Joshua commanded the sun to stand still; Psalm 109 speaks of God's unmoving foundation of the earth (cf. Gingerich and Voelkel, "Tycho and Kepler," 86).

12. *Harriot as Copernican*: He did not write about it, so it is not clear whether he became a Copernican at Oxford, but as we shall see, he was one of the first to recognize the significance of Kepler's adaptation of Copernicus's model.

13. *Copernicus, Recorde, Digges*: Johnson, "The Influence of Thomas Digges," 394–400. *British especially receptive to Copernicus*: Johnson, 390, 395.

14. *Harriot teaching parallax*: In addition to facilitating a triangulated measure of stellar distances, parallax affects the observed altitude of the sun taken by mariners when finding their latitude; this type of parallax arises because sailors take local measurements rather than measuring from the center of the earth (see also the next two chapters). Harriot mentioned this as a minor issue for short voyages: BL Add. MS 6789 ff486–9; Pepper, "Harriot's Earlier Work on Mathematical Navigation," 60.

15. *Tycho, Digges, and Dee's analysis of parallax of the 1572 "new" star; messages from God*: Pumfrey, "'Your Astronomers and Ours,'" 42–43, 33, 37; Stephen Clucas, ed. *John Dee: Interdisciplinary Studies in English Renaissance Thought*, Springer, 2006, 12–14, 44–53 (Goulding), 75–80 (Johnston).

16. *Harriot's religious views*: Many scholars have sifted through Harriot's writings in order to find out what he believed. He was a Protestant, but opinions vary as to his level of orthodoxy. See overviews in Mandelbrote, "The Religion of Thomas Harriot," and Jacquot, "Thomas Harriot's Reputation for Impiety."

17. *Dee on optics and spiritual rays*: Clulee, "Astrology, Magic, and Optics"; see also Woolley, *The Queen's Conjuror*, 168.

18. *Harriot's astrology*: There seems to be only one extant horoscope in Harriot's manuscripts, that of Queen Elizabeth, cast in 1596 (Gatti, "The Natural Philosophy of Thomas Harriot," 75).

19. *Chemistry and physics experiments with some substances used by alchemists (e.g., sal gemma)*: See, e.g., Harriot to Kepler, 1606, reproduced in Lohne, "Essays on Thomas Harriot," 310.

20. *Henry VIII's astronomer*: Nicholas Kratzer had been appointed in 1519.

21. *Ralegh and Dee*: Dee's diary entries, April 18 and July 31, 1583, quoted in Thompson, *Sir Walter Ralegh*, 69.

22. *Dee's magic mirror* was a large concave mirror that apparently focused an image of the viewer *in the air* between the mirror and the viewer—presumably somewhat akin to a mirage. For brief details, see Goulding, "Thomas Harriot's Optics," 158–159.

3: THE SCIENCE OF SEA AND SKY

1. *Harriot taught "many" captains*: Richard Hakluyt's 1587 preface to *De orbe novo*, quoted in Shirley, *Renaissance Scientist*, 18. *Amadas and Barlowe*: Quinn, *Set Fair for Roanoke*, 21–22.

2. *Christianity, Islam, and the decline of learning in Alexandria*: For a short, informative account—including the murder of Hypatia—see Greenblatt, *The Swerve*, 86–94, 283n.

3. *Baghdad, new Alexandria*: See, e.g., Boyer, *History of Mathematics*, 227; Scott L. Montgomery, "The Era of Translation into Arabic," *Isis*, 109, No. 2 (June 2018), 313–319.

4. *Carolingian Renaissance*: See the various works by Bruce S. Eastwood and reviews of his *Ordering the Heavens* by Zuccato, by Caudano, and by Lozovsky.

5. *Latin translations*: e.g. D. N. Hasse, "The Social Conditions of the Arabic-(Hebrew-)Latin Translation Movements in Medieval Spain..." in Speer, A. and Wegener, L, eds. *Wissen über Grenzen. Arabisches Wissen und lateineisches Mittelalter*, Gruyter, Berlin, 2006: 68–86. *Ptolemy's translator-correctors*: See Montgomery, *Science in Translation* (and Montgomery's summary in *Nature* 409 (February 8, 2001): 667), and Ragep, "Islamic Response," 121–134. Note that chief among Aristotle's translator-correctors was Gerard's Moorish contemporary Averroës.

6. *Sunset*: Interestingly, the earliest winter sunset and latest winter sunrise happen, respectively, a few days before and after the solstice, and vice versa in summer. This is because the earth moves a little way along the ecliptic during each daily rotation, and so it has to rotate for slightly longer to catch up to the sun and to realign the "solar day" (the time from one noon to the next, where noon is when the sun is at its highest for the day). For more detail, see "It's Going to Be a Long Summer's Day, Seriously," *The Conversation*, December 21, 2017, http://theconversation.com/its-going-to-be-a-long-summers-day-today-seriously-89491 and relevant links therein.

7. *Ancient astronomy?* Hayden and Villeneuve ("Astronomy in the Upper Palaeolithic") suggest an Upper Paleolithic date is possible, although all ancient observations were for ritualistic purposes. But Emília Pásztor ("Prehistoric Astronomers?") suggests that concrete evidence for human interest in the solstices and equinoxes is rare even into the

Bronze Age. On an 11,000(?)-year-old Australian stone circle tracking the directions of sunset: Norris et al., "Wurdi Youang." Stonehenge and solstices: Heggie, *Megalithic Science*, 151; Ruggles, "Astronomy and Stonehenge," 218–220. Both authors give critical analyses of the degree of astronomy practiced; but see Sims. On Chinese (and early Babylonian) astronomy and its political purpose: Steele, "A Comparison of Astronomical Terminology, Methods and Concepts," 253–254.

8. *Kepler, God, sun*: Gingerich and Voelkel, "Tycho and Kepler," 90.

9. *Simon Forman*: Kassell, *Medicine and Magic in Elizabethan London*, 75ff, 131. Cerasano, "Philip Henslowe, Simon Forman, and the Theatrical Community of the 1590s." He shows that Ben Jonson referenced Forman in several plays and conjectures that Forman may also have been a model for Marlowe's Faustus.

10. *The terms "Mesopotamian" and "Greek"*: The term "Mesopotamian" (or "Babylonian") refers to various peoples who lived, over the ages, in the region roughly corresponding to modern Iraq. Similarly, the term "ancient Greek" refers to Greek-speaking people in the various territories dominated by Greek-speaking rulers such as Alexander the Great. For instance, the city named after Alexander, Alexandria, is in Egypt; Hipparchus's birthplace, Nicaea, is in Turkey; and so on.

11. *Mesopotamians and the zodiac*: Gauquelin, *Cosmic Clocks*, 15–27. See also Jones and Steele, "A New Discovery." It is always risky to claim priority based on ancient evidence, but it seems that earlier peoples did not know of this band and that it played only a minor role in contemporaneous Chinese astronomy, although it has been suggested that it appears in Indian Vedic literature earlier than 700 BCE. Cf. the following: Hayden and Villeneuve, "Astronomy in the Upper Palaeolithic?" 340–341; *China*: Steele, "A Comparison of Astronomical Terminology, Methods and Concepts," 255; *India*: Kak, "Babylonian and Indian Astronomy," who also argues that Indian astronomy may have influenced the Mesopotamians, but see Pingree, "Astronomy and Astrology in India and Iran" (whom Kak critiques).

12. *Figure 1*: My diagram is inspired partly by F. Wright, *Celestial Navigation*, 4. Note that this and all my diagrams are not to scale.

13. *Babylonian predictions and mathematical methods*: Aaboe, "Observation and Theory in Babylonian Astronomy"; Steele, "Eclipse Prediction,"; Britton, "An Early Function"; Resnikoff and Wells Jr., *Mathematics in Civilization*, 84–85, 325–326.

14. *Mesopotamian eclipse predictions*: For details, see Steele, "Eclipse Prediction," 452–453.

15. *Mayan arithmetical Venus predictions*: Aldana, "Discovering Discovery." *Aboriginal solar eclipse stories* had the moon god covering the sun goddess (or Sun-woman) during sex, while lunar eclipses were explained as the Sun-woman chasing the moon-man: Ray Norris, "In Search of Aboriginal Astronomy," CSIRO Australia Telescope National Facility, www.atnf .csiro.au/people/rnorris/papers/n217_preprint.pdf.

16. *Eudoxus and Democritus quotes*: Boyer, *History of Mathematics*, 82, 62. *Phaethon*: Son of the sun god Helios. Phaethon's dream was to drive his father's chariot through the sky, even if just for a day. When Helios allowed him to do so, he could not control the fiery chariot and veered dangerously toward earth. Eudoxus wasn't interested in the thrill of chariot racing: he just wanted to *know* the sun.

17. *Greek geometric, causal models for eclipses*: Resnikoff and Wells Jr., *Mathematics in Civilization*, 326–327. *On Eudoxus's geometric model*: Boyer, *History of Mathematics*, 92; Goldstein and Bowen, "A New View of Early Greek Astronomy."

18. *Aristotle-Eudoxus*: Boyer, *History of Mathematics*, 92.

19. *Evolution of 360-degree circle (and 60 minutes in an hour)*: Earlier Greek astronomers spoke of fractions of a circle rather than numerical angles. But the Mesopotamians may have divided the zodiac circle into 360 parts (or degrees), since their number system was based on powers of 60. (By contrast, our decimal system is based on powers of 10: for instance, our number 125 is $1 \times 100 + 2 \times 10 + 5 \times 1$.) This apparent Mesopotamian heritage is also evident in our division of hours into 60 minutes, and minutes into 60 seconds.

20. *Contributions to saving* Almagest *by Syriac, Arabic, and Latin translators*: Montgomery, *Science in Translation*, 19–20.

21. *Astronomy in other cultures*: In 2017, the International Astronomical Union acknowledged the contributions of a number of ancient cultures by including eighty-six indigenous star names in a new star naming system: Duane Hamacher, "The Stories behind Aboriginal Star Names Now Recognised by the World's Astronomical Body," *The Conversation*, January 14, 2018, http://theconversation.com/the-stories-behind-aboriginal-star-names-now-recognised -by-the-worlds-astronomical-body-87617? For some Australian examples of indigenous astronomy, see Hamacher, "Stories from the Sky: Astronomy in Indigenous Knowledge," *The Conversation*, December 1, 2014, and "Stars That Vary in Brightness Shine in the Oral Traditions of Aboriginal Australians," *The Conversation*, November 9, 2017.

22. *Ptolemy's star catalog*: *Almagest*, ed. Toomer, 341–399.

23. *Tamerlane and Tamburlaine* are anglicized forms of the Turkic name, Timur Lenk.

24. *Star catalogs of Ptolemy, al-Sufi, Ulugh Beg*: Schaefer, "The Thousand Star Magnitudes in the Catalogues," 47–A97. See also Montgomery, *Science in Translation*.

25. *Molyneux globes*: Wallis, "A Newly Discovered Molyneux Globe," 78. Note that Portuguese mariners had been using model globes since the 1530s, as had Continental astronomers before them (Leitao and Gaspar, "Globes, Rhumb Tables, and the Pre-History of the Mercator Projection," 183).

26. *Flat earth idea widespread*: For Europe, see later in this chapter for Acosta's reference. In 1595, a learned Jesuit missionary (Ricci) reported that the Chinese still accepted a flat earth (Cullen, "A Chinese Eratosthenes," 109).

27. *Ptolemy's arguments for spherical earth*: *Almagest* 1.4.

28. *Eratosthenes*: Some of his assumptions were flawed and his measurements rough, but they gave a good guesstimate via an ingenious application of simple geometry and the length of the midsummer noonday shadows at two different locations on earth. For a critique, and an account of later Arabic tests of Eratosthenes's result, see Ragep, "Islamic Response."

29. *Tilted axis of earth*: Copernicus introduced this, but he still imagined that the earth was fixed to a crystalline sphere as it revolved around the sun; therefore, he had to assume the earth's tilted axis *rotates* (to keep pointing in the same direction), rather than remaining *parallel to itself* as in my figure 7. Kepler, Galileo, and Harriot (presumably, given he agreed with Kepler's elliptical orbits and with Gilbert on the different distances of different stars) would do away with crystalline orbs and Copernicus's modification, thus holding to the modern picture.

30. *Kepler's orbits*: Goldstein and Hon, "Kepler's Move from *Orbs* to *Orbits*," 74–111.

31. *Ptolemy claimed the earth does not move*: *Almagest* 1.7. Another reason heliocentrism did not take off was the apparent lack of stellar parallax, at a time when there was no appreciation of the size of the universe to account for this apparent lack.

32. *Father Acosta's views on the celestial sphere*: Acosta, *Natural and Moral History*, 5–9.

33. *From Copernicus's dedication to the pope*, quoted in Rosen, *The Intellectual Background*, 6.

34. *Ptolemy's method of measuring the ecliptic*: *Almagest* 1.12. (In contrast to earlier critics, Britton concludes it is probable Ptolemy did do as he said.) To measure his altitudes, Ptolemy measured the angle of the midday shadow cast by a horizontal cylindrical pin or rod that was fixed to the vertical face of a block of wood, on which a scale was inscribed like that on a protractor (as in figure 8a). It was a similar process to observing the way your own shadow falls: the angle between the shadow and the imaginary line from the end of the shadow to the top of your head is the altitude of the sun, whose rays are slanting so that they graze the top of your head while your body blocks out the rays behind you. Measuring such an angle precisely is another matter, though—which is why Ptolemy used his simple but effective instrument. (The use of a shadow cast by a pin or rod or column—collectively known as gnomons—was an ancient technique for taking astronomical measurements, and it was known in both the East and the West; see, e.g., Cullen, "A Chinese Eratosthenes.")

35. *Harriot's measurements of the obliquity of the ecliptic*: He used a cross-staff, of which more in the next chapter. The obliquity is necessary for making tables of the sun's declination, which Harriot certainly did in the early 1590s. But Pepper, the leading expert on Harriot's navigational manuscripts, suggests (in Shirley, *Renaissance Scientist*, 55) that Harriot

probably also did this in the early 1580s. Note that Edward Wright discussed his own method of measuring the obliquity of the ecliptic (more fully than Harriot did in his surviving manuscripts) in his *Certain Errors in Navigation* (1610), 304–308, and especially 236–237, where he shows that although his instruments were different from Ptolemy's (as were Harriot's), the underlying method and calculations were the same. Wright's observations were from the mid-1590s.

36. *More on calculating the angle between the ecliptic and the equator:* At the equinoxes—when the direction of the sunset is due west, halfway between the most northerly and southerly sunset points (the solstices)—the sun is at the intersection of the ecliptic and the celestial equator, as in figure 9. (You can see why this is so from my earlier bowling-ball analogy: the sun would always set due west if the sun "traveled" around the celestial equator rather than the ecliptic, so at the equinox it must be "on" the equator, and so the ecliptic must meet the equator at that point.) So Ptolemy concluded that the angle between these two intersecting curves must be half the difference between the summer and winter maximum solar altitudes (or equivalently the zenith angles shown in Figure 8), which arise from the same symmetry. His reasoning is illustrated in figure 8b, expanding Britton ("Ptolemy's Determination"). Ptolemy gave only the bare outline of his argument. Note that for an observer at the equator, the obliquity is simply half the angle about due west made by the setting sun at the summer and winter solstices. Normally, this angle depends on latitude (as Harriot's formula for magnetic variation will show, in figure 12), but latitude is zero at the equator.

37. *Harriot's lost* Arcticon *and lost fame/commercial secrets:* Pumfrey, "John Dee." *Harriot ensuring Ralegh's navigators the best trained:* Quinn, *Set Fair for Roanoke,* 27; Pepper, "Harriot's Earlier Work on Mathematical Navigation," 75. In the absence of the *Arcticon* manuscript, I have followed Pepper in my selections of Harriot's contributions to the Barlowe-Amadas voyage; such topics were discussed, albeit generally in less detail, in contemporary practical manuals such as Bourne's.

38. *Armillary spheres:* For Tycho and the 1577 comet, see Gingerich and Voelkel, "Tycho and Kepler," 79. Ptolemy's description of the armillary sphere is in *Almagest* 5.1.

39. *Barlowe's and Amadas's backgrounds:* Quinn, *Roanoke Voyages* 1: 78; Quinn, *Set Fair for Roanoke,* 21–22.

4: PRACTICAL NAVIGATION (AND WHY THE WINDS BLOW)

1. *Harriot, "the elevation of the pole":* BL Add MS 6788 f437r. These instructions are from later classes Harriot gave, but in them he refers to his *Arcticon,* which he was writing in 1584.

2. *Harriot on Polaris inaccuracies:* BL Add MS 6788 f484*; quoted in Pepper, "Harriot's Earlier Work on Mathematical Navigation," 74, 69. Note that Harriot referred to Cortes, Digges, and Pedro Nunes. Harriot obtained the information about "the King of Spain's cosmographers" on his 1585 voyage to America, a year later than his *Arcticon. Degree of latitude:* Because the earth is actually slightly flattened at the poles, a distance of 1 degree along a meridian ranges from 68.7 miles at the equator to 69.4 miles at the poles.

3. *Harriot's calculations on the Pole Star:* These were given in full in his *Arcticon,* according to a later reference he made (BL Add. MS 6788 f 484). My summary is also drawn from Pepper, "Harriot's Earlier Work on Mathematical Navigation," 69–71, and Cotter, *A History of Nautical Astronomy,* 130–136.

4. *Changing star positions:* Over very long time periods, these stars' relative positions *and* their declinations do change slightly, because of precession (of which more later), and because the stars *are* actually moving at different rates, albeit very slowly, relative to the sun—a fact that Newton's mentor and disciple Edmond Halley would discover in 1718. Also, some of the closer stars appear to change position as the earth revolves around the sun. See, e.g., Smart, *Textbook on Spherical Astronomy,* 249ff.

5. *Ancient Polynesian and Arab navigators using the stars:* See, e.g., Cotter, *A History of Nautical Astronomy,* 129–130. But see the following (and references therein) for a debate about the difference between deliberate and accidental reaching of destinations in order that star patterns could guide an *initial* (as opposed to a return) voyage: Andrew Sharp, "David

Lewis on Polynesian Navigation," *Journal of the Polynesian Society* 74, no. 1 (March 1965): 75–76; Ben Finney, "Rediscovering Polynesian Navigation through Experimental Voyaging," *Journal of Navigation* 46, no. 3 (1993): 383–394. *Meriam (Torres Strait) navigation:* Anna Salleh's report (www.abc.net.au, September 20, 2016) on a collaboration between the Meriam elder A. Tapin and the physicist D. Hamacher.

6. *Calculating tables of solar declination:* Ptolemy had shown (*Almagest*, 1.14) how to use trigonometry to calculate the sun's declination every day. Harriot and his sixteenth-century colleagues used a more modern trigonometric formula; cf. Pepper, "Harriot's Earlier Work on Mathematical Navigation," 65.

7. *Harriot's table ("flowchart") for calculating latitude from noon sun altitude and declination:* BL Add MS 6788, f474, transcribed in Pepper, "Harriot's Earlier Work on Mathematical Navigation," 66.

8. *Dee introducing the cross-staff, Harriot's favorite instrument:* Roche, "Harriot's 'Regiment of the Sun,'" 247.

9. *Figure 11:* See also Cotter, *A History of Nautical Astronomy*, 67.

10. *Harriot on how to observe the sun "safely":* BL Add MS 6788 f489; see also Pepper, "Harriot's Earlier Work on Mathematical Navigation," 63.

11. *Thales, sun's diameter:* Wasserstein, "Thales' Determination of the Diameters of the Sun and Moon." To time the rising sun, Thales may have used a dripping "water clock," a sandglass, or his pulse.

12. *Harriot on ocular parallax:* BL Add MS 6788, f486r; error of 1.5 degrees: Cotter, *A History of Nautical Astronomy*, 66, but see Harriot's examples in Pepper, "Harriot's Earlier Work on Mathematical Navigation," 62, table 2.

13. *Solving the "dip" of the horizon:* See Cotter, *A History of Nautical Astronomy*, 111–114, for a detailed mathematical outline of the required analysis, such as Harriot would have used. Harriot's table is given in Pepper, "Harriot's Earlier Work on Mathematical Navigation," 61.

14. *All Harriot's corrections to instruments:* BL Add MS 6788 ff485–8; see also Pepper, "Harriot's Earlier Work on Mathematical Navigation," 59–63, for a summary and additional explanation. *Harriot's loyal friend:* William Lower, of whom more later. Note: Wright's tables were more accurate (Cotter, *A History of Nautical Astronomy*, 115), but Harriot intended to correct for refraction later (Pepper, 60).

15. *John Dee on clocks:* In his preface to the 1570 English translation of Euclid's *Elements*.

16. *Rate of knots for sixteenth-century sailing ships:* Pepper, "Harriot's Earlier Work on Mathematical Navigation," 56.

17. *Choosing a site:* Quinn, *Set Fair for Roanoke*, 7–9; Quinn, *Roanoke Voyages* 1:77. *Carleill's plans* (and his argument for independent English trading bases): Hakluyt, *Principal Navigations* 8:134–147.

18. *Latitude:* In 1585, Hakluyt the elder wrote a pamphlet called "Inducements to the liking of the voyage intended towards Virginia in 40 and 42 degrees latitude"; the Outer Banks are at latitude 35.6 degrees north.

19. *Acosta on trade winds and westerlies: The Natural and Moral History of the East and West Indies,* 138–140. He got a bit tied up and contradictory with his rotating-wheel analogy, but the idea was right. *Aristotle still widely considered authoritative:* See, e.g., McConica, "Humanism and Aristotle in Tudor Oxford."

20. *Modern explanation of the trade winds and westerlies:* Today we say that the air above the equator does indeed move faster than air at higher latitudes, as Acosta suggested, and it is also "lighter" in the sense that warm air expands and rises upward, causing a low pressure area near the ground. Unlike Acosta, however, today we know that the earth *and* its atmosphere rotate from west to east (making the "celestial sphere" appear to rotate from east to west); the atmosphere is "anchored" to the earth by gravity. Nevertheless, winds do form within the rotating atmosphere, because of differences in air pressure, which are due to differences in surface temperatures—and broadly speaking, these air movements flow

to the north or south, because warm air at the equator expands and pushes upward, creating more pressure in the upper air. This pressure drives the air toward the poles.

At about 30 degrees north (or south) of the equator, however, some of this equatorial air has cooled sufficiently for it to condense and become "heavier" in the sense that the increased pressure above pushes it downward. This cooler, lower air begins to blow back toward the equator, while the remaining air continues on toward the poles. Air above the equator is traveling with the earth at the fastest possible rate; as it expands and flows north (or south), this easterly velocity component remains unchanged (according to Newton's first law of motion). But a person at a latitude north (or south) of the equator is traveling east (with the earth) more slowly than the wind that arrives from the equator; this extra relative motion to the east makes it appear as though the wind is moving east as well as north (or south) from the equator. In other words, it appears to be blowing from the southwest (or from the northwest in the southern hemisphere). These are the midlatitude "westerlies."

The Coriolis effect is in the opposite direction for the cooler air that begins blowing back *toward* the equator at latitudes around 30 degrees. This air began its southward journey at a latitude where the earth is traveling east more slowly than it is at a lower latitude (a latitude nearer the equator), so a person on the ground at a lower latitude will be traveling east *faster* than the incoming wind. Relatively speaking, then, this wind will appear to be turned to the west as it flows south (or north in the southern hemisphere) toward the equator. These are the trade winds, and they are most reliable at latitudes from about 5 degrees to 30 degrees north or south of the equator.

21. *Galileo and the trade winds (and Newton, Hadley, Coriolis)*: Burstyn, "Galileo's Attempt to Prove That the Earth Moves." Most of this paper concerns Galileo's analysis of the tides. Although it failed in its ultimate purpose, Burstyn shows the ingenuity behind Galileo's thinking. For Galileo's anticipating the Coriolis effect re the trade winds, see 168–169, 175–176; his failing to take account of heat differentials, 176. Galileo's analyses are from his *Dialogue concerning the Two Chief World Systems*.

22. *Galileo, Inquisition, proving earth moves*: See, e.g., Burstyn, "Galileo's Attempt to Prove That the Earth Moves." See also Hutchinson, "Galileo, Sunspots, and the Orbit of the Earth," 68–74. But John Henry has pointed out that to prove his point, Galileo used assumptions he knew were incorrect—notably, the uniform, circular motion of the earth and planets that even Ptolemy knew was wrong ("Why Harriot Was *Not* the English Galileo," 134–135).

23. *The principle of Galilean relativity*, from his *Two Chief World Systems:* Galileo gave the example of being shut in a cabin below deck in a ship sailing smoothly at a constant speed. If you had some butterflies and a bowl of goldfish with you in the cabin, you would notice that they continued to fly and swim in exactly the same way as they did when the ship was stationary. Similarly, if you threw a ball to a friend, it would require no more effort than it did when the ship was docked. So there is no way to tell from inside the cabin that you are moving relative to the shore. Only if you were able to look out a window could you notice the constant relative motion between ship and shore, although you wouldn't be able to tell, from any observation or feeling, whether the ship or the receding shoreline was "really" moving. The only way you could decide that it is the ship that is moving is by noticing the effects of its initial acceleration from rest to its constant sailing speed (or other acceleration effects, such as if the wind changed).

Incidentally, Kosso shows that Galileo's attempt to explain the earth's rotation by the tides *conflicts* with this relativistic assumption ("And Yet it Moves," 219). This helps explain why he did not succeed with his tidal analysis. *Einsteinian "special" relativity* was designed to understand what happens when two systems' constant relative velocity is close to the speed of light. It takes us into the far more sophisticated territory of post-Newtonian physics.

24. *Harriot reading della Porta*: Gatti, "The Natural Philosophy of Thomas Harriot," 75.

25. *Galileo missing temperature variation*: Burstyn, "Galileo's Attempt to Prove That the Earth Moves," 176.

26. *Della Porta, Benedetti, and al-Kindi on winds*: Arianna Borelli, "The Weatherglass and Its Observers in the Early Seventeenth Century," 67–130, esp. 84ff.

27. *For a transcript of the patent document* see Quinn, *Roanoke Voyages* 1:82–89.

28. *Spanish Crown's policy on spoils of conquest*: Batchelder and Sanchez, *The Encomienda and the Optimizing Imperialist*, 10.

29. *Did Harriot sail 1584?* Quinn says yes (*Set Fair for Roanoke*, 24), and others have agreed. Pepper, however, claims an annotation by Harriot dated August 1584 that makes it almost impossible he could have sailed on the first voyage ("Thomas Harriot: A Biography"); the annotation is in the Oslo Witelo, but Lohne's account of it suggests the last digit of the date is unclear, so it is not necessarily 1584 ("Thomas Harriot [1560–1621]," 114).

5: AMERICA AT LAST

1. *Amadas-Barlowe voyages, including Barlowe's quotes*: Based on Barlowe's report, which was published by Hakluyt in 1589 and republished, e.g., *Principal Navigations* 8:297–310. The report is also reprinted in Quinn, *Roanoke Voyages* 1:91–115. *Sketchy details of the reconnaissance fleet* are from Quinn, *Set Fair for Roanoke*, 24–25. *Harriot did use "year of the lord"* when helping Ralegh date biblical events for his *History of the World* (e.g., MS HMC 1 f65), but I noticed no such reference in his notes or his published treatise on America.

2. *Some said Columbus had supernatural powers*: Jackson, *Columbus*.

3. *Columbus's pledge, "neglect sleep"*: Jackson, *Columbus*.

4. *Hawkins, slave trade*: The slave trade was already widespread, but the first Englishman to engage in it was John Hawkins.

5. *Davis's acknowledgment*: See also Quinn and Shirley, "A Contemporary List of Harriot References," 15. On a voyage to the East Indies in 1605, Davis was killed by Japanese pirates (*Chambers Biographical Dictionary*).

6. *Hakluyt on Harriot*: Preface to 1587 edition of *De orbe novo petri martyris anglerii mediolanensis*, quoted in Shirley, *Renaissance Scientist*, 18.

7. *"Exemplary skill"* is the judgment of the Roanoke Voyages expert D. B. Quinn (*Set Fair for Roanoke*, 27).

8. *The island where Barlowe and Amadas made their first contact with North America* was probably "some two miles east of Cedar Point on Bodie Island, nearly a mile out to sea," according to Quinn (*Roanoke Voyages* 1:95n1).

9. *Value of Elizabethan money, then and now*: Singman, *Daily Life in Elizabethan England*, tables 2.2 and 2.3.

10. *Harriot's description of the Algonquian method of making canoes* is from his caption to White's drawing in de Bry, *America*.

11. *Barlowe as a persuasive promulgator*: There is little doubt that Barlowe's report was designed to promote Ralegh's colonizing venture; for a hardnosed analysis, see Moran, "A Fantasy-Theme Analysis." Nevertheless, as Moran acknowledges, Barlowe's report serves as a vital firsthand ethnographic record of first contact in North America. For more general critiques of racist stereotypes, see, e.g., Robert F. Berkhofer Jr.'s classic *The White Man's Indian: Images of the American Indian from Columbus to the Present* (New York: Knopf, 1978).

12. *Roanokes*: For current information about the Roanoke and Croatoan (Hatteras) tribes, see the website maintained by North Carolina Algonquian descendants, www.ncalgonquians.com.

13. *Amadas/Fernandes privateering en route home*: Moran, "A Fantasy-Theme Analysis," 34.

14. *On Shane O'Neill's visit*: Coote, *A Play of Passion*, 43; Salter, *Elizabeth I and Her Reign*, chapter 1, document 9a (from a report by the early seventeenth-century historian William Camden).

15. *Frobisher's Inuit*: Salter, *Elizabeth I and Her Reign*, chapter 3, document 1a. In 1576 and 1577, Frobisher brought two others, who also died (Vaughan, "Sir Walter Ralegh's Indian Interpreters," 344). See this reference also for the first Inuit in England—and also James A. Williamson, *The Cabot Voyages and Bristol Discovery under Henry VII*, works issued by the Hakluyt Society, 2nd ser., no. 120 (New York: Cambridge University Press, 1962), 128.

16. *"Childish and silly figure"*: Lupold von Wedel, quoted in Quinn, *Roanoke Voyages* 1:116n6.

17. *Carolina Algonquian*: Unfortunately, most Algonquian languages were lost by the late nineteenth century. Today, Harriot's work, along with that of other pioneers, is being used to reclaim the language; see Dawson, "The Vocabulary of Croatoan Algonquian," and Rudes, "Giving Voice to Powhatan's People."

18. *Harriot as teacher-student*: There is no specific evidence of Harriot teaching Manteo and Wanchese English, but since the evidence for his learning Algonquian is clear (cf. his phonetic alphabet and his *Brief and True Report*, to be discussed later), most scholars make the reasonable deduction that he was the primary player in the English-Algonquian exchange; e.g., Quinn, "Thomas Harriot and the Problem of America," 14–15.

19. *Ralegh's linguistic-cultural program, loyalty*: Vaughan, "Sir Walter Ralegh's Indian Interpreters," 341–376. *Spanish account of good treatment*: Interrogation of Hernando de Altamirano, June 1585, reprinted (with further acknowledgments) in Quinn, *Roanoke Voyages* 2:741.

20. *Harriot's Latin and Greek*: Harriot's elegant prose is the opinion of Fantazzani, "Harriot's Latin," 232–235.

21. *Harriot, Hart, phonetics*: Salmon, "Thomas Harriot and the English Origins of Algonkian Linguistics." Note the alternative spelling of "Algonquian"; a third spelling is "Algonquin." *Hart on spelling reform*: Quoted in Bushnell, "Early Education," 496.

22. *Harriot's systematic phonetics, loops, sounds formed in the mouth*: Cf. Salmon, "Thomas Harriot and the English Origins of Algonkian Linguistics," and the detailed analysis in Stedall, "Symbolism, Combinations, and Visual Imagery," 381–384. Stedall credits Alec Wallace, John Shirley, and Vivian Salmon with pioneering the study of Harriot's alphabet, Wallace being the first to have realized it was phonetic.

23. *"An universal alphabet"*: The original of Harriot's alphabet is in the Westminster School, London, and also in Harriot's manuscript BL Add MS 6782 f337. Reproductions are available in, e.g., Salmon, "Thomas Harriot and the English Origins of Algonkian Linguistics," and Stedall, "Symbolism, Combinations, and Visual Imagery." Harriot's heading in my text is from the Westminster School manuscript, as deciphered by Salmon. For a detailed account of phonetics with regard to Algonquian languages, see James A. Geary, "The Language of the Carolina Algonkian Tribes," in Quinn, *Roanoke Voyages* 2:873–883.

6: PREPARING FOR VIRGINIA

1. *Litigious Elizabethan businessmen*: For some of the court cases mounted by various investors in Ralegh's and Gilbert's voyages, see summaries and transcripts in Quinn, *Roanoke Voyages* 1:67–70, 234–242 (Amyas Preston case), 480–488, 2:596–598, 623–712 (this referred to the 1590 voyage led by John White; cf. chapter 8). *Gilbert's case, and Ralegh as witness*: Quinn, *Set Fair for Roanoke*, 6. Most of these cases dragged on for years, and even when the plaintiff was awarded compensation, as in Preston's case, it was often contested and not always paid. *Preston and piracy*: Quinn, *Set Fair for Roanoke*, 130–131.

2. *Acosta on the magnetic compass*: *Natural and Moral History of the East and West Indies*, 53–55.

3. *Harriot's illustrative examples on magnetic variation*: I have adapted the example in my narrative from BL MS 6789 ff534r–537. Note that these manuscripts form part of the notes that Harriot prepared in 1595 for another of Ralegh's ventures, but Pepper assumes that the original work formed part of Harriot's *Arcticon* ("Harriot's Earlier Work on Mathematical Navigation," 54). Harriot would certainly have had sufficient knowledge to do this at that time. In the absence of clear evidence of when he did what in regard to his early navigational research, I have chosen to include magnetism here, to break up the navigational narrative in earlier chapters.

4. *Borough on magnetic variation*: Pepper, "Harriot's Earlier Work on Mathematical Navigation," 71–72 (Borough's formula given on 71).

5. *My figure 12a* is an amplified, explicated version of Pepper's figure 4.9 ("Harriot's Earlier Work on Mathematical Navigation," 73). Pepper has based his diagram on an extract from Harriot's manuscript (Petworth House/Leconfield HMC 241 VIb f18), which shows the four different quantities for which sines were needed to construct his table of theoretical sunrise directions. Harriot listed these four sines I, II, III, IIII, but his meaning must have

been I: II = III: IIII. The extant table that Harriot constructed, and the examples he gave on how to use it, are from BL MS 6789 ff534r–537.

6. *Sunrise declination*: This required taking the tabulated declination for noon and estimating the declination at sunrise, using the "first difference" given in the table. Harriot would have taught his students to do this when he taught them about using tables of declination. See Pepper, "Harriot's Earlier Work on Mathematical Navigation," 72.

7. *The usual, less accurate method of finding magnetic compass variation at the time* is given in Bourne's *Regiment for the Sea*, 31–32 (1584 ed.).

8. *On Robert Hues*: Batho, "Thomas Harriot's Manuscripts," esp. 297; Fantazzi, "Harriot's Latin," esp. 233–235; Shirley, *Biography*, 373–375 Tanner, "Henry Stevens and the Associates of Thomas Harriot," 98.

9. *Gilbert's acknowledgment of Harriot*: De magnete (1600) 1.7. *Harriot's note on this*: Quinn and Shirley, "A Contemporary List of Harriot References," 16.

10. *Gilbert's discovery and "dip"*: Robert Norman was the one who discovered magnetic dip, but Gilbert realized its significance. For details, see Henry, "Why Harriot Was *Not* the English Galileo," 123–124; Henry, "Animism and Empiricism" 104–105.

11. *Gilbert's magnetic "soul"*: Discussed and quoted in Boner, "Life in the Liquid Fields," esp. 287. *On "agency" in matter*: See, e.g., Riskin, *The Restless Clock*.

12. *Search for evidence the earth moves*: Kosso, "And Yet It Moves" (on evidence for rotation). See also Hutchinson, "Galileo, Sunspots, and the Orbit of the Earth," and Burstyn, "Galileo's Attempt to Prove That the Earth Moves," on Galileo's attempts (tides, winds). As mentioned, the Coriolis effect was a credible inference of the earth's rotation; the first physical evidence of its revolution was stellar aberration, of which more later.

13. *Richard Mulcaster; Elizabethan English*: Richard Mulcaster, *The First Part of the Elementary* (1582) excerpted in Kinney, *Elizabethan and Jacobean England*, 540. For more on the Elizabethan battle for English—and its fluidity, changes, and influences—see McCrum, Cran, and MacNeil, *The Story of English*.

14. *Voltaire on Shakespeare*: Lettres philosophiques.

15. *Visiting German duke, English antiforeigner*: Extracted in Salter, *Elizabeth I and Her Reign*, chapter 5, document 1a.

16. *Harriot reading Acosta in Spanish*: Letter from Harriot to Robert Cecil, quoted in Rosen, "Harriot's Science," 12.

17. *Acosta on "barbarous blindness" (as his English translator put it)*: Natural and Moral History of the East and West Indies, 339.

18. *English writers' use of "savage"*: See also Kupperman, *Settling with the Indians*, 111–112. For a different perspective, see Parsons, "Wildness without Wilderness."

19. *Racism*: Actually, "the early modern period could be fluid in its thinking about matters of race and gender in ways that later ages were not" (Scott-Warren, "What can we learn from early modern drama?" 557). This is also supported by many of the earliest first contact reports, such as those by Barlowe and Harriot. See also Kupperman, *Settling with the Indians*, viii: many Englishmen saw the "Indians" as exemplars of a better society than that of an England-in-flux. On the other hand, many of Shakespeare's plays, including *Othello, Tempest*, and *Merchant of Venice*, contain racist stereotypes that present or interrogate racist views that evidently existed; see also Scott-Warren, esp. 557. For the role of Christianity in racism, see Quinn, *Set Fair for Roanoke*, 205ff.

20. *Saving souls*: Spreading Christianity was the primary justification in the bill, passed by the House of Commons, confirming Ralegh's "patent" to colonize America; reprinted in Quinn, *Roanoke Voyages* 1:126–129.

21. *Columbus, "love and friendship"*: Jackson, *Columbus*, locations 354, 378.

22. *Mayan tribute system*: Chamberlain, *The Pre-Conquest Tribute and Service System of the Maya*.

23. *On the* encomienda *system in the New World*: For a detailed account of the complexities and sophistries in balancing Christian ideals with greed/survival, see Simpson, *The Encomienda in New Spain*; see also Pierce, "The Mission," 243–249; Batchelder and Sanchez, *The Encomienda and the Optimizing Imperialist*.

24. *Spanish debates, Aristotle on slavery*: Pierce, "The Mission," 243–244; Quinn, *Set Fair for Roanoke*," 208.

25. *Roman colonies, Machiavelli, More*: Quinn, "Rennaissance Influences in English Colonization," 73–93. In the early 1580s, some pamphleteers were using versions of More's idea of "waste" to justify the idea of establishing of English colonies. For instance, in 1583, Peckham (*Western Planting*, 97ff.) used a biblical justification to bolster the right of colonists to unused land. By Ralegh's time, English debate about America focused mostly on trade and the practicalities of colonization in the face of increasing Anglo-Spanish tension, although it seems that Ralegh implicitly accepted the idea that Indians would willingly cede "unused" territory to colonists. Quinn also points out humanist influences in the desire of some theorists/promoters to provide settlers with a better life and an upward social mobility unattainable in England. For example, this idea, combined with More's concept of waste, was used by the elder Richard Hakluyt (cousin of Reverend Richard Hakluyt, the geographer): not only were "unused" land, timber, and other New World resources considered to be "wasted"; so, too, overpopulation at home produced unemployed, impoverished "waste people," who might find "great relief" as laborers or merchants in the New World; see Sweet, *American Georgics*, 14–22.

26. *Las Casas*: "Of the Island of Hispaniola" (1542), English extract in *Documents in United States History*, CD-ROM (Upper Saddle River, NJ: Prentice Hall, 2004), document 1-5. Note, though, that in order to save the indigenous people from annihilation, Las Casas suggested importing black slaves; hence the market for slaves (cf. Hawkins's later attempt at slave-trading).

27. *Hakluyt on Las Casas*: Coote, *A Play of Passion*, 91. Hakluyt's document for Ralegh was *Discourse of Western Planting*; the quotation is from paragraph 20, point 18.

28. *Report for Ralegh on a colonial code of laws*: Reprinted in Quinn, *Roanoke Voyages* 1: 138–139. The author of the report is unknown, although Quinn suggests Thomas Digges (then working on military problems) as one possible candidate (21).

29. *Mendoza to Philip II on Ralegh's plans*: Reprinted in Quinn, *Roanoke Voyages* 2:728–729.

30. *Hakluyt to Walshingham*: Reprinted in Quinn, *Roanoke Voyages* 1:155.

31. *Dependent on Spain; no legal trade in the Indies*: See the elder Richard Hakluyt (a lawyer, and cousin of the Rev. Hakluyt) on the urgency of finding a trade route to the East, excerpted in Salter, *Elizabeth I and Her Reign*, chapter 3, document 1c; and Mendoza to Philip II, excerpted in Quinn, *Roanoke Voyages* 2:759.

32. *Ralegh and privateering*: E.g., Grenville's return trip, 1585; cf. Spanish report from Mendoza in Quinn, *Roanoke Voyages* 2: 743–744, 1:169ff.

33. *Walsingham investing in Ralegh's Virginia*: Coote, *A Play of Passion*, 93.

34. *Grenville leading the expedition instead of Ralegh*: Rowse, *Ralegh and the Throckmortons*, 143.

35. *The Stile/Tucker case (Amadas's temper)*: Extracts from the case are reprinted in Quinn, *Roanoke Voyages* 1:139–144. The verdict in the case is not known.

36. *Harriot teaching new captains and masters*: Quinn, "Thomas Harriot and the Problem of America," 15. *Barlowe*: From Holinshed's *Chronicles*, reprinted in Quinn, *Roanoke Voyages* 1, esp. 175; also see 25.

37. *Personnel on voyage*: A Plymouth mayoral report suggested six hundred altogether; the number of crew and soldiers is conjectured by Quinn, *Roanoke Voyages* 1:173n5.

7: ROANOKE ISLAND

1. *Ships' tonnage*: For cargo ships, this refers to carrying capacity—and this can run to hundreds of thousands of tons; for passenger ships, it refers to the "weight" of the volume of the enclosed space in the ship. Note that Henry VIII's famous *Mary Rose* was 600–800 tons; it was still in Elizabeth's fleet in 1590 (Quinn, *Roanoke Voyages* 2:620n7).

2. *Harriot on size of Tiger*: Brioist, "Thomas Harriot and the Mariner's Culture," 186–187. But note that Harriot didn't employ decimal point notation (149.5 tons), as it was not then generally in use; this is discussed further in chapter 12. Quinn says that the *Tiger* was "officially" rated as either 160 tons or 200 tons—presumably depending on the definition

of tonnage used (*Roanoke Voyages* 1:178n6); the keeper of the *Tiger's* journal estimated it at 140 tons. *Harriot's more precise formula*: Shirley, *Biography*, 101–102.

3. *More crowded than* Mayflower: Brioist, "Thomas Harriot and the Mariner's Culture," 187.

4. *Harriot as navigational consultant on the voyage*: Quinn, *Roanoke Voyages* 1: 37.

5. *Harriot's notes on ships and sailors' culture*: BL Add MS 6788 ff1–48. These notes were made two decades later, revealing Harriot's remarkable memory and also his subsequent work on the topics. Important analyses of these manuscripts are given in Brioist, "Thomas Harriot and the Mariner's Culture," and Stedall, "Notes Made…on Ships and Ship Building," 325–327. Note that Quinn conjectures that before the voyage Harriot might have learned from seamen on the Thames ("Thomas Harriot and the Problem of America," 12).

6. *Seamen versus navigational theorists*: Bourne's preface to *Regiment for the Sea*. See also Bennett, "Instruments, Mathematics, Natural Knowledge," 144; Pumfrey, "Patronizing, Publishing and Perishing," 152, and see also 151; Brioist, "Thomas Harriot and the Mariner's Culture," 185. For alternative/general accounts, see Marrioli, "A Fruitful Exchange/Conflict"; Bennett, "The Mechanics' Philosophy and the Mechanical Philosophy"; Johnston, "Mathematical Practitioners and Instruments in Elizabethan England." *Wright as captain with Drake*: Pumfrey, "Patronizing," 156.

7. *Harriot on shipboard hierarchy*: BL Add MS 6788 ff32, 21; Brioist, "Thomas Harriot and the Mariner's Culture," 188. These notes were written in 1608 but were evidently informed by his own experiences in 1585, and perhaps on his later voyage to Ireland. The shares listed were for military expeditions, but presumably the hierarchy was similar to that in the 1585 expedition, in which Lane was a military officer in command of several hundred soldiers and, ultimately, of the whole colony.

8. *Examples of nautical terms recorded by Harriot before their appearance in print* are given in Stedall, "Notes Made…on Ships and Ship Building," 325. *Harriot's record of mariners' language*: I have drawn on the Harriot quotations, and commentary by the authors, in that Stedall article and Brioist, "Thomas Harriot and the Mariner's Culture."

9. *What happened to the other ships?* For more details, see Quinn's summary in *Roanoke Voyages* 1:164–166; for men dumped on Croatoan, Quinn, *Set Fair for Roanoke*, 64.

10. *Spanish prisoner's report on Grenville's company*: English transcript of what appears to be notes taken during an interrogation of Hernando de Altamirano, reprinted in Quinn, *Roanoke Voyages* 2: 740–743.

11. *The English-Spanish encounter at Puerto Rico, governor's report*: Letter from Governor Diego Hernández de Quinones to Philip II, June 12/22, 1585 (English translation acknowledged and reprinted in Quinn, *Roanoke Voyages* 2:733–738).

12. *Wococon* is near the island of present-day Ocracoke (Quinn, "Thomas Harriot and the New World," 41). The *Tiger* journal is reprinted in Hakluyt, *Principal Navigations* 8: 310–317, and annotated in Quinn, *Roanoke Voyages* 1:178–193.

13. *The struggle to save the* Tiger: This description is from a letter from Lane to Walsingham, in Quinn, *Roanoke Voyages* 1: 201.

14. *Harriot on damage to seed grain*: *Brief and True Report*, reprinted in Hakluyt, *Principal Navigations*, 363, and Quinn, *Roanoke Voyages* 1:344.

15. *Lane against Grenville and his supporters*: Letters from Lane to Walshingham, August 12 and September 8, 1585, reprinted in Quinn, *Roanoke Voyages* 1:199–204, 210–214.

16. *Vengeance re the stolen silver cup*: *Tiger* journal, in Hakluyt, *Principal Voyages* 8:316; Quinn, *Roanoke Voyages* 1:191. The journal entry does not name Amadas here—this is Quinn's annotation—and it is not clear whether "the Admiral" of the venture (Amadas) specifically gave such a reprisal order: the journal just says "we" burned the corn and town. Given his authority and his earlier behavior, it seems likely Amadas was a ringleader, at least, which Quinn also assumes (*Set Fair for Roanoke*, 72).

17. *Making contact with Wingina*: *Tiger* journal, footnote 3, Quinn, *Roanoke Voyages* 1:189.

18. *Wingina's invitation to settle*: This is a probable conjecture by Quinn, *Roanoke Voyages* 1:192n2. Oberg suggests that Granganimeo's interest in trading led to the invitation ("Gods and Men," 378).

19. *"Fort"*: Quinn, *Set Fair for Roanoke*, 78–82 and, on archaeological evidence, chapter 20.

20. *Lane on "lung diseases"*: See next note.

21. *Lane's patriotic letter to Walsingham*: August 12, 1585, reprinted in Quinn, *Roanoke Voyages* 1:202–204.

22. *Lane's letters; "only with savages"*: Letters to Sidney, August 12, 1585, and to Walsingham, September 12, 1585, reprinted in Quinn, *Roanoke Voyages* 1:204–205, 213, respectively. Lane did acknowledge that the "savages" were "courteous, and very desirous to have [woolen] clothes." But his recipient noted the mercantile intent, writing in the margin of the letter "commodities fit to carry to Virginia." Barlowe had already noted that the Algonquians were interested in trade: letter from Lane to Hakluyt the Elder (cousin of the geographer), September 3, signed "from the fort" (209; Hakluyt's marginal inscription is in footnote 4). Lane's later report did show some respect for Manteo and one or two others who were friendly to the English, notably Menatonon, who was, "for a savage, a very grave and wise man." (259). But Lane was too insensitive to diplomatic relations, with disastrous consequences, as will become clear below.

23. *Harriot's Roanoke friends, especially Wingina*: Harriot, *Brief and True Report*, in Quinn, *Roanoke Voyages* 1:378.)

24. *Harriot's descriptions of Algonquian life and culture*: All Harriot quotations pertaining to America, in this chapter and beyond, are from his *Brief and True Report* (1588)—reprinted in Hakluyt, *Principal Voyages* 8:348–386; also reprinted and annotated in Quinn, *Roanoke Voyages* 1: 317–387—or from his captions to engravings of White's drawings in de Bry, *America*.

25. *Indigenous names*: Harriot's was the most detailed report, but Acosta, for example, would also give a brief and fascinating account of native food plants in the Indies, using some of the local names, while Cartier, too, had given a (very) few names. Harriot's list is much more extensive than these, and it appears he *thought* in terms of the Algonquian names.

26. *Harriot on the winter expedition*: There is no direct evidence of personnel on this trip, but Harriot says that "often in the time of winter, our lodging was in the open air upon the ground" (*Brief and True Report*, in Quinn, *Roanoke Voyages* 1:384). Further indication of Harriot's presence (and White's, too) is the surveyed maps produced during the expedition. See Quinn, "Thomas Harriot and the problem of America," 21.

27. *Lavish Elizabethan "informal supper"*: Batho, "Thomas Harriot and the Northumberland Household," 28.

28. *Galen still current*: In 1595 and 1597, the mathematician-physician Thomas Hood failed to gain his medical license because he failed his exams on Galen! For details, see Johnston, "Mathematical Practitioners and Instruments in Elizabethan England," 333.

29. *Circulation of blood*: Harvey may have got his inspiration from Harriot's friend Warner; Jacquot, "Harriot, Hill, Warner and the New Philosophy," 117.

30. *Pipe smoking fashionable*: Quinn notes that the first English book specifically on tobacco would not be printed until 1595 and did not mention Ralegh or Virginia (*Roanoke Voyages* 1:345–346n3). It was the pipes, not tobacco, that Ralegh's men introduced to England. In his *Brief Report* (Hakluyt, *Principal Navigations* 8:364; Quinn 1:345–346), Harriot was referring to the use of Algonquian pipes ("sucking"). *Medicinal*: Beer quotes a contemporary account of Ralegh curing a fever and headache with tobacco ("Thomas Harriot and Sir Walter Ralegh's Wife," 163–164).

31. *Less hospitable villages*: Harriot wrote, "There was no town where we had any subtle device practiced against us, we leaving it unpunished or not revenged (because we sought by all means possible to win them by gentleness), but that within a few days after our departure from every such town, the people began to die very fast." *Brief and True Report*, in Hakluyt, *Principal Navigations* 8:380.

32. *Infectious diseases brought by the English*: Lane's earlier letter noted that his men's "lung diseases especially of rheums" were cleared up in the wholesome air; Quinn says, "Rheums were colds, catarrh, bronchitis, pulmonary tuberculosis...which would often be helped by dry weather" (*Roanoke Voyages* 1:202n6). But he suggests that the mystery illness was likely

to be measles or the common cold (378n2); later, he concludes it was colds or flu, because of the apparently short incubation time (*Set Fair for Roanoke*, 228).

33. *Digges telescope*: Whitaker, "The Digges-Bourne Telescope Revisited," 64–65; Johnson, "The Influence of Thomas Digges on the Progress of Modern Astronomy in Sixteenth-Century England."

34. *Distorting mirrors (cf. kaleidoscope)*: Cohen, "Tudor Technology in Transition," 98. In 1990, Quinn also advocated for the idea that Harriot's "perspective glass" was such an arrangement of mirrors ("Thomas Harriot and the Problem of America," 19). *Reasons for land-based telescope interpretation*: a) Digges had described his father's telescope, and had been associated with Dee and therefore Gilbert and Ralegh; b) the rest of Harriot's list of technical wonders is entirely practical. On the other hand, the English did bring toys such as dolls for gifts, and Harriot's wording "strange sights" could suggest distorting mirrors.

35. *Acosta's idol of sand*: *Natural and Moral History of the East and West Indies*, 340–341.

36. *Harriot not judgmental*: The closest he comes to judgment is in the last sentences in two captions to White's illustrations (nos. 45 and 46) in the de Bry edition of his report; cf. Quinn, *Roanoke Voyages* 1:433, 435. But it is quite possible that Reverend Hakluyt introduced this judgmental orthodoxy when he translated Harriot's Latin captions into English, because it is completely at odds with the rest of Harriot's report and captions. See also the next note.

37. *Hakluyt on religion (colonization advice); Harriot's agenda*: Hakluyt quoted in Coote, *A Play of Passion*, 90–91. *Harriot*: Note that Quinn suggests that in contrast to Hakluyt's mission, and that stated in official documents authorizing Ralegh's colony, Harriot subverted the orthodox procedure by suggesting the way to win over the Algonquians was to start with education and friendship, not proselytizing (*Roanoke Voyages* 1:372n1).

38. *Harriot's religion*: He generally kept his religious beliefs to himself and chose his friends for who they were as people, regardless of their religion. For a sober scholarly assessment, see Mandelbrote, "The Religion of Thomas Harriot," 246–279. *Kissing the Bible to gain "power"*: Oberg, "Gods and Men," 381.

39. *Postcolonial debates about Harriot, science, and colonization*: Some modern scholars have accused Harriot of deliberately "appropriating" the Algonquian language in a way that ultimately led to the destruction of the people he was "studying and objectifying." The most famous accusation is Stephen Greenblatt's "Invisible Bullets," his 1981 "New Historicist" critique of Shakespeare, in which he also accused Harriot of willfully deceiving and exploiting the Algonquians. By the 1990s a number of scholars had provided trenchant criticisms of Greenblatt's thesis and sought to position Harriot in his own times rather than in ours. After all, today *we* know that indigenous knowledge was later appropriated and denigrated, but Harriot did not know that: in his *Brief and True Report*, he did imply the superiority of European technology, but he was also respectful of indigenous knowledge, and he spoke of colonists living *with* the original inhabitants, not wiping them out. See, e.g., McAlindon, "Testing the New Historicism"; Sokol, "Invisible Evidence" and "The Problem of Assessing Thomas Harriot's 'A Briefe and True Report' of His Discoveries in America"; Kupperman, *Settling with the Indians*.

The imperialist, masculine "penetration of nature" analogy became a popular trope among postcolonialist and feminist analysts of science itself in the 1980s and '90s (e.g. Alexander, "The Imperialist Space of Elizabethan Mathematics"); it certainly provided food for thought, but I do not think it stands up to scrutiny from the point of view of a working mathematician or scientist (cf. later critical analysis by North, "Stars and Atoms," 194; Mandelbrote, "The Religion of Thomas Harriot," 259n50.) It's true that Francis Bacon applied imperialist and penetration imagery to scientific discovery, but that does not prove that Harriot and his scientific peers saw (even subconsciously) every topic they touched in terms of an imperialist carving up of the globe or a masculine sexual penetration of nature. Besides, it is pertinent that Harriot's most significant mathematical and scientific works were carried out precisely at the time he became free of his patron's imperialist/commercial demands.

Other recent analyses of Harriot's work or context tend to acknowledge *both* the imperialist/commercial driver of much scientific research *and* more "scientific" motivations. For instance, Pumfrey ("Harriot's Maps of the Moon," 163–168) extends another early analysis by Alexander ("Lunar Maps and Coastal Outlines"), adding an alternative context and likely motivation: the simple desire to understand and know nature. (I will discuss Harriot's moon maps in a later chapter.) For analyses from other perspectives, see, e.g., Young, "Narrating Colonial Violence and Representing New World Difference" (an analysis of the objectifying effect of the glossy de Bry edition in contrast to Harriot's original *Report*); Booth, "Thomas Harriot's Translations" (for a possible influence of Algonquian language on Harriot's thinking); Oberg, "Gods and Men" (for a perspective on the Algonquians' reaction to the English); and Sweet, *American Georgics*, 19, who considers Harriot to have been "the most environmentally sensitive of all early promoters of American colonization" (see also Kupperman, 91, for an example of this).

This is just a sample of the scholarly papers on this issue. The tragedy of colonization weighs on us today, but when Harriot prepared for Virginia, he did not foresee what was about to unfold. Perhaps he should have done. It is a difficult balance for modern readers, and a problematic story to tell from a non-indigenous perspective.

40. *Harriot on sharing knowledge: Brief and True Report*, in Quinn, *Roanoke Voyages* 1:372. For discussion see Quinn, *Set Fair for Roanoke*, 222–223. *Chilling in hindsight*: Although Harriot wrote his report as propaganda for Ralegh, he could not hide his own very different interests. As Quinn says, (168), often "we can see the intellectual and linguist dominating the publicist"—and I would add that we also often see the human being. Quinn speaks of Harriot's honesty, and the unique "freshness and critical authority" of his report (173).

41. *Wingina versus the English*: In the following account, the details of the breakdown in relations are from Lane's report, reprinted in Quinn, *Roanoke Voyages* 1:255–294. I have also drawn on Quinn's editorial notes and on Oberg's analysis in "Gods and Men," 382–390.

42. *Abandoning Roanoke for hunting?* Quinn, *Set Fair for Roanoke*, 120.

43. *"Knock my brains out"; "better sort"*: From Lane's report, reprinted in Quinn, *Roanoke Voyages*, 1:282. My details of Lane's actions, including subsequent quotations from him, are from this report (see note 41).

44. *Drake's voyage, and planned Spanish attack on First Colony; Drake's relations with the indigenous people*: Quinn, *Set Fair for Roanoke*, 131–134, 210.

45. *Hakluyt on the "hand of God"*: *Principall Navigations* (1589), reprinted in Quinn, *Roanoke Voyages* 1:478.

46. *Three men left behind*: From Drake's report of his voyage, reprinted in Quinn, *Roanoke Voyages* 1:307. See also Quinn, *Set Fair for Roanoke*, 141.

47. *Loss of notes, maps, drawings, and pearls*: Harriot's report is the source for the pearls ("and many things else") (Quinn, *Roanoke Voyages* 1:334), and Lane's for the books, maps, etc. (293).

48. *Ralegh's intentions?* Quinn emphasizes this point (*Set Fair for Roanoke*, 143); but see 414 for his definite later intention not to displace the "Native American polity."

49. *Jesuits, Chesapeakes*: Quinn, *Set Fair for Roanoke*, 208–209; Karliana Sakas, "The Indigenous Authorship [...] Jesuit Mission of Ajacán," *Journal of Iberian Studies*, 19, 2011, 511–524.

50. *Grenville's privateering* was sometimes challenged in court; transcripts of court cases are in Quinn, *Roanoke Voyages* 1:480–488. *Grenville's cruelty*: Quinn, *Set Fair for Roanoke*, 146; the Spaniards' complaints—in a request to be ransomed via a prisoner exchange—are reprinted in Quinn, *Roanoke Voyages* 2:770–771. *His 1585 raids repaying investors*: Quinn, *Set Fair for Roanoke*, 85–86.

8: AFTER ROANOKE

1. *Harriot and Wright*: In Harriot's 1596 letter to Robert Cecil about Guiana (reprinted in full in Stevens, *Thomas Hariot*, and British History Online, http://www.british-history.ac.uk/cal-cecil-papers/vol6/pp255-272), he mentions Acosta's book, "which you had from Wright, and which I have seen." This sounds as though Wright was no stranger to Harriot.

Roche also suggests that Harriot may have influenced Wright, e.g., through the Durham House circle ("Harriot's 'Regiment of the Sun,'" 258).

2. *Mercator, Dee*: Sailing due north or south, or due east or west, means sailing along a well-understood great circle; sailing in other directions gives rise to the bizarre properties of rhumb lines. Mercator's was the first map of the world to use such a projection, although an earlier "map" appeared on the case of a sundial made in 1511 by Erhard Etzlaub, according to Pepper, "Harriot's Earlier Work on Mathematical Navigation." 84. *On Nunes and rhumb lines*: Leitao and Gaspar, "Globes, Rhumb Tables, and the Pre-History of the Mercator Projection." *Dee on Nunes*: Letter to Mercator, quoted in Almeida, "On the Origins of Dee's Mathematical Programme," 460. *Portuguese model globes*: Randles, "Pedro Nunes' Discovery of the Loxodromic Curve," 86, 92–93.

3. *Mercator savvy, or was his method "obvious"?* It is true that trade secrets were often guarded: Ralegh guarded Harriot's work, and Galileo and Thomas Bedwell are just two further examples (discussed in Johnston, "Mathematical Practitioners and Instruments," 328). But Gaspar and Leitao ("Squaring the Circle") suggest Mercator used a table of rhumbs, such as Dee and Nunes had already constructed (although the authors rule out these latter tables as Mercator's source). If so, this was certainly not the complete theoretical solution later offered by Harriot, or even the accurate tables constructed by Harriot and by Wright.

4. *Nunes on loxodrome at the poles; not carrying out calculations; promising further work*: Leitao and Gaspar, "Globes, Rhumb Tables, and the Pre-History of the Mercator Projection," 188–190. A Latin edition of Nunes's collected works was published in 1566. *Nunes's critics*: Da Sà, quoted in Randles, "Pedro Nunes' Discovery of the Loxodromic Curve," 91–92. (Da Sà was half right in his attack on Nunes's attempt to turn rhumb lines into great-circle arcs, of which more later.) *Nunes abandoning mathematics*: Unpublished manuscript, quoted in Randles, 92.

5. *Harriot on Pedro Nunes*: BL Add MS 6788 f485r.

6. *Provenance of the name "loxodrome"*: Harriot's "helical line" is in his *Doctrine of Nauticall Triangles Compendious*, Petworth House/Leconfield HMC 241 VIb f2. Jon Pepper is the leading expert on this manuscript; see Pepper, "Harriot's Earlier Work on Mathematical Navigation," 74–83. *On Snell*: Randles, "Pedro Nunes' Discovery of the Loxodromic Curve," 88.

7. *Mercator map's 400th anniversary* was celebrated at NASA in 1969, as a new navigational milestone was reached: the moon landing (Resnikoff and Wells, *Mathematics in Civilization* 155–156, figure 6.2).

8. *Figure 13:* This derivation is designed to fit with the diagrams in figures 2 and 14. It is also the same as that given by Rickey and Tuchinsky ("An Application of Geography to Mathematics," 163), who point out that Wright used a slightly different diagram that focused on the 2-D face of the wedge shown here.

9. *Figure 14b* is adapted and expanded from Pepper's figure 4.11 in "Harriot's Earlier Work on Mathematical Navigation," 76, and Smart, *Textbook on Spherical Astronomy*, 326–327.

10. *Harriot and Wright's tables of "meridional parts"*: Wright used the term "parts of the meridian" for the modern "meridional parts"; see extract given in Rickey and Tuchinsky, "An Application of Geography to Mathematics," 164. Pepper believes that Harriot had produced his tables using additions of secants in the 1580s ("Some Clarifications of Harriot's Solutions of Mercator's Problem," 235), but this is conjecture. What survives is Harriot's formula, and a reference saying he had the corresponding tables in another manuscript, which is now lost. Wright's tables apparently dated from 1589 (Pumfrey, "Patronizing, Publishing and Perishing," 158). Almeida notes that the similarity between methods suggests Wright may have benefited from discussions with Harriot or Dee ("On the Origins of Dee's Mathematical Programme," 466n30). In another context, however, Wright said he learned Mercator's secret from no one (quoted in Pumfrey, 159).

11. *Early trigonometric tables*: Ptolemy, *Almagest* 1.10. *Ptolemy and trig. identities*: Boyer, *A History of Mathematics*, 165–166, 238. *Viète's tables*: Boyer, 308. *Clavius*: Sadler, "Nautical Triangles Compendious," 142. *Viète's multiple-angle formula*: Boyer, 310. *Abu'l-Wefa*: Boyer, 238.

12. *Al-Khwarizmi*: Boyer, *A History of Mathematics*, 228, 231–233. "Algorithm" initially referred simply to the Hindu numerals that al-Khwarizmi popularized. *Geometric trigonometry of the Greeks, Indians, and Arabs*: Boyer, 166. *Sines from India*: Boyer, 209, 215, 252. *Arabic intuition of algebraic geometry*: Boyer, 241–242, e.g., Omar Khayyam; however, they lacked the symbolism and generality that enabled Harriot and Descartes to begin to formulate analytic geometry.

13. *Harriot's analytic and algebraic geometry*: His manuscripts, including the one shown, are listed in the "Analytic Geometry" box on the ECHO website. Stedall ("Symbolism, Combinations, and Visual Imagery," 386–388) discusses a beautiful example in which Harriot rewrote entirely algebraically a proof done geometrically by Viète. *Descartes's analytic geometry*: Descartes did not actually systematically employ the coordinate grid that was later named "Cartesian" in his honor, and he emphasized the geometrical aspect far more than we do today, where we tend to think purely symbolically as Harriot did (cf. Boyer, *A History of Mathematics*, 336–346). *Predecessors* who glimpsed the connection include Apollonius and Oreseme. *Harriot's algebraic trigonometry*: Stedall ("Notes Made by Thomas Harriot on the Treatises of François Viète," 190–191) points out that Harriot's usage of his trigonometric symbols suggests he was intuitively using the idea of trigonometric *functions*, because he delineated the angular argument using commas equivalent to modern parentheses. Stedall says to her knowledge no one did this for another century.

14. *Harriot's early trigonometric formulae*: He listed these proportions, or ratios, in a chart covering all the trigonometric relationships in two different right-angled triangles: one giving the changing distance and latitude along a right-angled triangle whose hypotenuse lies on the rhumb line of compass bearing angle a (as in triangle aAB in the left-hand diagram in figure 14b), and the other between distance and longitude as represented on the equivalent triangle on a Mercator map (as in triangle a′AB′ on the right-hand side in figure 14b). Harriot's table of proportions is at Petworth House/Leconfield HMC 241 VIb ff5–6. See also Pepper, "Harriot's Earlier Work on Mathematical Navigation," 78–79.

15. *Harriot's interpolation instructions*: His rule was given in words, reproduced in Pepper, "Harriot's Earlier Work on Mathematical Navigation," 80. His verbal rule is not very clear (and appears to be incorrect in the last term, according to Pepper's interpretation of it), but certainly Harriot later found correct, algebraic rules in his *Doctrine of Triangular Numbers*, and he seems to have been the first to have begun the process of generalization of exact algebraic interpolation (Stillwell, *Mathematics and Its History*, 123; Edwards, *Pascal's Arithmetical Triangle*, 11–14). *Mesopotamians' linear interpolation*: Boyer, *A History of Mathematics*, 30.

16. *Bürgi*: Waldvogel, "Jost Bürgi's Artificium of 1586 in Modern View." *Brahmagupta*: Edwards, *Pascal's Arithmetical Triangle*, 16; Krishnachandran, "On Finite Differences, Interpolation Methods and Power Series Expansions in Indian Mathematics."

17. *Dee's Mercator tables (his* Canon gubernauticus)*: Leitao and Gaspar, "Globes, Rhumb Tables, and the Pre-History of the Mercator Projection," 181; Pepper, "Harriot's Earlier Work on Mathematical Navigation," 84–85; Almeida, "On the Origins of Dee's Mathematical Programme," 466.

18. *Harriot and Dee friendly*: In 1590, Dee inscribed a book given to him by Harriot as being from "my friend Thomas Harriot" (Quinn and Shirley, "A Contemporary List of Harriot References," 15–16). In 1594, Harriot used some of Dee's astronomical observations (cited in Pepper, "Harriot's Earlier Work on Mathematical Navigation," 75).

19. *Publishing Wright's tables*: More details of this saga—including quotes here and in the preceding and the following paragraph—are in Pumfrey, "Patronizing, Publishing and Perishing," 157–162.

20. *Ralegh: son of farmer, hard-working MP; Devonshire accent*: Coote, *A Play of Passion*, 95–97, 57 (from Aubrey's *Brief Lives*).

21. *Vacant land at Chesapeake Bay*: Quinn, "Thomas Harriot and the Problem of America," 21. *Colony not intending to displace the original inhabitants*: Quinn, *Set Fair for Roanoke*, 414. Note that Menendez de Avilés—he of the 1565 St. Augustine massacre—had reconnoitered

Chesapeake Bay (called the Bay of Madre de Dios by the Spanish) in 1572. Later, Menendez warned Philip II that Spain should colonize North America before the Protestant English or French, because "they and the Indians are nearly of one faith," so they would "very easily make friends," after which it would be impossible to wrest the territory from them (as he had done when he destroyed the Huguenot settlement St. Augustine); Menendez's letter is excerpted in Quinn, *Roanoke Voyages* 2:772.

22. *Disappointed merchants*: See, e.g., Thomas Harvey, transcript in Quinn, *Roanoke Voyages* 1:232–234 and footnote 2.

23. *Interrogation*: Nicholas Burgoigon told Hakluyt and Harriot that the Spanish were searching for Virginia (reprinted in Quinn, *Roanoke Voyages* 2:763–766).

24. *Harriot and colonialism*: In addition to his *Report* supporting Ralegh's venture, he also wrote some very brief notes on the kinds of corporations involved in colonization (BL Add MS 6789 f523, reprinted in Quinn, *Roanoke Voyages* 1:389).

25. *Towaye*: Vaughan, "Sir Walter Ralegh's Indian Interpreters," 352. *Manteo; list of colonists*: Quinn, *Roanoke Voyages* 2:502, 539–42, respectively.

26. *Harriot-White maps used till 1650*: Quinn, *Roanoke Voyages* 1:58. *Satellite images*: Quinn, "Thomas Harriot and the Problem of America," 21.

27. *Details of the 1587 colony* are from White's report, published in Hakluyt's *Principal Navigations* 8:386–403, reprinted in Quinn, *Roanoke Voyages* 2:515–538. For a detailed analysis of what went wrong, see Quinn, *Set Fair for Roanoke*, 279ff.

9: WAR, AND A NEW CALENDAR

1. *All ships to stay at port*: Quinn, *Roanoke Voyages* 2:554–555.

2. *Bells and bonfires*: Arthur Throckmorton's diary, quoted in Rowse, *Ralegh and the Throckmortons*, 111.

3. *Elizabeth, James, and the death of the Queen of Scots*: Mueller, "The Correspondence of Queen Elizabeth I and King James VI"; G. Batho, "The Execution of Mary, Queen of Scots"; Perry, *Elizabeth I: The Word of a Prince*, 260–277; Salter, *Elizabeth I and Her Reign*, chapter 7, document 2c.

4. *"Give God the vomit"*: Quoted in Perry, *Elizabeth I: The Word of a Prince*, 262; for Cardinal Allen, etc., 279–280.

5. *Ambassadors*: Giovanni Mocenigo and Giovanni Gritti, respectively; Venetian State Papers, August 20, 1588, quoted in Hart, *Battle of the Spanish Armada*, 58.

6. *The Armada, including Ralegh's advice, Philip taking London, Tilbury speech, fire ships*: Hart, *Battle of the Spanish Armada*; Perry, *Elizabeth I: The Word of a Prince*, 278–292; Coote, *A Play of Passion*, 144–147.

7. *Piracy on White's voyage*: Quinn, *Roanoke Voyages* 2:565n1, and White's report, 568. Litigation often ensued in such cases; see 1:480–488.

8. *White's report of his aborted 1588 voyage*: Reprinted in Quinn, *Roanoke Voyages* 2: 562–569; for ships' tonnage, etc., 555.

9. *Lost Colonists' destination: Chesapeake Bay versus Weapemeoc, Croatoan, etc.*: The primary evidence for the Chesapeake Bay/Powhatan massacre theory is William Strachey's reference to it in his *History of Travel into Virginia Britannia*, published in 1623, although probably circulated earlier (Strachey had arrived in Jamestown in 1610) and Samuel Purchas's 1625 *Purchas His Pilgrimage* (Purchas had been advised by John Smith, confidant of Powhatan). Quinn gives a detailed analysis of the Lost Colonists' likely fate and the contemporary sources stating Powhatan's admission to Smith. He reconstructs what he considers the probable story, including Powhatan's motivation for the massacre (*Set Fair for Roanoke*, chapter 19).

Other scholars suggest that the Jamestown reports were garbled interpretations that conflated two different massacres in intertribal warfare, and that the Powhatans had nothing to do with the massacre of the Lost Colonists—which makes sense given the apparently warm welcome Powhatan gave John Smith just a few weeks after the supposed massacre. Parramore gives a good overview of this point of view and makes his case for the

Lost Colonists perishing when their hosts, the Weapemeocs, were attacked by other hostile tribes ("The 'Lost Colony' Found").

Archaeological digs and DNA testing for possible descendants have been under way for some years. See, e.g., Tanya Basu's *National Geographic* report on the First Colony Foundation's project, November 2013, at http://news.nationalgeographic.com/news/ 2013/12/131208-roanoke-lost-colony-discovery-history-raleigh/. For the legal implications of recognizing mixed-race survivors, see Padget, "The Lost Indians of the Lost Colony," who argues that the Lumbee Indians of North Carolina are descended from the Lost Colonists.

10. *White's 1590 voyage*: White's report is reprinted in Quinn, *Roanoke Voyages* 2:598–622; see also Quinn's summary, 579–598.

11. *Harriot's captions*: His authorship is revealed in the caption to "Their sitting at meat," where he says maize tastes good, "as I described it in the former treatise" (Quinn, *Roanoke Voyages* 1:430; cf. 414n5).

12. *De Bry's* America: This is a beautiful book, but de Bry Europeanized the Algonquians' features, in contrast to White's realistic representations. For a postcolonial account of the political implications of this publication, see Young, "Narrating Colonial Violence and Representing New World Difference"; see also Sokol, "Invisible Evidence." For *Harriot's note "4 languages"*: Quinn and Shirley, "A Contemporary List of Harriot References," 11.

13. *Essex, gossip*: Rowse, *Ralegh and the Throckmortons*, 154.

14. *On government business*: Quinn, *Roanoke Voyages* 2:557. *Ralegh's poetry*: At a time when most poetry circulated in manuscript form only, dates and indeed attributions of surviving manuscripts are problematic. But Spenser's 1590 reference to a poem of Ralegh's suggests he began writing about Cynthia around 1589; see Rudick's edition of *The Poems of Sir Walter Ralegh*, xlviii–xlix. Note, though, that the only extant sections of the *Ocean* cycle are book 21 and part of book 22.

15. *Ralegh's great poem* was not yet finished, as we shall see. In our own times, Seamus Heaney recalled Ralegh's disgraceful role in Ireland with a poem entitled *Ocean's Love to Ireland*, in which Ralegh's (somewhat overhyped) contemporary reputation as a womanizer is turned to devastating effect as a metaphor for the ravishment of Ireland. It is telling, though, that Ralegh's one reputed illegitimate child was conceived in Ireland. (I'm indebted to Nicholls ("Last Act?" 181) for reference to Heaney's poem.)

16. *Ralegh jesting*: See, e.g., letter to George Carew, quoted in Beer, *Bess*, 4.

17. *Ralegh in* Faerie Queene: Rowse, *Ralegh and the Throckmortons*, 154–159; Coote, *A Play of Passion*, 158–159.

18. *Essex, gossip, Spenser*: Rowse, *Ralegh and the Throckmortons*, 154. See Hadfield, "Another Look at Serena and Irena," for analysis of some of the critical (and sexist) irony in *The Faerie Queene*. For a different take on Spenser's irony, see Coote, *A Play of Passion*, 158–159.

19. *Ralegh's Irish colonists*: Coote, *A Play of Passion*, 151, 223. *Harriot in Ireland*: Shirley, "Sir Walter Ralegh and Thomas Harriot," 20–22; note that Ralegh would remain active in Ireland until the bloody battle at Kilcolman in 1598, during which Spenser was killed. Note too that Coote suggests Ralegh's enemies had worked to thwart his Munster colony (223–224).

20. *Literal choice*: In a sense, Dee had made the same choice, turning from mathematics to the summoning of angels. *Dr. Faustus* opened "c.1589," according to Ousby, *The Cambridge Guide to Literature in English*, 262.

21. *Irish fishing*: *Brief and True Report*, in Quinn, *Roanoke Voyages* 1:360.

22. *Correct angle of obliquity of the ecliptic in Ptolemy's time and Harriot's*: My figures are, respectively, from Britton ("Ptolemy's Determination of the Obliquity of the Ecliptic," 30) and Roche ("Harriot's 'Regiment of the Sun,'" 251). See this latter reference for Harriot's value being more accurate than Tycho's—but note that Roche points out that, following Ptolemy and subsequent astronomers (including Copernicus and Tycho; cf. Gingerich and Voelkel, "Tycho and Kepler," 80), Harriot then added an incorrect correction for solar parallax. This meant his tables were not as accurate as they should have been, given his fine observation of the obliquity. For today's value, the formula for calculating the obliquity for

any year is given in Smart, *Textbook on Spherical Astronomy,* 238; but note there is also a periodic change in obliquity due to nutation (234, 247).

23. *Other ancient knowledge of precession?* E.g., Grofe, "Measuring Deep Time," suggests the Mayans might have known of it. Such claims are difficult to assess; see also Neugebauer, "The Alleged Babylonian Discovery of the Precession of the Equinoxes," 247–249.

24. *Hipparchus knew the rate of precession:* The current estimate is fifty seconds of arc a year; Hipparchus had calculated the rate of precession as thirty-six seconds per year. See Smart, *Textbook on Spherical Astronomy,* 226.

25. *Hipparchus and precession:* Described in *Almagest* 7.3.

26. *Precession, nutation, and changing obliquity:* There is a little more to it than I have described, especially when it comes to the change in the angle of the tilt of the earth's axis. For a detailed but reasonably accessible technical account, see Smart, *Textbook on Spherical Astronomy,* 226–238. For values of the periodically changing obliquity, see Guy Worthey, "Astronomy: Precession of Earth," Washington State University, http://astro.wsu.edu/worthey/astro/html/lec-precession.html.

27. *Harriot knew rate of precession:* See Pepper's analysis of his manuscript in "Harriot's Earlier Work on Mathematical Navigation," 69.

28. *Twenty-six thousand years:* The axis takes this long to rotate 360 degrees, but the vernal equinox takes about 21,500 years because of the eclipticial precession as well.

29. *"If the world do last so long," quoting Ptolemy:* Harriot, BL Add MS 6788 f480r. Most of the handwriting in Harriot's manuscripts is very difficult to decipher, and Pepper ("Harriot's Earlier Work on Mathematical Navigation," 69) has given valuable clarifications of the text here; indeed, throughout this chapter, I am indebted to his overview of Harriot's early navigation work (54–83). Here as elsewhere, in trying to present Harriot's work in a context for the lay reader, my narrative sometimes draws from these sparse manuscripts and contextualizes what Harriot would have known (such as the effect of precession when in 1595 he told his students about Polaris's changing position).

30. *Other factors affecting spin axis, including climate:* See, e.g., Adhikari and Ivins, "Climate-Driven Polar Motion." Some scientists (cf. the still-disputed 1938 Milankovitch Theory) have considered the possible feedback relationship between long-term climate change and changing axial spin.

31. *Nile floods and Sirius:* Gauquelin, *The Cosmic Clocks,* 9; Boyer, *A History of Mathematics,* 11.

32. *Acosta on the Mexican and Peruvian calendars, pictures/writing, "foolish and ignorant zeal":* Natural and Moral History of the East and West Indies, 434, 436–437, 444, 445. For more on the Mayan calendar, including illustrations of their "pictures" denoting the months, see Duncan, *The Calendar,* 24–27.

33. *Mayan tropical year:* Grofe discusses Mayan records that may suggest an awareness of the numerical difference between these two years ("Measuirng Deep Time"). Aldana discusses similar corrections to the Mayan ritual Venus-based year ("Discovering Discovery").

34. *Year according to the calendar, and according to the equinoxes:* Smart, *Textbook on Spherical Astronomy,* 144–145, 131–132. Ptolemy (*Almagest* 3.1) discusses the difference between the sidereal and tropical year.

35. *Official date of equinox according to the Church:* Duncan, *The Calendar,* 5. *Actual date according to Harriot:* BL Add MS 6788 f472, ff205–10. See Pepper, "Harriot's Earlier Work on Mathematical Navigation," 65, and Roche, "Harriot's 'Regiment of the Sun,'" 255. Note that Harriot's date was according to the Julian calendar.

36. *Ten "lost" days:* Duncan, *The Calendar.* For Protestant opposition, see 293–296; on Dee's advice to Elizabeth, and Archbishop Grindal's objections, 302–306; Roger Bacon, 1–9; dates of Protestant and other countries adopting Gregorian calendar, 318–319. See 321–333 for further intricacies in calendar making and timekeeping, such as the use of atomic clocks, and the fact that the earth's rotation—the marker of our "day"—is slowing down because of the weakening gravitational pull of the moon. This is because the moon is moving slowly away from the earth. It's been doing this for a billion years—and it takes about seventy-four thousand years for an earth day to increase by one second—but the lunar

orbit should eventually stabilize (Ian Sample, "The Days Are Getting Longer—but Very Slowly," *Guardian,* June 5, 2018).

37. *Further reform:* See, e.g., Phillip Orchard, "The Geopolitics of the Gregorian Calendar," *Sratfor Worldview,* 2014, https://worldview.stratfor.com/article/geopolitics-gregorian-calendar.

38. *Modern calculations on the accuracy of Harriot's solar declination tables:* Roche, "Harriot's Regiment of the Sun,'" 255–257.

39. *"My own experiment":* BL Add MS 6788 ff469, 480. Harriot's 12-foot instrument was presumably a very large (astronomer's rather than navigator's) cross-staff. For history of the instrument, see Roche, "Harriot's Regiment of the Sun,'" 247. For Harriot's work and accuracy, see Pepper, "Harriot's Earlier Work on Mathematical Navigation," 64.

40. *White abandoning colony to God:* Letter to Hakluyt, February 4, 1593, reprinted in Quinn, *Roanoke Voyages* 2:716.

41. *1602 search for Lost Colonists:* Documents quoted in Quinn, "Thomas Hariot and the Virginia Voyages of 1602." Note that Ralegh probably had a commercial rather than an altruistic interest in establishing the continuity of his Lost Colony (276). He also used the expedition to harvest timber for sale in England.

10: NEW CHANCES

1. *Suicide to protect the Percy line:* Thomas Wilson, *The State of England, Anno Dom. 1600,* reprinted in Kinney, *Elizabethan and Jacobean England,* 3–32, esp. 5.

2. *Percy's (i.e., Northumberland's) claim to the throne:* From the 1600 report by Wilson (previous note), 5. *Ralegh as parvenu:* Ralegh actually came from an old and not undistinguished family, although recent generations had not done so well as earlier ones; Rowse, *Ralegh and the Throckmortons,* 129–130.

3. *Northumberland's dissolute youth:* Henry Percy's *Advice to His Son,* 1609 (published, e.g., G. B. Harrison, ed. [London: E. Benn, 1930]).

4. *Northumberland's household accounts re Ralegh:* Alnwick Castle records, quoted by Shirley, *Renaissance Scientist,* 23. For Northumberland's lavish spending, see transcripts in Batho, *The Household Papers of Henry Percy.*

.5. *Northumberland's interest in science:* In 1591, the Latin edition of G. B. della Porta's *De furtivis literarum notis vulgo de Ziferis libri IIII* was dedicated to him by the translator; Batho, "Thomas Harriot and the Northumberland Household," 31.

6. *Harriot buying tobacco:* From Northumberland's accounts; Batho, "Thomas Harriot and the Northumberland Household," 31. *Harriot gambling (and winning):* Shirley, *Biography,* 218; Mandelbrote, "The Religion of Thomas Harriot," 262. *Northumberland's wins and losses at cards and dice:* Batho, *Household Papers of Henry Percy,* 62–63. *Elizabethan currency:* There were twenty shillings (s.) in a pound, and twelve pennies (d.) in a shilling.

7. *Dice:* Silva, "On Mathematical Games," 80. *Date of* I Ching: *I Ching, or Book of Changes,* Richard Wilhelm translation, rendered into English by Cary F. Baynes (London: Routledge & Keagan Paul, 1951, rpt. 1975), xiv.

8. *History of gambling math:* Lists of dice possibilities first appeared in Europe, around the thirteenth century, but no further progress was made until the sixteenth century; Edwards, *Pascal's Arithmetical Triangle,* 5, 36. *Combinations of letters; humors (Europe in 1603), number mysticism:* Rabinovitch, "Rabbi Levi Ben Gershon and the Origins of Mathematical Induction"; Maclean, "Harriot on Combinations," 83–84. *Combinations of tastes and syllables (in India):* Edwards, 27.

9. *Beginning of modern probability theory:* Sixteenth and seventeenth centuries (Cardan and Pascal); Boyer, *A History of Mathematics,* 363.

10. *Harriot on a shilling per toss:* Harriot's scrawled table is at the bottom right of BL Add MS 6782 f185; I owe the basic financial interpretation to the online editors of Harriot's manuscripts at ECHO (European Cultural Heritage Online).

11. *Gift of velvet to Bess; Proclamation on Apparel:* Alnwick household accounts, quoted in Shirley, *Renaissance Scientist,* 23; the Proclamation on Apparel is reprinted in Kinney, *Elizabethan and Jacobean England,* esp. 319—but note (297) that it was not necessarily obeyed!

12. *Nicholas as "wise and expert"*: Earl of Bedford, quoted in Rowse, *Ralegh and the Throckmortons*, 33. *Elizabeth's fury*: Quoted in Beer, *Bess*, 20. *Saving Mary*: Rowse, 102.

13. *Female literacy rate*: Beer, *Bess*, 27. Note that Bess's spelling was idiosyncratically phonetic—thereby preserving the speech of upper-class Elizabethans. To take just a couple of examples, she wrote "hit" instead of "it" and "will" instead of "well."

14. *Bess's inheritance case*: Beer, *Bess*, 238; see also 52, 99–101, 110–111. Recovering the money was another matter (and Rowse, *Ralegh and the Throckmortons*, 199, says she never got it). *Bess's spirit*: For detailed analyses through Bess's letters, see Beer, "Ralegh's History of Her World"; Robertson, "Negotiating Favour."

15. *Alice Goold (or Gould or Gold)*: Ralegh's letter to Judge Goold, in *Letters of Sir Walter Ralegh*, 379. See also Beer, *Bess*, 4, 113.

16. *Elizabethans' sex life*: Using astrologer Simon Forman's diaries of his patients, Rowse paints a colorful picture (*Elizabethan Renaissance*, 145ff.). *Essex*: Beer, *Bess*, 9. NB: A 1588 Ralegh marriage date has been conjectured but late 1591 is more likely: Paul Hammer, *The Polarisation of Elizabethan Politics*, CUP, 1999, 116n23.

17. *Undivided loyalties*: I agree with Susan Doran's view in her 1996 book *Monarch and Matrimony: The Courtships of Elizabeth I*, quoted in Beer, *Bess*, 69.

18. *Bess the chosen one*: Letter from Ralegh to Bess, December 4–8, 1603, in *Letters of Sir Walter Ralegh*, 263–265; in Beer, *Bess*, 269–271, and quoted later in my narrative, in chapter 14.

19. *Essex circle, war, etc.*: Beer, *Bess*, 7; Rowse, *Ralegh and the Throckmortons*, 126–127.

20. *Essex keeping Ralegh's secret?* Beer suggests he did (*Bess*, 7); Rowse thinks it "hardly conceivable" that he kept quiet about his rival and suggests this may be the reason Bess and her brother were estranged for a few months (*Ralegh and the Throckmortons*, 159).

21. *August 7, to the Tower*: Arthur Throckmorton's diary, cited in Coote, *A Play of Passion*, 201. Arthur's fascinating, intimate diary was published as *The Diary of an Elizabethan Gentleman*. *Queen waiting for remorse; Ralegh offended*: Rowse, *Ralegh and the Throckmortons*, 162; Beer, *Bess*, 68.

22. *Bess's letter from the Tower*: Reprinted (with her phonetic spelling) in Rowse, *Ralegh and the Throckmortons*, 164.

23. *Ralegh's prison poems*: I have modernized the spelling in my excerpts from poems 25–27 ("The 'Cynthia' holographs," 26–27, are books 21 and 22 of *Ocean's Love to Cynthia*) in Rudick's edition of *Poems of Sir Walter Ralegh*.

24. *Fate of Ralegh's* Ocean's Love to Cynthia: Ralegh had given his Tower outpourings to Robert Cecil to pass on to the queen, but the only extant copy was found among Cecil's papers in the nineteenth century: Coote, *A Play of Passion*, 207; Beer, *Bess*, 71.

25. *Harriot's calculations of* Madre de Dios *size*: Batho, "Thomas Harriot's Manuscripts," 295.

26. *The* Madre de Dios, *and Ralegh*: Coote, *A Play of Passion*, 208–211; Rowse, Ralegh and the Throckmortons, 166–168; Beer, *Bess*, 73–74.

27. *Ralegh's rapturous welcome* was recorded in Cecil's report; Coote, *A Play of Passion*, 210.

28. *Division of* Madre de Dios *spoils, and Ralegh's complaint* (which shows just how much planning and action, as well as money, he had put into it): Rowse, *Ralegh and the Throckmortons*, 168.

29. *"Toiling terribly"*: Cecil's report, quoted in Coote, *A Play of Passion*, 210.

30. *Ralegh's desperate plans and moping*: For details, see his letters, e.g., Beer, *Bess*, 80–81; Coote, *A Play of Passion*, 212ff., including excerpts of his speeches to Parliament (217–218).

31. *Warner and Harriot*: Jacquot, "Harriot, Hill, Warner and the New Philosophy," 108–109, Shirley, *Biography*, 66. Conflicting estimates of Warner's age appear in the literature, but he died in the early 1640s. *Warner's intellectual standing*: Prins considers him a minor philosopher (*Walter Warner*, viii), but Warner's achievements include tables of logarithms and refraction (following Harriot, from whom he learned the law of refraction), and speculations on natural philosophy and the circulation of the blood: discussed in Jacquot, 117–125, who sees him as a significant transitional figure (109).

32. *Northumberland on original scholarship*: From his 1609 advice to his son, quoted in Gatti, "Giordano Bruno," 64. *Northumberland's gift to Harriot*: Batho, "Thomas Harriot and the Northumberland Household," 32.

33. *Harriot's students:* A grateful student recorded his thanks in a book given to Harriot, which still survives (Batho, "Thomas Harriot and the Northumberland Household," 32).

11: SETBACK

1. *Religious "tolerance":* It is interesting to compare the religious path Elizabeth had been trying to steer with that of the Umayyad caliphs in tenth-century Andalusia. This is seen today as a relatively tolerant era, because the caliphs had allowed those of the nondominant faith to keep their religion. Elizabeth had done this, too. Nevertheless, like Elizabethan Catholics, Spanish Christians and Jews had been forbidden from building houses of worship, or from displaying their faiths in public, and religious rebels were executed; Hinkle, "Medieval Islamic Spain (al-Andalus) as a Civilizational Bridge between Later Antiquity and Early Modernity," 90.

2. *Responsio a bestseller:* Coote, *A Play of Passion,* 185. Most scholars think the author was Robert Parsons (e.g., Strathmann, *Sir Walter Ralegh,* 174); Jacquot says the author of *Responsio* was either Parsons or Thomas Creswell ("Thomas Harriot's Reputation for Impiety," 166n7). Quinn and Shirley say the pamphlet was "said to be" an extract of the *Responsio,* by Parsons ("A Contemporary List of Harriot References"). *Quotations* are given in Kargon, "Thomas Harriot, the Northumberland Circle, and Early Atomism in England," 132; Shirley, "Sir Walter Ralegh and Thomas Harriot," 23. *Parsons and Shakespeare:* John Finnis suggests that Parsons's theory of good government, notably that it was justifiable to rebel against an unjust government, influenced Shakespeare—who was "probably a Catholic"—in, e.g., *Titus Andronicus* ("Shakespeare, Identity and Religion, *Philosophers' Zone,* RN, ABC, May 12, 2012).

3. *"Learned conjuring":* Strathmann, *Sir Walter Ralegh,* 174.

4. *On Harriot and accusations of atheism:* Harriot listed Parsons's and Nashe's works among those he believed contained references to himself. Excerpts in Quinn and Shirley, "A Contemporary List of Harriot References," 19–21; on Dee's claim, see 21n28. *Scholarly papers on Harriot's supposed atheism:* Jacquot, "Thomas Harriot's Reputation for Impiety"; Mandelbrote, "The Religion of Thomas Harriot." Both Jacquot (174) and Mandelbrote (267–270) discuss ways in which Harriot's empathy with Algonquian values in his *Brief and True Report* might have fostered atheism charges.

5. *Harriot on submission to God:* In *Brief and True Report,* when telling Wingina that God would not listen to the requests of mere mortals, and specifically his letters, as we'll see: a letter to his doctor (in Mandelbrote, "The Religion of Thomas Harriot," 247), and a letter from Lower, February 10, 1610 (in Jacquot, "Thomas Harriot's Reputation for Impiety," 169. *Neighborly love:* Harriot preached this in Virginia. He also made notes on St. Augustine's analysis of the Ten Commandments, most of which concern neighborly love (BL Add MS 6789 f463; cf. Jacquot, 174). In his *Brief and True Report* he showed not only empathy for the Algonquians but also what some saw as a scandalous pragmatism in his savvy comment that the Algonquians' belief in heaven and hell led them to be remarkably law abiding. See also Jacquot; Mandelbrote 267ff.; and Maclean, "Harriot on Combinations," 84.

6. *Carew, Ralegh, and Ironside at dinner:* Quotations from Ironside's notes (BL MS Harleian 6849 ff183–190) are given in Coote, *A Play of Passion,* 224–226, and (with additional analysis) Strathmann, *Sir Walter Ralegh,* 139–147.

7. *Ralegh's religious views,* History of the World *and other writings, love of debate:* Strathmann, *Sir Walter Ralegh,* esp. chapters 3 and 4.

8. *Unorthodox debate, skeptical interlocutor:* This evidence only came to light when the notes of the conversation were part of a bundle of manuscripts auctioned by Sotheby's in 1986. The manuscript owner had entitled it, in Latin, "Notes from the discussions of Thomas Harriot…concerning God, the first cause and many other things—so I believe." Stephen Clucas noticed the manuscript on display at the British Library in 1992, and my account is based on his summary and analysis of the discussion in his "Thomas Harriot and the Field of Knowledge in the English Renaissance," 129–135. Mandelbrote has studied the manuscript, too, and cannot definitely link the writer to anyone with ties to Harriot, but

he suggests a date of between 1594 and 1603, in both of which years "Harriot's name was a byword for impiety" ("The Religion of Thomas Harriot," 261).

9. *Harriot reading Trithemius for codes, magic, or his alphabet?* Codes: Seaton, p113; but note that Seaton decoded Harriot's alphabet before his papers regarding Algonquian were found, so she assumed it was for a secret code. Gatti (Fox (ed.)) p74 discusses Trithemius, the teacher of the famous magus Agrippa, and notes there seems to be no reference in Harriot's MSS to Agrippa or other Renaissance magicians; this reinforces the idea that Harriot's interest in Trithemius was *not* primarily magical.

10. *Torporley's notes* are reproduced in Jacquot, "Thomas Harriot's Reputation for Impiety," 183–186. *On Harriot's skepticism:* Almost all of Harriot's manuscripts are devoted to mathematics and physics, but in BL Add MS 6789 f460, he discusses ancient Skepticism and the philosophy of doubt. Gatti notes that this passage suggests (in its brief, unresolved way) an almost Cartesian philosophy of doubt and clear conceptualization ("The Natural Philosophy of Thomas Harriot"). See also Clucas, "'Noble virtue in extremes,'" 135.

11. *Atomists, ex nihilo:* Jacquot, "Thomas Harriot's Reputation for Impiety," 179. Most of Harriot's contemporaries would not have associated *ex nihilo* with atomism, just with rejection of Creationism; cf. Henry, "Thomas Harriot and Atomism," 272. *Not only atomists:* North says atomists were not the only ones to consider *ex nihilo* ("Stars and Atoms," 203).

12. *Harriot and Harvey:* Excerpts from Harriot's notes are in Quinn and Shirley, "A Contemporary List of Harriot References," 19–21.

13. *Harriot's biblical chronology:* BL Add MS 6789 ff463–474.

14. "In the beginning": BL Add MS 6789 f494v, decoded and reproduced in Seaton, "Thomas Harriot's Secret Script," figure 4.

15. *Harriot, Hood, and Forman:* Quinn and Shirley, "A Contemporary List of Harriot References," 22; Kassell, *Medicine and Magic in Elizabethan London*, 43.

16. *Harriot as Durham House's key atheist:* See chapter 14 for Justice Popham's tirade.

17. *Baines against Marlowe:* Harleian MS 6848 ff185–6.

18. *Marlowe and Harriot:* A letter from Thomas Kyd to Sir John Puckering claims that Marlowe had at least one conversation with Harriot and Warner, cited in Jarrett, "Algebra and the Art of War," 29. Jarrett also discusses here Roche's claim ("Harriot, Oxford, and Twentieth-Century Historiography," 236) that the evidence for a connection is slight. But note that Kyd briefly shared lodgings with Marlowe and was also accused, no doubt unjustly, of atheism, and he tried vehemently to dissociate himself from Marlowe and his beliefs. Kyd was possibly tortured, but see Owens, "Thomas Kyd and the Letters to Puckering," on this and its implication for the authenticity of Kyd's letter.

19. *Marlowe's death:* For an insightful discussion, see Hammer, "A Reckoning Reframed."

20. *Cerne Abbas:* Shirley, "Sir Walter Ralegh and Thomas Harriot," 24; Coote, *A Play of Passion*, 228.

21. *Witnesses against Harriot:* Extracts from Harleian MSS listed in Shirley, "Sir Walter Ralegh and Thomas Harriot," 24–25.

22. *Hues on Harriot's tract: Tractatus de globis* (1594), 111.

23. *Ralegh hunting Jesuits; Cornelius:* Coote, *A Play of Passion*, 229.

24. *The English had already briefly visited Guiana:* Hawkins's account of his 1567–68 voyage to Guiana is published in Kinney, *Elizabethan and Jacobean England*, 429.

25. *Help from Harriot:* Shirley, "Sir Walter Ralegh and Thomas Harriot," 25. *Help from Bess:* Letter from Bess to Cecil, in Rowse, *Ralegh and the Throckmortons*, 182.

26. "My Bess": Letter from Ralegh to Cecil, in *Letters of Sir Walter Ralegh*, 119. *Separation between Bess and Wat:* Possibly Wat was still being wet-nursed, and hence the separation (120n6). Or perhaps Bess thought it safest, in view of Damerei's death, that she should send Wat further away, while she stayed close enough to keep an eye on Sherborne (Beer, *Bess*, 94).

27. *Lord Mayors of London on plays:* Excerpted in Ashley, *Elizabethan Popular Culture*, 167.

28. *Updating* Arcticon: BL Add MS 6788, ff205–20, 323, 468–91; BL Add MS 6789 ff534–7; cf. Pepper, "Harriot's Earlier Work on Mathematical Navigation," 57ff.

29. *Not the norm to publish:* Pumfrey, "Patronizing, Publishing and Perishing," 143.

30. *Sanderson:* Original accounts of the dispute with Ralegh, and other details on Sanderson, are given in Shirley, "Sir Walter Raleigh's Guiana Finances," and in McIntyre, "William Sanderson."

31. *Harriot attending Ralegh, doing Northumberland's accounts, too*: Shirley, "Sir Walter Ralegh and Thomas Harriot," 25.
32. *Northumberland's lawsuits*: Batho, *The Household Papers of Henry Percy*, e.g., 60.
33. *Northumberland's wealth*: Batho, "Thomas Harriot and the Northumberland Household," 34. *Harriot's leases*: Pepper says Harriot had leases in various lands or manors from Northumberland and Ralegh ("Thomas Harriot: A Biography," 213).
34. *Northumberland's experiments*: BL Birch MS 4458 ff4–5; Lohne, "Essays on Thomas Harriot," 226–227.
35. *Ralegh's quote*: From excerpt in Kinney, *Elizabethan and Jacobean England*, 442.
36. *Ralegh's quote*: See previous note. *Columbus and syphilis*: Charles Q. Choi, "Case Closed? Columbus Introduced Syphilis to Europe," *Scientific American*, LiveScience, December 27, 2011, https://www.scientificamerican.com/article/case-closed-columbus/.
37. *Ralegh in Guiana (and "a lady in England")*: Coote, *A Play of Passion*, chapter 12 (and 245). *Ralegh in awe of Bess*: He feared her complaints (e.g., letter 81 in *Letters of Sir Walter Ralegh*), but many of his letters have little messages or asides about Bess, as though he enjoyed her feisty opinions, and as though they mattered: e.g., "Bess remembers herself to your lordship and says your breach of promise shall make you fare accordingly!" (letter 141); see also, e.g., letters 139, 140, 145, 149, 159. See chapter 14 for Cecil and Howard on Bess.
38. *Ralegh's plan for Guiana*: Thompson, *Sir Walter Ralegh*, 103.
39. *Harriot to Cecil on Acosta*: Shirley, "Sir Walter Ralegh and Thomas Harriot," 12.
40. *Harriot to Cecil*: Quoted in Batho, "Thomas Harriot's Manuscripts," 296. Harriot's complete letter is in Stevens, *Thomas Hariot*, and British History Online, http://www.british-history.ac.uk/cal-cecil-papers/vol6/pp255-272.
41. *Arthur's diary re Harriot*: Rowse, *Ralegh and the Throckmortons*, 191–192. *Ralegh's 1597 will*: Shirley, "Sir Walter Ralegh and Thomas Harriot," 25; Beer, *Bess*, 114. *Women and property*: Beer, 165. Ralegh urged Bess to remarry for her own security if anything happened to him; Ralegh to Bess, c. July 27, 1603, in *Letters of Sir Walter Ralegh*, letter 72. *Trusting Harriot as kin*: Beer, "Thomas Harriot and Sir Walter Ralegh's Wife," 12.

12: ROYAL REFRACTION

1. *"Quantitative relationship"*: Ptolemy, *Optics* 5.31, quoted in Smith, "Ptolemy's Search for a Law of Refraction," 231. Smith's paper shows the intellectual magnitude of Ptolemy's achievement, given the incorrect prevailing conceptual framework. *Ptolemy's measurements of refraction*: His experimental design is described in Smith, 222n2. It is similar to Harriot's later experiments with his astrolabe (to be described later in this chapter), and Smith says that it could have yielded an experimental determination of the sine law if Ptolemy did not have a preconceived notion of what he was looking for—a notion arising from his use of a particular kind of geometric optics that carried more weight for him than experiment. *Data massaged (and why)*: Smith, 232–235. See also Jean-Luc Godet, "A Short Recall about...Refractive Index," http://okina.univ-angers.fr/publications/ua13816/1/refrindexstor.pdf.
2. *Al-Haytham and refraction*: There is no law of refraction even in al-Haytham's masterly *Book of Optics (Kitab-al-manazir)*; cf. al-Haytham expert Mark Smith, endorsed and referenced in Goulding, "Thomas Harriot's Optics," 144n21. Al-Haytham did make an attempt to explain *why* refraction changes the direction of a light ray: he likened it to encountering a resistance at the interface of the denser medium. It was not until the midseventeenth century, however, that Christiaan Huygens developed a fledgling wave theory of light and suggested that light slows down in a denser medium. The *actual* slowing down of light in a denser medium was experimentally verified in the nineteenth century. Then, in the early twentieth century, Albert Einstein introduced the idea that light speed is constant, but this really means that it is constant in any given medium. It is highest in a vacuum, and this vacuum speed is the legendary c of $E = mc^2$ fame, the speed beyond which light in our universe cannot pass (according to the theory of relativity).

3. *Ibn Sahl's manuscripts on lenses* were discovered and reconstructed by Rashed, "A Pioneer in Anaclastics." On 478, Rashed shows that a geometric ratio in Ibn Sahl's diagram (reproduced on 467) is equivalent, in hindsight, to the modern sine law, which he claims is the basis of Ibn Sahl's analysis of lenses. Overall, it is an impressive reconstruction of an impressive early geometric analysis of lenses. One pertinent point is that Ibn Sahl realized that when reversing the direction of view, the angles of incidence and refraction are simply reversed (478). This is an important insight, but it does not depend on a *particular* law of refraction, and Robert Goulding noted that an accurate law of refraction is not actually necessary for such geometric analyses: al-Haytham and his commentator Witelo developed geometric optics even further than Ibn Sahl, and they developed sophisticated theorems for lenses and other optical phenomenon using only a qualitative understanding of refraction ("Thomas Harriot's Optics," 147).

    Indeed, Ibn Sahl gave no account of where his ratio came from. Andrew Young suggests that he likely began with mathematical conic surfaces of revolution and worked backward to find the light path that makes these lenses suitable for burning ("Discovery of the Law of Refraction," https://aty.sdsu.edu/explain/optics/discovery.html). On the available evidence, Young does not think Ibn Sahl discovered the *law* of refraction. A similar analysis and conclusion is given in Jean-Luc Godet, "A Short Recall about…the Refractive Index," http://okina.univangers.fr/publications/ua13816/1/refrindexstor.pdf.

    For a more detailed critique of Ibn Sahl's analysis, and a gentle suggestion that Rashed exaggerated Ibn Sahl's importance regarding the law of refraction, see Sabra, review of Rashed's "*Géometrie et dioptique au Xe siècle*", 685. But in his "A Pioneer in Anaclastics" Rashed does clearly acknowledge (491) that rather than deducing a law of nature from experiment, Ibn Sahl's was a purely geometric exercise devoted to lenses. In fact, Rashed's main point here is that Ibn Sahl was a forgotten pioneer in geometric optics, and especially the study of lenses, and I certainly agree that he deserves recognition for this.

4. *Pre-seventeenth-century laws of nature.* Cf. Penrose's choice, according to his definition of what qualifies as a "superb" theory (*The Emperor's New Mind*, 197–198). Kepler spoke of the inverse square law of light; Gingerich and Voelkel, "Tycho and Kepler," 92.

5. *Kepler and Tycho on atmospheric refraction.* Péoux, "Atmospheric Refraction and the Ramus Circle"; Goulding, "*Chymicorum in morem*," 34.

6. *Measuring astronomical refraction.* Tycho followed the method outlined by Witelo, and he used the calculations embodied in an armillary sphere; Péoux, "Atmospheric Refraction and the Ramus Circle," 464, 471. For an example regarding the geometry of the celestial sphere, see Smart, *Textbook on Spherical Astronomy*, 46–48, 65–70, for a mathematical derivation of theoretical rising and setting times of celestial bodies; these can be compared with observations, to show the effect of refraction.

7. *Refraction: sunrise, sunset.* Christophe Rothmann and Tycho, for example, knew this: Péoux, "Atmospheric Refraction and the Ramus Circle," 472, 470. *Twinkling: Torres Strait Islanders.* Anna Salleh's report (www.abc.net.au, September 20, 2016) on a collaboration between the Meriam elder A. Tapin and the physicist D. Hamacher. *Planets don't twinkle* because they are so much closer to earth than stars that they appear relatively large; atmospheric refraction does not change their apparent direction by more than their apparent diameter, so they do not twinkle.

8. *Refraction: heavens versus atmosphere.* Jean Pena was notable for conjecturing there was no difference: Péoux, "Atmospheric Refraction and the Ramus Circle," 468; see also 470 for Rothmann.

9. *Heavenly matter.* Kepler would dispense with physical planetary orbs (creating the idea of orbits in space instead), but he still believed in the existence of the ether.

10. *Harriot on refraction in navigation.* Quoted in Pepper, "Harriot's Earlier Work on Mathematical Navigation," 60.

11. *Al-Haytham, Witelo, reading Risner editions.* Péoux, "Atmospheric Refraction and the Ramus Circle," 461–462. See my final chapter for Lohne's thrilling discovery that Harriot read Risner.

12. *Harriot, Kepler, and della Porta.* Goulding, "Thomas Harriot's Optics," 144–149. Goulding reproduces della Porta's constructions and shows the refraction "law" they imply. He

points out that Harriot was one of the few readers of della Porta's tract, and surely the only one to test it! *Dedication to Northumberland*: In 1591; Batho, "Thomas Harriot and the Northumberland Household," 31 Clucas, "Thomas Harriot and the Field of Knowledge," 128.

13. *Witelo copied Ptolemy's tables*: Goulding, "Thomas Harriot's Optics," 170.

14. *Harriot's method*: This is my "modern" interpretation of his BL Add MS f407v: in the top row he has measured *cd* as 860 units, then converted to 10,000 units (the usual approach to trigonometry in those days before decimal point notation was used). Then he recorded his measurements of *bc* for angles 70°, 60°, 50°, etc., then converted them all to proportions of 10,000. Tangents *bc/cd* then give the angles he has written in the table. In the far right-hand column he has a couple of measurements for *bf*, apparently as a check on his initial angles: e.g., in the 60° row, for *bf* = 9815, his *bc/bf* does indeed give sine 60°.

15. *Status of al-Haytham's experiments*: Ruling out Ptolemy's law: Boyer, *The Rainbow*, 81. As far as studying refraction further, in "Le 'De aspectibus' d'Alhazen" Mark Smith explored al-Haytham's most famous optical experimental design and deduced that it would have been technologically impossible to carry out such an experiment at the time; Smith concluded that al-Haytham was in the conceptual tradition of ancient thought experiments. Einstein (and Galileo) famously showed the value of such thought experiments, but unlike his ancient and medieval forerunners, Einstein knew that such "experiments" need to make predictions so that they can be tested with physical experiments. See also King, "Mediaeval Thought Experiments." *Roger Bacon*: Duncan, *The Calendar*, 2, 8. Francis Bacon also did some experiments, and he died because of one of them: he was stuffing a dead chicken with snow in order to experiment on refrigeration, and caught a lethal cold in the process (*Chambers Biographical Dictionary*).

16. *Islamic science*: Nasr believes the lack of quantitative results by Arab scholars was due to a religious respect for unity, leading them to seek God beyond the heavens, not rend them asunder as in the West (*Islamic Science*). On the other hand, Ragep suggests that Islamic work in observational astronomy and geography points to a change in Islamic culture, under the influence of Greek thought but, under Caliph Ma'mun, passing beyond it to "testing" hypotheses ("Islamic Response to Ptolemy's Imprecisions," esp. 132ff.). Sufis such as the tenth-century century Persian Abu Said emphasized wholism rather than reductionism or scholasticism or, indeed, rigid theology: to those who seek truth in conventional religion, Abu Said wrote, "Until college and minaret have crumbled, this holy work of ours will not be done." And speaking of the Aristotelian Arab philosopher Avicenna (Ibn Sina), Abu Said said, "What I see, he knows," and Avicenna said, "What I know, he sees." Shah, *The Way of the Sufi*, 239, 240. *Christianity*: The thirteenth-century friar Roger Bacon's Franciscan order had frowned upon the idea of experimentally discovering nature's secrets. In its reliance on the biblical "word of God" for its truths, the order was suspicious of "novelties" uncovered by mere humans, and Bacon was eventually imprisoned, his work censored; Duncan, *The Calendar*, chapter 1. Bacon was influenced by Sufi thought and wore Arab dress (Shah, 18, 40).

17. *Francis Bacon's science*: Martin, *Francis Bacon, the State, and the Reform of Natural Philosophy*, 5.

18. *Francis Bacon, "already inclined"*: Quoted in Gatti, "Giordano Bruno," 63. *Harriot not reading him*: Gatti, "The Natural Philosophy of Thomas Harriot," 75.

19. *Kepler using Ptolemy/Witelo tables*: Goulding, "*Chymicorum in morem*," 34–35; Lohne, "Thomas Harriot (1560–1621)," 114. *Kepler a Neoplatonic Christian*: Lindberg, "Continuity and Discontinuity in the History of Optics," 436.

20. *Harriot's astrolabe measurements*: Lohne, "Essays on Thomas Harriot", 283. *Harriot's diagrams*: See, e.g., BL Add MSS 6785 f1, 6789 f320. Most laws of physics are derived from a combination of experiment and intuition: nature is never an exact fit with an idealized mathematical law. Harriot's experimental data did *not* fit the idealized law *exactly*, as his own tables tell us when he lists measurements alongside results obtained "by calculation."

21. *Harriot's sine ratios*: From figure 16b, consider triangles *bcf* and *bcd*: $\sin r = \dfrac{bc}{bf}$, and $\sin i = \dfrac{bc}{bd}$. So $\dfrac{\sin i}{\sin r} = \dfrac{bf}{bd}$. If *bf* and *bd* are the radii of concentric circles, as in Harriot's astrolabe experiment, then the sine law is generalized.

22. *Harriot's famous* Regium *is at* BL Add MS 6789 f320. Even this was a work in progress, though: he'd added to it to prove that the shape of the refracted image of the circular bottom of the jar in which he'd immersed his astrolabe was an ellipse.

23. *Kepler on hidden ratio:* Quoted in Goulding, "*Chymicorum in morem*," 35.

24. *Fame:* As a young man, Kepler realized he was "obsessively interested in fame," according to Rublack, "The Astronomer and the Witch."

25. *Descartes plagiarizing Snell?* Boyer, *The Rainbow*, 203.

26. *Harriot invented his own symbols for the trigonometric functions* sine, secant, and tangent, and delineated the angular argument using commas. See Stedall, "Notes Made by Thomas Harriot on the Treatises of François Viète," 191.

27. *Ibn Sahl, al-Haytham, reversibility:* Rashed, "A Pioneer in Anaclastics," 478; Alex Small, "Ibn al-Haytham on Refraction," *Physicist at Large* blog, September 13, 2016.

28. *Harriot's tables of refraction* are drawn up at BL Add MS 6789 ff88–90, for air and glass, air and water, water and crystal. At the top he writes, e.g., from air to water, and at the bottom he shows the reversibility by writing from water to air.

    Note that textbooks generally give the index of refraction of a substance with respect to a vacuum, whose refractive index is said to be 1. The refractive index of air is then 1.0003. Textbooks (such as Robert Resnick and David Halliday, *Physics* [New York: John Wiley & Sons, 1966] 1014) often give the index of refraction of water as 1.33 with respect to a vacuum; multiplying Harriot's value for water with respect to air (1.335194) by 1.0003 gives 1.3355945, which is 1.34 rounded to two decimal places.

29. *Such a calculation is in,* e.g., BL Add MS 6789, f161r. Lohne was the first to uncover Harriot's discovery of the law of refraction ("Thomas Harriot [1560–1621]").

30. *Harriot's sine law:* Finding "sines" in Harriot's papers is not entirely straightforward, because he used his own special symbol to designate sines. In addition to Harriot's many pages of experimental notes and sine law calculations, Goulding suggests he tested his sine law in his analysis of the geometric optics of curved mirrors; Goulding analyzes these papers in detail ("Thomas Harriot's Optics").

31. *Accuracy of Harriot's arithmetic calculations:* Sadler checked these for Harriot's later tables of meridional parts ("Nautical Triangles Compendious," 141, 143), but the point is relevant to all his calculations.

13: SPIRALS AND TURMOIL

1. *Harriot's shopping lists; scrawled note:* BL Add MSS 6789 ff514–5, 6787 f204r.

2. *Elizabeth's angry letters to Essex:* Excerpted in Perry, *Elizabeth I: The Word of a Prince*, 305–308.

3. *Guianans with Ralegh:* Vaughan, "Ralegh's Indian Interpreters," esp. 342–345, 371, 375 (but see 370, Spanish account); Beer, *Bess*, 123–124.

4. *Visiting Hardwick, Knole:* Beer, *Bess*, 115.

5. *Daughter in Jersey?* Beer, *Bess*, 113; Coote, *A Play of Passion*, 282.

6. *Bad harvests in late 1590s:* Coote, *A Play of Passion*, 263.

7. *"Blackamoors" with Ralegh; 1596 proclamation:* Beer, *Bess*, 124. *"Blackamoor" with Percy (Northumberland)* (in 1586): Batho, "Thomas Harriot and the Northumberland Household," 31.

8. *Monopolies:* Rowse says Ralegh blushed when the playing cards were mentioned but went on to tell Parliament that since he had had his tin patent (as Lord Warden of the Stannaries), the tinners were assured of an improved wage, no matter the price of tin (*Ralegh and the Throckmortons*, 225).

9. *The Essex Rising (Rebellion):* In Elizabethan documents the date is 1600 because the new year did not start until March, but as with all my dates, unless I quote directly from a dated manuscript, I am following that of scholars cited, who usually take the year as beginning on January 1. On Essex's rebellion: Hammer, "Shakespeare's *Richard II*, the Play of 7 February 1601, and the Essex Rising" (2008) (Hammer challenges the legend that Essex's staging of Shakespeare's *Richard II* was an integral part of his treasonous plan);

Kinney, "The Essex Rebellion"; Salter, *Elizabeth I and Her Reign*, chapter 5, documents 5a, 5b.

10. *Rumors of Ralegh's supposed plan* may have been deliberately circulated to panic Essex into revealing his hand: Hammer, "Shakespeare's *Richard II*, the Play of 7 February 1601, and the Essex Rising," 15.

11. *Rumor and supposition:* Kinney, "The Essex Rebellion," 296–297, including n2.

12. *"L of E ready"*: BL Add MS 6787, f204r as deciphered by Gatti, "The Natural Philosophy of Thomas Harriot," 67. It is such a cryptic (and barely legible) note that it is only conjecture that it refers to Essex and his execution. But it's a highly probable one, given the poignant ambivalence of Harriot's note and Essex's close relationship to both Ralegh and Northumberland. It would not be the last preexecution note Harriot would make, as we shall see. And if it does not refer to the execution, Harriot would still have been caught up in the drama surrounding the Essex Rebellion, given the danger it posed to Ralegh.

13. *Harriot reading Bacon's apologia:* This is a conjecture based on a "note to self" in Harriot's manuscripts: "Mr Willis book, Bacon" (BL Add MS 6789 f514r); Gatti, "The Natural Philosophy of Thomas Harriot," 75 (also citing Shirley, *Biography*).

14. *Northumberland on Ralegh:* Quoted in Rowse, *Ralegh and the Throckmorton*, 232.

15. A *"great circle"*—such as the meridians of longitude, or the equator, or any circle encompassing the sphere and lying in a plane through the center of the sphere—has a radius equal to the radius of the sphere. Other lines of latitude have a smaller radius, and so they are not great circles. Neither are rhumb lines, of course: they are spirals, and in his earlier work on the Mercator map, Harriot designated triangles bounded by a meridian, a parallel, and the arc of a rhumb line as "nautical" rather than spherical triangles.

16. *Nunes, spherical triangles:* Randles, "Pedro Nunes' Discovery of the Loxodromic Curve," 88–91.

17. *Hues on Nunes's great circles: Tractatus de globis*, e.g., English edition (1659), 161.

18. *Shadows, lines on sphere:* Booth, "Thomas Harriot's Translations," 352–353. Note that Booth considers Harriot's work in connection with conceptual patterns Harriot may have absorbed from his study of Algonquian grammar (353ff).

19. *Stereographic projection conformal:* Harriot's working is at BL Add MS 6789 ff.17–18. Pepper has reproduced and explicated Harriot's proof ("Harriot's Calculation of the Meridional Parts," 366–367, 411–413) . *For Harriot's apparent priority (incl. Ptolemy did not have it):* Pepper, "Harriot's Calculation," 367n44; Pepper, "Thomas Harriot and the Great Mathematical Tradition," 18; Berggren, "Ptolemy's Maps of Earth and the Heavens," 134–135, 138–139, including note 9: the earliest extant proof of the circle-preserving property (not angles in general) was by the ninth-century scholar al-Farghani. The Jesuit scholar Clavius also asserted the conformality of circles in stereographic projection in his 1593 treatise on the astrolabe, but Derek Whiteside pointed out that this is not a general proof of conformality: Clavius referred only to angles made by intersecting circles, not to the angles between *any* lines as in Harriot's general proof; Whiteside quoted in Pepper, "Harriot's Calculation," 367n44.

20. *Date of Harriot's spiral:* Pepper includes this with Harriot's early work ("Harriot's Earlier Work on Mathematical Navigation," 74). He certainly had all his results on the equiangular spiral, and on the conformality of stereographic projection, well before 1614 when he completed his work on the Mercator problem. In "Harriot's Calculation of the Meridional Parts," 365–366, Pepper again says he probably began it before 1600.

21. *Descartes, Torricelli:* Boyer, *A History of Mathematics*, 342–343. The spiral was what Descartes called a "mechanical" curve and we call a "transcendental" one; he thought that unlike an algebraic curve, it was not geometric. Boyer points out that despite its brilliance the spiral result did not disprove Descartes on *algebraic* curves, and that the spiral actually never reaches its pole.

22. *Spiral equation:* θ is not the constant angle of the compass bearing on the loxodrome but the angle between the x axis and a radial line from the origin to a point on the spiral.

23. *Harriot's mnemonic for (what we call) pi*: BL Add MS 6788 f547v; cf. Sokol, "Thomas Harriot...on Population," 205. *Stevin, decimals*: Biggs, "Thomas Harriot on Continuous Compounding," 67. Decimal point notation: Boyer, *A History of Mathematics*, 303.

24. *Emmy Noether* is the most famous such female mathematician: she unified earlier work on mathematical symmetries in her famous theorems.

25. *Harriot as forerunner, precalculus*: Pepper, "Thomas Harriot and the Great Mathematical Tradition," 21. Others soon followed independently: for Kepler's precalculus, see Boyer, *A History of Mathematics*, 325 (including area of ellipse, which Archimedes had done in a manuscript not then extant), and for Fermat, Gregory of St. Vincent, and other pre-Newtonian/Leibnizian pioneers, see 348ff. For slightly earlier forerunners, see Boyer, *The History of the Calculus and Its Conceptual Development*: e.g., unlike Harriot, Stevin followed Archimedes, stopping short of summing a geometric series to infinity (102); Stifel and Viète regarded the circle as made of an infinite-sided polygon (93), but Kepler was the one who applied this concept to other problems than the circle (108)—and this seems to have been a little *after* Harriot's work on the length of the spiral.

26. *Sum of an infinite geometric series* is also now taught in high school, but it had not been published at that time and was not yet common knowledge Pepper ("Harriot's Calculation of the Meridional Parts," 371 and n50. *Euclid's formula: Elements* 9.35. *Harriot's working*: See, e.g., MS HMC 240 II ff211, ff221, 224, 227, BL Add MS 6784 f430; analyses in Lohne, "Essays on Thomas Harriot," and Pepper, "Harriot's Calculation." *Harriot first to use* n^{th}-*term expressions*: Beery and Stedall, "Thomas Harriot's Doctrine," 10.

27. *Harriot had general results*: For instance, he wrote that "every helical [equiangular spiral length] is equal to the secant of his angle" (MS HMC 240 II f231, transcribed in Pepper, "Harriot's Calculation of the Meridional Parts," 403). His proof was remarkable for his times but incomplete by modern standards. But then, so were some of Newton's!

28. *Harriot's spherical triangles*: See, e.g., BL MS 6787 ff.204–210, 106–110. (f106 has the date September 18, 1603). See also Pepper, "Harriot's Calculation of the Meridional Parts," 367; Pepper, "Thomas Harriot and the Great Mathematical Tradition," 21; and Lohne ("Essays on Thomas Harriot"), 299–301. Albert Girard was the first to publish the area formula, in 1629, although Lohne says his proof is not as elegant as Harriot's (300).

29. *Briggs praising Harriot*: My quotation is Pepper's paraphrase ("Harriot's Calculation of the Meridional Parts," 367n41), from a letter from Briggs to George Hakewill, who published it in 1630. It is possible Briggs and Harriot knew each other, although Stedall suggests it is almost certain that it was Torporley who passed on Harriot's results to Briggs (Stedall, "Notes Made by Thomas Harriot on the Treatises of François Viète," 191).

30. *Nine months of no work*: Batho, "Thomas Harriot and the Northumberland Household," 35. This is based on Harriot's dated manuscripts, so it certainly seems that he did no experimental work, at least.

31. *Harriot's list*: BL Add MS 6787 f111v.

14: CHANGING OF THE GUARD

1. *Arthur's diary*: Quoted in Beer, *Bess*, 122.

2. *Feisty Bess*: Sometimes Bess would rail against her husband during periods of stress, leaving him "weary by her crying and bewailing," as he put it; Thompson, *Sir Walter Ralegh*, 228.

3. *Howard and Cecil*: Quoted in Coote, *A Play of Passion*, 295, 297. *Opposing Lady Ralegh*: Beer, *Bess*, 133, 144–145; Rowse, *Ralegh and the Throckmortons*, 228. *Will Cecil*: Beer, 127.

4. *James' procession*: Shirley, "Sir Walter Ralegh and Thomas Harriot," 26. *Percy/Northumberland's advice*: Letter from Northumberland to James VI of Scotland, and James's reply, reprinted in Kinney, *Elizabethan and Jacobean England*, 137–139.

5. *James I's speeches* are reprinted in Kinney, *Elizabethan and Jacobean England*, 141–148, 157ff.

6. *Wine patent*: Coote, *A Play of Passion*, 321.

7. *Moving from Durham House*: Rowse, *Ralegh and the Throckmortons*, 233–234.

8. *Would Ralegh betray Cobham?* Near-contemporary Aubrey said so, and Rowse found this convincing (*Ralegh and the Throckmortons*, 236–237); so did Coote (*A Play of Passion*, 301). *Ralegh confessing his suspicions:* Quoted in Nicholls, "Sir Walter Ralegh's Treason."

9. *Cecil and Howard trapping Cobham, Ralegh:* Beer, *Bess*, 142–143; Mandelbrote, "The Religion of Thomas Harriot," 262–263. Nicholls says that Ralegh was probably not brought down by Cecil's machinations, but he did little to save him ("Last Act?" 169).

10. *Ralegh's 1603 trial:* A transcript is available in Oldys and Birch, *Lives*, 648–690. Note that one witness (Dyer) *was* sworn (682), but his evidence was extremely thin. *No presumption of innocence:* Rowse, *Ralegh and the Throckmortons*, 234. *Impact of Ralegh's trial today:* See, e.g., Boyer, "The Trial of Sir Walter Ralegh." *Other current discussions:* A plethora of books touch on the subject, but for a succinct account of the prosecution case, see Nicholls, "Sir Walter Ralegh's Treason." For an interesting brief legal analysis—which argues that neither Spain nor Ralegh could have been seriously involved in such a plan as Cobham wildly confessed to—see the 2007 articles on the Main Plot by the feisty free speech advocate Anthony J. H. Morris, QC, on his *Lex Scripta* website.

11. *Ralegh's careless words* and their legal implication: Nicholls, "Last Act?" 168.

12. *Two witnesses:* Patterson, *The Trial of Nicholas Throckmorton*, 47; Beer, *Bess*, 17; L. Hill, "The Two-Witness Rule in English Treason Trials." *12 Am. J. Legal Hist.* 95 (1968), 95–111.

13. *Harriot searching the Bible for Ralegh's trial:* Mandelbrote, "The Religion of Thomas Harriot," 264; cf. BL Add MS 6787 f119v.

14. *"Damnable atheist":* Trial transcript in Oldys and Birch, *Lives*, 686.

15. *Popham's speech:* In the published version by Oldys and Birch, the words "Harriot nor any Doctor" have been replaced by "any devil"; Harriot's name appears in a manuscript version (Jacquot, "Thomas Harriot's Reputation for Impiety," 167). Apparently there were several copies of the document: Thompson's version has "any devil Harriot, nor any such doctor" (*Sir Walter Ralegh*, 197). Batho cites a manuscript version that suggests a misprint in Oldys and Birch, whose full quote is "You know what men said of Harpool…Let not any devil persuade you to think there is no eternity in hellfire": Batho's version replaces "Harpool" with "Harriot," and the MS reference is given in his footnotes ("Thomas Harriot and the Northumberland Household," 35).

   Note that the accusation that Ralegh was willing to spy for the Spanish in return for money was baseless; back in the 1580s, however, when Mary Queen of Scots was being investigated via Walsingham's double agents, Ralegh had dreamed up a foolhardy scheme whereby he would act as a double agent, in the hope of gaining information on the movements of Spanish treasure fleets. He bungled the whole thing! Coote, *A Play of Passion*, 103–105.

16. *James's mock executions:* Coote, *A Play of Passion*, 316–317; Rowse, *Ralegh and the Throckmortons*, 240–241; Thompson, *Sir Walter Ralegh*, 206ff.; Beer, *Bess*, 159.

17. *Bess's letter to Cecil:* Reproduced in Beer, *Bess*, 155, with all its phonetic spelling.

18. *Ralegh's farewell letter to Bess:* December 4–9, 1603, reprinted in Beer, *Bess*, 269–271, and Thompson, *Sir Walter Ralegh*, 206–207. My spelling is slightly more modernized than Beer's, and slightly less so than Thompson's.

19. *Ralegh spared scaffold charade:* Beer, *Bess*, 159.

20. *Cecil:* Coote, *A Play of Passion*, 321 (helping Bess); Thompson, *Sir Walter Ralegh*, 212 (saving Ralegh's life?).

21. *Reactions to Ralegh's trial:* Quoted in Coote, *A Play of Passion*, 315, and Rowse, *Ralegh and the Throckmortons*, 238 (both referring also to Edwards's *The Life of Sir Walter Ralegh*, vol.1, e.g., 410, 415, 432–433). *More on Justice Gaudy (the judge quoted), Coke:* Thompson, *Sir Walter Ralegh*, 198–199.

22. *Pensions to Howard and Cecil:* Thompson, *Sir Walter Ralegh*, 201; Rowse suggests Ralegh and Cobham had been "gaping" for one, too (*Ralegh and the Throckmortons*, 245). *Pension policy:* Carter, "Gondomar: Ambassador to James I," *Hist. J.*, 7, No. 2 (1964), 189–208.

23. Sejanus, Measure, *and Ralegh:* I owe this possible connection to Beer, *Bess*, 160.

24. *Chapman to Harriot:* Excerpted in Quinn and Shirley, "A Contemporary List of Harriot References," 23. Note that Goulding sees possible allusions to alchemy in Chapman's ode ("*Chymicorum in morem,*" 50–51).

15: ALGEBRA, RAINBOWS, AND AN INFAMOUS PLOT

1. *Specific weight/gravity* is the weight of a given volume of a substance divided by the weight of the same volume of a reference medium such as water. To determine this value, Harriot used Archimedes's proposition 7 from book 1 of *On Floating Bodies,* which says that if you weigh an object in air, then weigh it again in a medium such as water, the object will weigh less when immersed by an amount equal to the weight of fluid it displaces. So, to measure his specific weights, Harriot weighed the body in air and in water, then calculated the difference between them; then he found the ratio of this difference to the weight in air. He did not use the modern terms "specific weight" or "density," but his diagrams and working show that he had the idea. See Schemmel, *The English Galileo* 1:135–139. Harriot's experiments on specific weight were incredibly meticulous. He even compensated for the porosity of his materials! Clucas, "Thomas Harriot and the Field of Knowledge in the English Renaissance," 122–123.

2. *Algebra and falling motion:* An example (BL Add MS 6788 f144v) is explained in Stedall, "Symbolism, Combinations, and Visual Imagery," 384, and Lohne, "Essays on Thomas Harriot," 244–245. *First fully symbolic algebra:* Cf. detailed analysis by Seltman, "Harriot's Algebra," 185.

3. *Index notation:* A fifteenth-century French manuscript (published only in the nineteenth century) by Nicolas Chuquet contained a notation analogous to that of Bombelli: the unknown was implied rather than stated, and he would have written n^2 as 1^2: Boyer, *A History of Mathematics,* 277.

4. *Torporley's index form* is discussed in Beery and Stedall, *Thomas Harriot's Doctrine of Triangular Numbers,* 23. *Algebraic notation predecessors:* Boyer, *A History of Mathematics,* 181, 304; Mazur, *Enlightening Symbols,* e.g., 81–83.

5. *Zero algebraic:* BL Add MS 6783 f187r, analyzed by Seltman, "Harriot's Algebra," 169.

6. *Harriot first to algebraically solve cubics:* Seltman, "Harriot's Algebra," 185. Several decades earlier, Cardan had published the first algorithmic solution for some cubic equations (Boyer, *A History of Mathematics,* 285–286). *Viète's lack of generality:* Boyer, 304.

7. *Harriot on complex (and negative) roots:* See, e.g., BL Add MS 6783 ff49r, 103, 157, 313. As noted, he did not always include them in his papers, which—together with the fact that the editors of his *Praxis* declared them "impossible"—has led some scholars to conclude that he rejected them outright. But see Seltman, "Harriot's Algebra," 167–170, for a detailed analysis of Harriot's work on this topic. It shows that Harriot did accept both positive and negative complex roots, and Seltman conjectures that the inconsistent cases may have been for pedagogical reasons, or possibly the development of Harriot's own thinking. *First to fully solve biquadratics:* Stedall, "Rob'd of Glories," 480. (Cardano was perhaps the first to publish a mention of complex numbers, in 1545; he did not understand them, describing them as "useless"; quoted in Flood and Wilson, *The Great Mathematicians,* 70, and Stillwell, 190; *Mathematics and Its History,* 190.) *Bombelli:* Stillwell, Flood and Wilson, 70.

   Harriot and the product of two negative numbers: Gatti reproduces a note to Warner suggesting that at one time, Harriot did not believe the answer was positive ("The Natural Philosophy of Thomas Harriot," 68); however, Tanner shows that Harriot correctly realized it was positive for real numbers and negative for imaginaries, although dating this correct conclusion is fraught ("The Ordered Regiment of the Minus Sign").

8. *Descartes and Harriot:* Stedall, "John Wallis and the French" (including open question, 276); Stedall, "Rob'd of Glories," 488 (visiting England), 489 (H and D: same predecessors); Stedall, "Reconstructing Thomas Harriot's Treatise on Equations" (including Beaugrand, 62). *Descartes's "lack of frankness"/amnesia on sources:* Boyer, *The Rainbow,* 203, 211.

   Wallis was not fair to Descartes (Stedall, "John Wallis"), but it is easy to see why he had his sword out for his unsung countryman: Descartes did not always use index notation and x's, but sometimes used lowercase letters from the beginning of the alphabet in just the same way as Harriot did, even writing *bb* instead of b². Also, Wallis apparently had seen some of Harriot's manuscripts, not just the *Praxis* (Jacqueline Stedall, *A Discourse concerning Algebra: English Algebra to 1685*, New York: Oxford University Press, 2002). In 1949, however, Paul Turán gave due recognition to Harriot in the title of a paper, "On Descartes-Harriot's Rule" (*Bulletin of the American Mathematical Society* 55, no.8 [1949]: 97–800), which discusses a rule for the number of solutions of polynomials of arbitrary degree.

9. *Harriot's symbolic innovations*: Seltman, "Harriot's Algebra," 155–157. *Recorde on equality sign*: Quoted in Boyer, *A History of Mathematics*, 269; see also 290 for sample from Recorde's 1557 book *Whetstone of Whitt*.

10. *Torporley to Harriot*: The 1586 date is conjectured by Pepper, "A Letter from Nathaniel Torporley to Thomas Harriot," 289; Torporley's letter is printed on 289–290.

11. *Viète's form of equations*: Flood and Wilson, *The Great Mathematicians*, 79; Stedall, "Rob'd of Glories," 465–466; Mazur, *Enlightening Symbols*, e.g., 134–137, 139, 140, 142, 162–163. As noted, Mazur also discusses symbolic attempts by Viète's precursors, including Bombelli and Stevin, while Boyer gives an early use of letters in algebra by Jordanus Nemorarius (*A History of Mathematics*, e.g., 258) and an even earlier use by Diophantus of Alexandria (181). But none of these had the generality of Harriot's symbolism.

12. *Wallis on Harriot*: Quoted in Booth, "Thomas Harriot's Translations," 349.

13. *Harriot's student* was William Lower, of whom more shortly. Letter printed in Rigaud, *Supplement to Dr Bradley's Miscellaneous Works*, 42–45, quote on 44. *Harriot rewriting Viète's propositions in symbolic notation*: Stedall, "Notes Made by Thomas Harriot on the Treatises of François Viète," 190–191.

14. *Pascal's triangle*: This triangle, and its link with integer (whole number) binomial powers, had been known in medieval China, India, the Middle East, and mid-sixteenth-century Europe; Pascal would develop the associated mathematics more abstractly and fully in 1654. The first two rows in the triangle at the top of Harriot's page are 1 and 1, 1, which correspond to the coefficients in $(a+b)^0(=1)$ and $(a+b)^1(=a+b)$. Then, considering $(a+b)^2=a^2+2ab+b^2$ and $(a+b)^3=a^3+3a^2b+3ab^2+1$, the numerical coefficients of the terms on the right-hand side are 1, 2, 1, and 1, 3, 3, 1 respectively, and these are the third and fourth rows in the triangle; and so on. The entries in each successive row can be routinely generated from the preceding row; for instance, the 3s in the fourth row can be routinely found from 1+2 and 2+1 by adding pairwise the entries in the third row. So the binomial coefficients for higher powers could be generated by repeating these operations, whereas Harriot's general form makes them more readily obtainable as well as more conceptual. For background history, see Edwards, *Pascal's Arithmetical Triangle*.

15. *Harriot on the binomial theorem*: The binomial coefficients in Pascal's triangle can be applied to whole numbers of combinations of cards or dice, say. But Harriot also wrote (e.g., BL Add MS 6782 ff142, 165) negative and fractional values of *n* in binomial coefficients, and applied them in his "Newton-Gregory" interpolation formulae, of which more in chapter 21. For examples, see Beery and Stedall, *Thomas Harriot's Doctrine*, 126–127 (and background theory: 66–67, 88–89, 108–109); see also Edwards, *Pascal's Arithmetical Triangle*, 12–13; Maclean, "Harriot on Combinations," 82.

   Indeed, looking at the binomial coefficients simply as formulae—not as "numbers of things," as in gambling and other concrete contexts—there seems no reason that *n* can't be a fraction, or a negative number. Newton is credited as the first to write out and apply binomial expansions for fractional powers (cf. Boyer, *A History of Mathematics*, 393); although Harriot had long ago applied them, Newton derived them in a more general context. (Today, a proper proof of the binomial theorem with fractional powers involves proof of convergence of the associated Taylor series.) Boyer also says (393) that Harriot's contemporaries Stevin and Girard "had suggested fractional powers but did not really use them"—unlike Harriot in his work on interpolation.

   It is difficult to know when Harriot did much of his work, but I am including the binomial theorem here because he would have had to have at least the integer version in hand well before he discovered his marvelous interpolation formula around 1611 (cf. chapter 21).

Harriot's algebraic advances: For more detail, see Stedall, "Symbolism, Combinations, and Visual Imagery," and Seltman, "Harriot's Algebra." *Binomial theorem context*: Edwards, *Pascal's Arithmetical Triangle*; Pepper, "Thomas Harriot and the Great Mathematical Tradition," 14.

16. *Harriot's symbols and commas for the trigonometric functions sine, secant, and tangent*: Stedall, "Notes Made by Thomas Harriot on the Treatises of François Viète," 190–191; as noted earlier, Stedall says no one to her knowledge treated trigonometric quantities in this way for another century.

17. *Briggs and Napier*: Boyer, *A History of Mathematics*, 314.

18. *Harriot's binary numbers*: Exploring binary arithmetic: BL Add. MS 6788 f244v (on a page of work on specific weights; the next folio, f245v, is on specific weights and is dated 25 July 1604); BL Add MS 6786 ff346v–347. Pepper discusses Harriot's binary decomposition in his work on the spiral ("Harriot's Calculation of the Meridional Parts," 371–372). *No known predecessor*: Rosen, "Harriot's Science," 10.

   Harriot's contemporary Francis Bacon—who rose to power under James I, attaining the positions of attorney general and lord chancellor—developed a simple binary code for his secret messages of state. He denoted the letter *a* by aaaaa, the letter *b* by aaaab, *c* by aaaba, *d* by aaabb, and so on. These correspond to the binary numbers, 00000, 00001, 00010, 00011, ... (See Heath, "Origins of the Binary Code," but note that Bacon published details of his code in 1623, after Harriot's death; Harriot's work was unknown to Heath back in 1972.) Bacon may or may not have understood the deeper nature of these numbers, but it is likely his code arose simply from arbitrary patterns. Certainly he did not display the insight into the general arithmetical properties of binary numbers that Harriot and Leibniz did.

19. *Harriot's color experiments*: See, e.g., BL Add MS 6789 ff.184–197, including dates of April 8, 9, 10, 11, 1605.

20. *Indigo*: David Topper ("Newton on the Number of Colors in the Spectrum," *Studies in History and Philosophy of Science Part A* 21, no. 2 [1990]: 269–279) shows that the musical octave analogy was embraced by Newton but was not the only reason for his conclusions. Alan Shapiro ("The Evolving Structure of Newton's Theory of White light and Color", *Isis*, 71, No. 2 (June 1980), 211–235) shows that Newton realized there are gradations in the spectral colors (which we now attribute to varying wavelengths).

21. *Dispersion*: Nowadays, we know that light has wavelike properties, and that the spectrum is made of bands of subtly changing colors, as the wavelengths change, rather than precise "rays" of color. *Harriot's dispersion diagrams and calculation of index of refraction for colors*: See, e.g., BL Add MS 6789 ff184r, 198–200, 209. See also Lohne, "Thomas Harriot (1560–1621)," 120–121. *Witelo, Kepler, Descartes refraction but not correct dispersion*: Boyer, *The Rainbow*, 105, 110ff, 182, 217, 240–241; and Boyer, "Kepler's Explanation of the Rainbow." Boyer shows (*Rainbow*, 240) they assumed that such things as the width of the sun caused incident rays to come in at different angles, producing corresponding single refracted rays at different angles. *Medieval work on raindrops*: Boyer, *Rainbow*, 209.

22. *Harriot's three colors*: Lohne, "Essays on Thomas Harriot ," 285, from Harriot's manuscript *De coloribus*. See also, e.g., BL Add MS 6789 ff190r, 198.

23. *Official report on Gunpowder Plot* from spymaster Robert Cecil, letter to Cornwallis, reprinted in Kinney, *Elizabethan and Jacobean England*, 148–152. Cecil refers to Fawkes by his alias Johnson.

24. *Sir William Lower*: Hunneyball, "Sir William Lower and the Harriot Circle." *Cecil's dislike of Northumberland*: Rowse, *Ralegh and the Throckmortons*, 253–254. *Cecil as spy at Sherborne; Throckmortons*: Beer, *Bess*, 173–175. *Did Cecil organize the Gunpowder Plot?* This theory has a long history—beginning with the wrongly accused Jesuit Father John Gerard's 1606 account—and is still being debated. A key modern proponent of the idea that Cecil organized the whole thing is Father Francis Edwards. For a brief outline of his case (in his review of Antonia Fraser's *The Gunpowder Plot*), see *Catholic Historical Review* 83, no. 2 (April 1997): 347–348. But the arguments in his own book are criticized in Pauline Croft's

review in a later issue of the same journal (96, no. 2 [April 2010]: 360–361). Other arguments against conspiracy: e.g. A. Okines, "Why was there so little government reaction to Gunpowder Plot?", *J. Ecclesiastical History*, 55, No. 2 (April 2004), 275–292; Nicholls, "Strategy and Motivation in the Gunpowder Plot." Nevertheless, Cecil used it to his advantage: Shirley, *Biography*, 327–8.

25. *Interrogation of Harriot; searching his rooms*: Original documents are cited and quoted in Batho, "Thomas Harriot and the Northumberland Household," 36, and in Shirley, "Sir Walter Ralegh and Thomas Harriot," 28, and (report to Cecil) Mandelbrote, "The Religion of Thomas Harriot," 248–249.

26. *Harriot's letter to the lords*: The original document is cited and fully transcribed (with the original spelling) in Shirley, "Sir Walter Ralegh and Thomas Harriot," 29.

16: SOLVING THE RAINBOW

1. *Harriot's concentrated rays*: See, e.g., BL Add. MS 6789 ff388, 289. The former is a geometric drawing;, the latter shows a cluster of angles of incidence with virtually the same angle (so they all make a cone within 4′ of the greatest angle, 41°46′). *Kepler*: Boyer, *The Rainbow*, 212–213. *On dates of Harriot's work*: He rarely dated his manuscripts, and so may have done his geometrical analysis of the rainbow before his work on color. By his 1606 correspondence with Kepler, however, he had all the pieces of the rainbow in hand.

2. *Harriot's 41°46′*: His method is at BL Add MS 6789 ff289, 292. See also Lohne, "Thomas Harriot (1560–1621)," 118–119. Note that in f289 Harriot's table (like Descartes's; cf. Boyer, *The Rainbow*, 214) shows that the *maximum* "arc of egress" is 20°52′, so the total angle of the cone is twice this (cf. Lohne's diagram), i.e., 41°46′. Harriot also went one step better than Descartes in the mathematical method of his deduction of the angle of incidence (59°17′) corresponding to the angle of maximum egress: Harriot used his value of (what we call) the index of refraction, his law of refraction, and his algebra applied to the geometry of the arc of maximum egress, to deduce that the required angle of incidence can be found from cosine (i) = 0.51079748. Descartes's approach was simply to compare experimental values and see which gave the largest egress (which Harriot also did as a preliminary). Like Descartes, though, Harriot did not specifically mention the *minimum* change of direction (or "deviation") of the incoming ray as we do today, but the two statements are equivalent: 2 times the arc of egress = 180° minus the deviation. (180° is the highest possible deviation, which occurs when a ray from the sun passes head-on into the raindrop (so the angle of incidence is zero), and is reflected straight back toward the sun.) *Least change of direction*: For more contemporary detail, see, e.g. Walker, "Multiple Rainbows from Single Drops of Water and Other Liquids," 422. *For Harriot's maximum egress and exact formula for finding 59°17′*: e.g., BL Add MS 6789 f292. Lohne shows how Harriot uses a result on the maximal intercept of the chord (such as the refracted ray through a raindrop) that gives the maximum angle of return ("Essays on Thomas Harriot," 302).

3. *Average size*: From the analysis in Walker, "Multiple Rainbows from Single Drops of Water and Other Liquids," 422, the cone of red rays has an angle of 42°22′, and blue has 40°39′. The average of these is 41°31′. Nevertheless, as shown in the previous note, Harriot calculated the relevant theoretical angle 59°17′ exactly; see also Lohne, "Thomas Harriot (1569–1621)," 119.

4. *Harriot on dispersion*: His methods and calculations, including his use of his sine law of refraction, are at BL Add MS 6789 ff184r–206r. Most of his experimental and mathematical work on dispersion focused on the difficult enough task of quantitatively differentiating between just two colors, usually red and yellow or orange. It is not clear if he realized red has the smallest index of refraction (e.g., f198): he didn't use modern terminology, and also, because *he* knew what he was doing, he didn't explain it fully during the process of recording and deriving his results. Either way, though, here and elsewhere (e.g., f190), he has shown there are different indices of refraction for different colors. *Do six different "rays" emerge on dispersion of a ray of white light?* Today we say there are six bands of color, each varying continuously in frequency/color.

5. *Descartes on the rainbow.* Boyer, *The Rainbow*, chapter 8. On 203 Boyer wonders if Descartes had somehow heard of Harriot's work and points out Descartes's general "lack of frankness" with regard to his sources; see also 211. *Descartes not correct on color.* Boyer, 217–218.

6. *Harriot's diagrams:* BL Add MS 6789 ff206r, 207. They show that when light enters a prism, the angles of refraction of two emerging rays are "swapped" if the ray undergoes two or three reflections inside the prism, rather than one. This is what happens with the rainbow: in the primary bow, red is on the top and violet is on the bottom, while in the secondary bow, the order is reversed. This is related to the fact that when a ray of sunlight is refracted, red rays have the smallest angle of refraction, and violet the largest: so if the eye sees rays with the angles reversed, it sees the colors reversed.

7. *Newton's first paper on color* (including different raindrops for different colors): Boyer, *The Rainbow*, 243. See also Alan Shapiro, ibid. (p322n20 above).

8. *Northumberland on the rainbow.* From his 1604–5 letter to his son, quoted in Clucas, "'Noble virtue in extremes,'" 270.

9. *Northumberland's trial:* The background is summarized in Nicholls, "The 'Wizard Earl' in Star Chamber." For Fawkes likely being tortured, see 178n55. The entire transcript of Northumberland's trial is given and annotated on 177–189. My details in this and the following two paragraphs are from the transcript and from Nicholls's summary, esp. 175 ("troubled...himself"). *Advice on toleration of Catholics*: Northumberland to James VI, in Kinney, *Elizabethan and Jacobean England*, 137–139.

17: CONVERSATIONS WITH KEPLER

1. *Northumberland as family man:* Batho, "A Difficult Father-in-Law," esp. 744–746. It was Dorothy who brought Northumberland the lease to Syon (744), which James gave to Northumberland outright soon after his accession to the throne (Shirley, "Sir Walter Ralegh and Thomas Harriot," 26).

2. *Still house:* Cordial, Rowse, *Ralegh and the Throckmortons*, 252, 243; recipes (including Bess, Guianans), Beer, *Bess*, 163. *Northumberland's still house in 1607*: Batho, "Thomas Harriot and the Northumberland Household," 42–43.

3. *Prince Henry: Ralegh:* Rowse, *Ralegh and the Throckmortons*, 260; *Harriot*: Tanner, "Henry Stevens and the Associates of Thomas Harriot," 96 (citing BL Add MSS 6782 f27r, 6783 f75r).

4. *Kepler's* Optics: Letter to Harriot, October 2, 1606, English extract in Goulding, "*Chymicorum in morem*," 36n26.

5. *Tycho in Prague.* Gingerich and Voelkel, "Tycho and Kepler," 97.

6. *Harriot-Kepler correspondence.* The extant letters are reprinted in J. Kepler, *Gesammelte Werke*, vol.15 (*Briefe: 1604–1607*), 348–352. They are not available in English translation as far as I know, so I have relied on paraphrases and English excerpts from a variety of sources, cited in the relevant endnotes below.

   Note that Goulding asserts that Harriot initiated the correspondence, via Erikson ("*Chymicorum in morem*"). It is true that Harriot knew of Kepler through his books, and Erikson certainly arrived armed with considerable detail of Harriot's work. On the other hand, it is more consistent with what is known of Harriot's and Kepler's personalities that the cautious Harriot was approached by the outgoing Kepler, who had no hesitation in sending his works out to people for comment. This, in fact, is how he came to meet Tycho (Gingerich and Voelkel, "Tycho and Kepler," 95).

7. *Kepler on astrology.* Quoted in Gauquelin, *The Cosmic Clocks*, 47. For Kepler's views on astrology, see also Grafton, "Kepler as a Reader," 570–572. For reference to Kepler's 1610 defense of astrology, see Boyer, *The Rainbow*, 365. *Unsealed letter.* English excerpt (and Latin) in Goulding, "*Chymicorum in morem*," 36 and n28.

8. *Kepler's concepts theological.* Jacquot, "Thomas Harriot's Reputation for Impiety," 180; Gingerich and Voelkel, "Tycho and Kepler," 90; Lindberg, "Continuity [...]", 436.

9. *Latin verses to the sun:* Gatti, "The Natural Philosophy of Thomas Harriot," 70. But I agree with Mandelbrote, who does not believe these verses and other "pagan" reading lists of

Harriot's tell us anything more concrete about his beliefs than that he read widely ("The Religion of Thomas Harriot," 251–252).

10. *Kepler's letter*: English quotes in this and the next paragraph are from Goulding, "*Chymicorum in morem,*" 37–38.

11. *Northumberland on alchemy*: Henry Percy's "Advices to His Son," reprinted in Clucas, "'Noble virtue in extremes,'" 108.

12. *Harriot, Turner free to visit Ralegh*: Shirley, "The Scientific Experiments of Sir Walter Ralegh, the Wizard Earl, and the Three Magi in the Tower," 53. Turner was retained by Northumberland and was probably Harriot's doctor (Trevor-Roper, "Harriot's Physician," 48). *Turner's recipes*: One is for making "spirit of wine"; the other appears to be for transmuting clay into pewter. They speak of "phlegm" and say that "common menstruum is water [but] salt water is best" ("menstruum" meant "solvent"). Transcribed in Seaton, "Thomas Harriot's Secret Script," 112–113.

13. *Harriot's phonetic script as amusement/doodles/practice*: To take two examples: his signature on the title page of his *Doctrine of Nautical Triangles* (BL Add MS f7 shown in plate 1); in one of his sheets of calculations of angles of refraction, he has put the heading in his script (BL Add MS 6789 f266). *Anagrams for fun*: Bennett, "Instruments, Mathematics, Natural Knowledge," 75.

14. *Harriot's alchemical experiments*: See, e.g., BL Add MS 6788 ff393r, 397r, 403r. Clucas has described and contextualized these ("Thomas Harriot and the Field of Knowledge," 123–126). He also reproduces BL Add MS f403r, which mentions water, earth, and hot ashes. Shirley, *Biography*, 268–287, gives more details.

15. *Kepler's letter*: English quotes from Goulding, "*Chymicorum in morem,*" 38, except I have inserted "mysteries of" in the last quote, following Boyer ("Kepler's Explanation of the Rainbow," 364), who also replaces "initiate" with "priest." Goulding gives the Latin in footnote 35: *O Excellens naturae mysta.*

16. *Nunes*: The manuscript with this quotation was discovered in 1949 (Randles, "Pedro Nunes' Discovery of the Loxodromic Curve," 91–92).

17. *Harriot's letter*: English quotes from Goulding, "*Chymicorum in morem,*" 40 (Latin in n39).

18. *Harriot to Kepler*: Summarized and quoted in Boyer, "Kepler's Explanation of the Rainbow," 364.

19. *Harriot on atoms causing refraction*: Letter to Kepler quoted in Jacquot, "Thomas Harriot's Reputation for Impiety," 181; see also Goulding's translation, "*Chymicorum in morem,*" 41, 45–46.

20. *Harriot's playful metaphor*: Jacquot translates part of Harriot's letter as telling Kepler that he is being "playful" with his theory of refraction ("Thomas Harriot's Reputation for Impiety," 181).

18: ATOMIC SPECULATIONS

1. *Torporley's summary and rejection of Harriot's views on atomism*: Reprinted in the appendix in Jacquot, "Thomas Harriot's Reputation for Impiety."

2. *Torporley on Harriot*: Quote from *Diclides* in Jacquot, "Thomas Harriot's Reputation for Impiety," 168. For an analysis of the *Diclides*, see Silverberg, "Nathaniel Torporley and his *Diclides coelometricae.*"

3. *"Anathema," "pseudo-philosophy"*: Torporley's *Corrector*, quoted in Rosen, "Harriot's Science," 5.

4. *Aristotle*: Strathmann, *Sir Walter Ralegh*, 105–106. *Aristotle on continuity and indivisibles*: *Physics* 6.1.

5. *Rhumb line paradox*: BL Add MS 6786 f349v.

6. *Harriot on Aristotle and Zeno*: BL Add MS 6782 f367.

7. *Finding when Achilles overtakes the tortoise*: BL Add MS 6782 f368. The tortoise has a head start of one unit, and Achilles runs ten times faster than the tortoise. Suppose d is the distance the tortoise has moved when Achilles catches it; then Achilles has moved $1 + d$ units in this time. But Achilles's speed is ten times the tortoise's, so in this amount of time, the ratio of the distances traveled is 10:1. In other words, $1 + d$: $d = 1$: $1/10$, or $(1 + d)/d = 10/1$, so $d = 1/9$ units.

8. *Harriot on infinities/infinitesimals*: BL Add MS 6782 f363. This page is the beginning of quite a number of pages that compose an unfinished draft treatise, *De infinitis* (On infinity), and include his work on the Zeno paradox. (On f362, Harriot says his intention is to try to make Aristotle's concept of the continuum clearer.)

9. *Different amounts of infinity?* "Countably infinite" sets, such as the rational numbers, are those that can be listed one at a time, by 1:1 correspondences with the natural numbers or integers (which are themselves countably infinite); "uncountably infinite" sets, such as the real numbers, are larger, as Georg Cantor proved in the late nineteenth century. *Harriot's contemporaries grappling with infinities* included Galileo; see Boyer, *The History of the Calculus and Its Conceptual development*, 70 (and also chapter 4).

10. *Harriot's infinity/"Much ado"*: BL Add MS 6785 f436. Gatti thinks that the "nothing" in the quote refers to vacuums between atoms or indivisibles ("The Natural Philosophy of Thomas Harriot," 79). Mandelbrote suggests it refers to "nothing comes from nothing" and to Harriot's attempt to clarify Aristotle's view of the continuum ("The Religion of Thomas Harriot," 258–259).

11. *"Fools"*: These two quotes are given in North, "Stars and Atoms," 200, 223, although North does not connect them with Harriot's verse.

12. *Harriot on "nothing"*: BL Add MS 6782, f374v; Jacquot, "Thomas Harriot's Reputation for Impiety," 178. North seems to suggest that atoms were not the only concept of "nothing": Aristotle's potentialities, or the related idea of continuous transformations, also related to *ex nihilo nihil fit* ("Stars and Atoms," 201–202).

13. *Warner on matter*. BL Add MS 4394 ff.396r–399v; cf. Jacquot, "Harriot, Hill, Warner and the New Philosophy," 118. Warner's deduction that "actual" matter cannot involve "potentialities" is logical enough on the face of it. Today, though, quantum theory invokes potentiality in the sense, for example, of Schrödinger's cat having *probabilities* of being dead and alive, until it is actually observed and its condition is known for certain.

14. *Northumberland on alchemical atomism*: Quoted in Clucas, "'Noble virtue in extremes,'" 108. *Warner*. Clucas, 29. *Galileo*: Henry, "Thomas Harriot and Atomism," 281. *Bruno*: Although Bruno spent the early 1580s in England, Northumberland's massive library did not contain his works on the subject (Henry, 275–276); Henry gives good evidence that Bruno did not influence Harriot and the Northumberland circle. However, Gatti disputes Henry and believes Bruno did influence Harriot ("The Natural Philosophy of Thomas Harriot," 77n41). Mandelbrote ("The Religion of Thomas Harriot," 250–251) and Clucas (100) agree with Henry!

15. *Harriot on stacking balls*: BL Add MS 6786 ff375v–376; cf. Harriot, *Thomas Harriot's Doctrine of Triangular Numbers*, 5. *On Harriot, Bhaskara, and Kepler's conjecture*: Hales, "An Overview of the Kepler Conjecture."

16. *Harriot's symbolism for algebraic products*: Instead of using parentheses as we do to indicate multiplication, he wrote our $n(n+1)/2$ or $\dfrac{n(n+1)}{2}$ with the factors one on top of the other, and then the divisor below them, as shown:

$$\dfrac{\begin{array}{c} n \\ n+1 \end{array}}{2}$$

He added a vertical line on the right-hand side bracketing all three rows to indicate multiplication and division rather than addition. It was a beautifully simple advance on the earlier rhetorical expression.

17. *Al-Kindi, Northumberland's library, Warner*. Northumberland owned a manuscript copy of *De radiis*; Clucas, "Thomas Harriot and the Field of Knowledge," 128–129. *Force, radiative virtue*. BL Add MS 6786 f428r; Clucas, "Thomas Harriot and the Field of Knowledge," 128.

18. *Torporley's objections*: His list is reprinted in Jacquot, "Thomas Harriot's Reputation for Impiety," 183–186; see also Goulding, "*Chymicorum in morem*," 44–45, including n47: the notes were a record of his and Harriot's ongoing debate.

19. *Kepler to Harriot*: Jacquot, "Thomas Harriot's Reputation for Impiety," 181; Goulding, "*Chymicorum in morem*," 47–48.

20. *Harriot to Kepler*: Quotation in Goulding, "*Chymicorum in morem*," 48; gold leaf: Jacquot, "Thomas Harriot's Reputation for Impiety," 182; Shirley, Biography, 387.

21. *Harriot and* Nicomachean Ethics: Discussed and paraphrased in Gatti, "The Natural Philosophy of Thomas Harriot," 90.

22. *Harriot to Kepler on Gilbert*: Jacquot, "Thomas Harriot's Reputation for Impiety," 181; Boyer, "Kepler's Explanation of the Rainbow," 365. Gilbert's *De mundo nostro sublunary philosophia nova* was not published until 1651, so Harriot (or Northumberland and his library) must have been connected with Gilbert's circle.

23. *Fighting for Sherborne*: Thompson, *Sir Walter Ralegh*, 228–231; Beer, *Bess*, 188–189; Rowse, *Ralegh and the Throckmortons*, 230–231. Young Prince Henry tried to take it over, presumably so that he could return it to his hero Ralegh, but he died in 1612 (May, *Sir Walter Ralegh*, 22).

24. *King of Denmark*: Thompson, *Sir Walter Ralegh*, 231–232.

25. *"Stuck in the mud"*: My version is a mix of snippets in Kargon, "Thomas Harriot, the Northumberland Circle, and Early Atomism in England," 131; Maclean, "Harriot on Combinations," 85–86; Rosen, "Harriot's Science," 4; and Mandelbrote, "The Religion of Thomas Harriot," 259; most of the original Latin version is given in Maclean's n70.

26. *Galileo to Kepler*: Rosen, "Harriot's Science," 4.

27. *Prohibited ideas for Catholics*: Alexander, *Infinitesimal*, 19, 147–148.

28. *Carew Ralegh*: Thompson, *Sir Walter Ralegh*, 229; Beer, *Bess*, 260. Bess's spirit: See Beer, *Bess* and "Ralegh's History of Her World," and Robertson, "Negotiating Favour," for Lady Ralegh's letters regarding her attempts to fight for her own rights and her husband's over the years.

29. *Kepler's last letter to Harriot*: Boyer, *The Rainbow*, 185; Shirley, *Biography*, 388.

30. *Kepler on packing, atomism*: Hales, "An Overview of the Kepler Conjecture," 5. A mathematical proof of Kepler's conjecture—that face-centered cubic packing is "the tightest possible, so that in no other arrangement could more pellets be stuffed into the same container"—was published only in 2017 (having been checked by scholars for a decade). Its lead author was Hales. Although this is known today as "the Kepler Conjecture," Hales assumes that Kepler took his lead from Harriot, following their correspondence in which Harriot introduced Kepler to his atomic model of matter. So does Chuanming Zong in "A Mathematical Theory for Random Solid Packings," arXiv preprint, 2014, https://arxiv.org/abs/1410.1102.

19: SEARCHING THE SKIES

1. *On William Lower*: I've drawn from Hunneyball, "Sir William Lower and the Harriot Circle."

2. Lower to Harriot, February 6, 1610, transcribed in Rigaud, *Supplement to Dr Bradley's Miscellaneous Works*, esp. 42–44 (quotes in this and the next paragraph). The date is probably 1610 New Style: Batho, "Thomas Harriot's Manuscripts," 287, and Chapman, "A New Perceived Reality," 1.30–31.

3. *Kepler and Gilbert*: Kepler was excited by Gilbert's *De magnete* because the idea that the earth is a magnet made Kepler's intuitive notion of a solar "force" more concrete. Much of his *Astronomia nova* is mathematical, but he sometimes spoke in the emotional, animistic analogies of della Porta and Gilbert. Gilbert believed the planets and sun "collaborated," with "good will," because through their magnetic souls the planets "perceived" the most orderly way of arranging themselves around the sun. Kepler was more inclined to see the sun and planets as having equally strong perceptive souls, so that they engaged in a "wrestling match." Quoted and discussed in Dana Jalobeanu, "'Borders,' 'leaps,' and 'orbs of virtue: Francis Bacon's extension-related concepts," 2017, http://philsci-archive.pitt.edu/

12318/1/Orbs%20of%20virtue%20submitted.pdf. Kepler had also used a more modern-sounding clockwork analogy, although it involved an animistic, alchemical notion of celestial generation and decay (Boner, "Life in the Liquid Fields," 287–290).

4. *Carolingian diagrams*: Eastwood, "Johannes Scottus Eriugena, Sun-Centered Planets, and Carolingian Astronomy."

5. *Kepler versus Tycho et al.*: Gingerich and Voelkel, "Tycho and Kepler," 90; Voelkel, *The Composition of Kepler's "Astronomia Nova." Kepler's attempt at religous reconciliation with Copernicanism*: Quoted in Rosen, "The Intellectual Background," 6-7. For instance, Kepler explained that although Joshua commanded God to stop the sun, God would have stopped the earth instead, because He knew that to humans on earth the sun did seem to move.

6. *Tycho's heirs; confidentiality*: Voelkel, *The Composition of Kepler's "Astronomia Nova,"* chapter 8; Gingerich and Voelkel, "Tycho and Kepler," 98.

7. *Kepler assumed an inverse linear force (not inverse square!)*: Gingerich and Voelkel, "Tycho and Kepler," 92, 102.

8. *Computational device*: Gingerich and Voelkel, "Tycho and Kepler," 102. *Kepler on area of ellipse*: Boyer, *A History of Mathematics*, 325.

9. *Kepler faithful to empirical evidence*: For a short, clear summary of Kepler's method, see Einstein, "Johannes Kepler," 262–266. For further analysis, see Gingerich and Voelkel, "Tycho and Kepler," e.g., 105. *Newton on Kepler's guess*: Gingerich and Voelkel. 106. *Kepler did initially guess*: Miller, "*O Male Factum*," 46; Einstein, 265; Whiteside, "The Mathematical Principles underlying Newton's 'Principia Mathematica,'" 130. Kepler's bizarre speculations included the idea that the planetary orbits fitted one inside the other according to the way the five Platonic solids—built from the five regular polyhedra beginning with cubes and tetrahedra—fit inside each other.

10. *Kepler's orbits*: Goldstein and Hon, "Kepler's Move from *Orbs* to *Orbits*," 83ff.

11. *Newton's theory*: In particular, while Kepler assumed the sun *pushed* the planets around because of the (undefined) power of its "orb of virtue"—to use a term made popular by Gilbert and della Porta—Newton would mathematically define the notion of a physical force. Then he proved (following Robert Hooke's inspired guess) that the required force for planetary motion is not an outward push but an inward force of "attraction," and that it is due to gravity, not magnetism or light. *Final step?* Yes, in the sense that Newton's theory was a physical model, not merely computational. Of course, astronomical physics has continued to progress as new technologies and new theories have opened up the universe beyond Newton's solar system. In particular, Einstein's relativistic notion of gravity has replaced Newton's inverse square law when dealing with the wider cosmos. Newton's theory still correctly describes most interactions in the solar system, which is what it was designed to explain.

12. *Copernicus still needed epicycles*, created by Ptolemy: To keep circular motion, planetary motions needed to be seen as circles rolling along the larger circular "orbit."

13. *Lower to Harriot*: February 6, 1610, published in Rigaud, *Supplement*, 43.

14. *Harriot on ellipses*: BL Add MS 6787 ff455–463, 511–514, 408–412, and many more folios.

15. *Elliptical orbits*: It is possible Harriot was initially inspired by Gilbert's manuscript *De mundo*, where Gilbert claimed that planetary orbits were not circular because of forces from other planets (cf. Pumfrey, "Harriot's Maps of the Moon," 166). It was a brilliant hunch but not a theory: Gilbert was not a mathematician, and he did not specify that the orbits were elliptical. He was arguing on purely intuitive grounds that are scientifically flawed, except he acknowledged that some sort of force was moving the planets (for him it was planetary magnetic souls and the sun's light) and that these forces would impact each other. It is another example of seemingly modern results being guessed at imaginatively and laying the groundwork for others to test and develop them properly.

16. *Elliptical orbits not widely accepted*: Whiteside, "Newton's Early Thoughts on Planetary Motion," 121.

17. *Kepler versus Harriot, Digges*: Jacquot, "Thomas Harriot's Reputation for Impiety," 12–15, including letter from Lower testifying to Harriot's theory; see also Johnson, "The Influence of Thomas Digges on the Progress of Modern Astronomy in Sixteenth-Century England," 406. Today physicists think the universe may well be infinite, although it may also curve in on itself so as to be finite in volume.

18. *Lower to Harriot*: February 6, 1610. Quotations here and in the next two paragraphs are from Rigaud, *Supplement*, 43, 45.

19. *Telescopes for sale, etc.*: Chapman, "A New Perceived Reality," 1.27, referring to a contemporary report by Pierre de l'Estiole; see also Bucciantini, Camerota, and Giudice, *Galileo's Telescope*, e.g. 12, 14, 29–30, 37, 39–40. Harriot made his first telescope? Shirley, *Biography*, 397.

20. *Harriot's moon-drawing priority*: Pumfrey, "Harriot's Maps of the Moon," 163.

21. *Gilbert and Harriot mapping the moon*: Lower described the shapes on the moon as like "coasts"; quoted in Chapman, "The Astronomical Work of Thomas Harriot," 102. See also Pumfrey, "Harriot's Maps of the Moon" and "The *Selenographia* of William Gilbert"; Chapman, "A New Perceived Reality," 1.31; Alexander, "Lunar Maps and Coastal Outlines." All these scholars have made a compelling case that Harriot was mapping the moon.

22. *History/comparison of moon drawings*: Chapman, "A New Perceived Reality," 1.31.

23. *Lunar libration*: In "Harriot's Maps of the Moon" and "The *Selenographia* of William Gilbert," Pumfrey makes a good case that following Gilbert, Harriot made the first intentional mapping of lunar libration, the phenomenon first mentioned in print by Galileo three decades later. Pumfrey describes Harriot's record of libration, as well as Gilbert's cosmological program, and gives a good explanation of the modern understanding of the cause of libration.

24. *Gilbert's deduction* from lunar libration of axis tilt is shown in the description on 187ff. and the diagram on 190 of the 1651 edition of his *De mundo*.

25. Starry Messenger *or* Message? Rosen, in his introduction to his translation of Kepler's *Conversation with the Starry Messenger*, xiv–xv, prefers *Message*; he explains why, and why Kepler's use of "Messenger" was controversial. Note that Chapman ("A New Perceived Reality," 1.30–31) shows that Harriot was spurred on but not influenced by Galileo's drawings in *Siderius nuncius*; he also points out that while Galileo was the better artist, Harriot was the better draftsman-surveyor. *Harriot reading Kepler's* Conversation: The evidence is discussed by Bucciantini, Camerota, and Giudice, *Galileo's Telescope*, 132.

26. *Kepler's quotation* is from Rosen's translation of *Conversation with the Starry Messenger*, 12; Rosen's 71n85 gives an excerpt from Galileo's grateful reply.

27. *1616 decree on heliocentrism*: The exact wording is given by Graney, "The Inquisition's Semicolon," 8–9.

28. *Lower to Harriot*: June 1610, published in Rigaud, *Supplement*, 26.

29. *Bessel, Halley's comet*: Chapman, "The Astronomical Work of Thomas Harriot" 101; Roche, "Harriot, Oxford, and Twentieth-Century Historiography," 240. Bessel knew of Harriot's observations because Zach (of whom more in the epilogue) had published them in the late eighteenth century: Schemmel, *English Galileo* 1:20.

30. *Galileo's anagram, Kepler, Harriot and friends*: Here and in the following paragraph I have drawn from North, "Thomas Harriot and the First Telescopic Observations of Sunspots," 137–138. North doesn't give the bawdy rearrangement of the anagram! The friends are named only by their initials, and North assumes Torporley and Warner are most likely. Kepler published the correct answer later: Bucciantini, Camerota, and Giudice, *Galileo's Telescope*, 145.

31. *First telescopic records of sunspots*: In "Thomas Harriot and the First Telescopic Observations of Sunspots," North reproduces Harriot's first observation (132 and plate 4) and notes Galileo's various, inconsistent, claims of his own priority (135–136). North accepts December 1610 as the earliest reasonable date for Galileo and discusses Harriot's priority/ independence (132, 135–136, 153, and for the moon 136). In this article, North also gives a detailed analysis of Harriot's sunspot methods and results.

32. *"Sight was dim"*: Harriot's notes transcribed by Rigaud, *Supplement*, 34.

33. *Armillary sphere:* Lower to Harriot, quoted in North, "Thomas Harriot and the First Telescopic Observations of Sunspots," 150 and n92. The sphere probably helped fix the ecliptic, which, as North points out, Harriot drew on all his sunspot drawings; it helped him realize the direction of the sun's rotation.

34. *Harriot's compilation and accuracy:* North, "Thomas Harriot and the First Telescopic Observations of Sunspots," 152–155, 139; Herr, "Solar Rotation Determined from Thomas Harriot's Sunspot Observations," 1079. Pepper suggests Harriot's probable priority on solar rotation in "Thomas Harriot: A Biography," 216.

35. *First to calculate orbits of Jupiter's moons:* Rigaud, *Supplement,* 31. *Lower, "great glass":* Rigaud, 28. See also North, "Thomas Harriot and the First Telescopic Observations of Sunspots," 137, 143.

36. *Harriot's accuracy:* Herr, "Solar Rotation Determined from Thomas Harriot's Sunspot Observations," 1079, and Lohne, "The Fair Fame of Thomas Harriott," 79 (sunspots); Chapman, "A New Perceived Reality," 1.32, citing J. J. Roche's entry in the *Oxford Dictionary of National Biography* (Jovian moons; but see critique of one of his satellite measurements in Rigaud, *Supplement,* 31); Chapman, 1.31–32 (lunar map).

37. *Recorde, Dee, Digges:* Johnson, "The Influence of Thomas Digges on the Progress of Modern Astronomy in Sixteenth-Century England."

38. *Harriot's critics:* See, e.g., North, "Thomas Harriot and the First Telescopic Observations of Sunspots," 155–157; Henry, "Why Harriot Was *Not* the English Galileo," 122–123. *Sunspot debate:* North, 134–135; "This Month in Physics History," *APS News* 24, no. 3 (March 2015). *Harriot agreed they were "spots in the sun":* North, 132. *Sunspots show both sun and earth rotate,* although Galileo tried, unsuccessfully, to use them to prove earth orbits, too: Hutchinson, "Galileo, Sunspots, and the Orbit of the Earth," 70. *Harriot and Galileo equally good observational astronomers:* North, 132, 147.

39. *Galileo, Harriot, and patronage:* Chapman, "A New Perceived Reality," 1.32; see also Pumfrey, "Patronizing, Publishing and Perishing," for an analysis of the effect of patronage on Harriot's publication record. *Galileo's demonstrating and promoting his telescope:* Sobel, *Galileo's Daughter,* 31; Bucciantini, Camerota, and Giudice, *Galileo's Telescope,* 38–39.

40. *Galileo censored, warned:* Mayer, "The Censoring of Galileo's *Sunspot Letters* and the First Phase of His Trial" and "The Roman Inquisition's Precept to Galileo."

41. *Galileo's sunspot argument analyzed:* See, e.g., Topper, "Galileo, Sunspots, and the Motions of the Earth"; Hutchinson, "Galileo, Sunspots, and the Orbit of the Earth." *Flaws:* Topper, 765ff. (i.e., Galileo did not account for precession); Hutchinson, 68–71. Galileo also adduced the phases of Venus as evidence that Venus moved around the sun as the moon moves around the earth.

42. *Transcripts of Galileo's trial* are reproduced in Sobel, *Galileo's Daughter,* e.g., 284, 290.

20: GRAVITY

1. *Lack of clarity about rates of changing speed:* Schemmel, *The English Galileo* 1:53.

2. *Harriot's experiments and calculations on free fall:* Schemmel gives a detailed analysis in *The English Galileo,* vol. 1, chapters 4 and 5. (The relevant manuscript pages are reproduced in vol. 2.) For falling 55.5 feet, etc., see, e.g., 130.

3. *Harriot's experiment:* Schemmel gives a modern analysis of what Harriot's balance experiment was actually measuring, and he replicated the experiment himself (*The English Galileo* 1:99–101, and cf. his appendix E). He shows that it could indeed determine that the rate of change of speed was proportional to time, as Harriot deduced.

4. *Harriot's hypothesis testing:* To take just one example, see Petworth House/Leconfield HMC 241 VI f1r, reproduced and discussed in Schemmel, *The English Galileo* 1:121.

5. *Galileo, Two New Sciences,* the Third Day.

6. *Three units in the second time interval:* Some earlier scholars had got this far, but it seems none of them continued the progression, let alone articulated the times-squared law; Lohne, "Essays on Thomas Harriot," 235–236.

7. *Galileo assuming constant acceleration: Two New Sciences,* the Third Day.

8. *Is gravitational acceleration constant?* It is essentially constant over the small distances covered in everyday falling motion, but Newton's (inverse square) law of gravity takes into account the way gravity diminishes with distance from the source. His law comes into its own in studying the effects of gravity in the solar system, where, for most purposes, it is an excellent approximation to Einsteinian gravity.

9. *Harriot on fall in a medium:* Schemmel, *The English Galileo,* chapter 6, esp. 1:137 incl. n2, and 1:151 for Harriot's meticulous experimental and mathematical analysis but flawed theory. *Weight and falling motion:* As Newton would show—and as is evident in everyday experience—the *force* of the impact of a freely falling body does depend on its mass. Newton defined weight itself as a force. It is equal to the body's mass times the gravitational acceleration. Finding adequate quantitative definitions that distinguished between intuitive concepts such as weight, force, acceleration, and impetus was a difficult task that largely eluded Galileo, Harriot, and their contemporaries. Even Newton had trouble articulating subtle concepts such as inertia and energy, which achieved their modern form only in the nineteenth century.

10. *Harriot law of free fall:* Schemmel, *The English Galileo,* e.g., 1:82–88, 130, 135; Lohne, "Essays on Thomas Harriot," 236. *More rigorous than Galileo:* Schemmel 1:238–240.

11. *Galileo on arrows and other projectiles:* Quoted in Lohne, "Essays on Thomas Harriot," 241.

12. *Northumberland in Low Countries:* Shirley, "The Scientific Experiments of Sir Walter Ralegh, the Wizard Earl, and the Three Magi in the Tower," 60, and *Biography,* 295. *Harriot on water flow at Syon:* His manuscripts are analyzed by Clucas, "Thomas Harriot and the Field of Knowledge," 111–117. *Northumberland and Archimedes:* Clucas, 112–113.

13. *Northumberland and Archimedes:* Clucas, "Thomas Harriot and the Field of Knowledge," 112–113. *Northumberland's tract (weight of water):* BL Birch MS 4458 ff4–5; Lohne, "Essays on Thomas Harriot," 226.

14. *Reading on peace of mind:* Clucas, "'Noble virtue in extremes.'" *Northumberland's war games, leaden soldiers, manuscript on war:* Batho, "Thomas Harriot and the Northumberland Household," 42.

15. *Harriot on military embankments and troop organization:* BL Add MS 6788 ff50r–51r, 55r–58r.

16. *Harriot on projectile motion:* For detailed analyses, see Lohne, "Essays on Thomas Harriot," and Schemmel, *The English Galileo,* vol. 1. *Galileo beginning with projectiles:* Schemmel 1:240 (cf. Harriot's path, 238). *Galileo on projectiles:* Schemmel 1:237.

17. *Harriot on precursor to law of inertia (Newton's first law):* BL Add MS 6789 f30r; see also Schemmel, *The English Galileo* 1:236.

18. *Oblique projectiles:* Galileo's flawed reasoning: Schemmel, *The English Galileo* 1:237; Damerow et al., *Exploring the Limits of Preclassical Mechanics,* e.g. 253, 264. Harriot's flawed reasoning: Schemmel 1:227 (cf. section 8.6 for detailed analysis). For a general analysis of the flaws and achievements in Harriot's and Galileo's approaches to falling motion, see Schemmel, e.g., 1:233–241.

19. *Harriot's superior algebraic prowess in analyzing physical problems:* Schemmel, *The English Galileo* 1:46, and 1:236–237 for Harriot's *parabolic trajectories* (more rigorous than Galileo, partly [1:46] because Galileo lacked Harriot's algebraic skill and symbolism). *For an example* of Harriot's correct use of mathematics despite flawed physics, see Stedall, "Symbolism, Combinations, and Visual Imagery," and Lohne, "Essays on Thomas Harriot."

20. *Was Harriot lesser than Galileo, Descartes (and Kepler and Tycho)?* This is the view especially of Henry in "Why Harriot Was *Not* the English Galileo." North concurred with regard to astronomy, saying that unlike Galileo and others, Harriot made no *causal* explanations of the phenomena he observed ("Thomas Harriot and the First Telescopic Observations of Sunspots"; North later qualified his position—see below). For such scholars, having a go seems to be key: they seem to gloss over the fact that almost all the early seventeenth-century speculations about causes were wrong or weak. Yet, as pointed out by Scott Hyslop in his review of Fox, *Thomas Harriot,* vol. 2 (*Aestiomatio* 10 [2013]: 293–313, esp. 302), Henry dismisses the causal importance of Harriot's atomism precisely because he thinks it

was weak! In fact, Derek Whiteside, in his review of Shirley, *Thomas Harriot: Renaissance Scientist* (*History of Science* 13 [1975]: 61–70, esp. 61), takes the opposite view of Harriot's attempted theory, speaking of his "physical explanation of [reflection and refraction at an interface founded] upon a sophisticated atomic substratum." Schemmel, too, points out that in his mechanics Harriot was not interested only in solving isolated practical problems, but sought to use general physical principles (*The English Galileo* 1:19). And in "Stars and Atoms," 189n8, North qualified his earlier position, saying that he "meant only that they [Harriot and Galileo] entered history on a different evidential footing, and not that one was intrinsically inferior to the other."

Bennett, however, consigns Harriot to a lower rank than Galileo, Kepler, and even Tycho ("Instruments, Mathematics, Natural Knowledge"), although he disagrees (e.g., 150) with those who say that Harriot did not combine mathematics with natural philosophy. Clucas shows Harriot as a brilliant man of his times and not a modern scientist ahead of his time ("Thomas Harriot and the Field of Knowledge"). But Bennett and Clucas are particularly concerned to highlight Harriot's considerable achievements, and to judge him according to his own times, not ours. Henry and North also give due praise to Harriot's achievements. Of course, the exercise of comparing Harriot with his better-known peers is fraught—and indeed all these recent critics were trying to counter earlier scholars (such as Shirley and Lohne) who, as the first to discover Harriot's astounding achievements, naturally sought to place him alongside his famous contemporaries. All of these critics make pertinent points, as of course did Lohne and Shirley.

This debate has been important for formulating a more nuanced assessment of Harriot's position in the scientific pantheon. His is a case of apples rather than oranges: as I've already indicated and will discuss further in the following chapters, he did not provide a speculative theoretical overview of cosmology or matter and motion, but he *did* provide a uniquely diverse body of individual results and methodologies, which could have had an equally important effect on the development of science and mathematics, if they had been published. Newton's ambition to formulate his own "system of the world" in the first place may have been indirectly derived from the speculative syntheses of Galileo and Kepler. But he hadn't read *Two New Sciences* when he wrote *Principia* (cf. *Dictionary of Scientific Biography* on Galileo). In fact, *Principia* was especially influenced by *individual* landmark results rather than grand syntheses—landmark results such as Harriot's might have been, had they been completed and published. Aside from Kepler's bravura planetary laws, Newton used other individual results such as the laws of free fall, collisions (on which Galileo did little), parabolic trajectories, and refraction, all of which Harriot pioneered. Newton also used some rediscovered mathematical techniques that Harriot had developed.

To sum up the debate over whether or not Harriot was "the English Galileo" as Schemmel suggests, while Galileo was arguably more "modern" than Harriot in his willingness to seek causes and unifying connections between phenomena, Harriot was more "modern" than Galileo in his use of experiment to test hypotheses. Galileo did experiments, of course, but he often relied on intuition to justify his conclusions, and he generally tested only preconceived ideas; Harriot was more rigorous and open in his appeal to experimental data and mathematical proof. It is not a matter of either/or: modern science combines both these approaches. And in the end, neither Galileo nor Harriot nor any of their peers was truly modern. But their work certainly pointed toward the way of the future, and it helps us to better understand and appreciate the concepts and processes we take for granted today.

21. *Harriot, "better ability"*: Quoted in Lohne, "Essays on Thomas Harriot," 205. This letter, and presumably (cf. Pepper below) his tract on collisions, is from 1619, but it is a collection of results made many years earlier (Harriot calls them his "ancient" results) (Pepper, "Harriot's Manuscript on the Theory of Impacts"). Pepper says it is "not at all unlikely" that the manuscript under discussion is the one referred to in the letter to Northumberland (132–133). *Harriot's first principles in projectile motion*: Schemmel, *The English Galileo* 1:25 and subsequent analysis.

22. *New World disease?* Pepper, "Harriot's Manuscript on the Theory of Impacts," 132n6. Harriot's vague, intermittent ill health as early as 1605, when he was forty-five, does not seem to be related to his later issues from his New World habit of smoking. Some of it was surely stress related. Between bouts of illness and drama, he seemed to enjoy reasonably good health, judging by his intellectual output.

21: MATHEMATICS, JAMESTOWN, GUIANA

1. *Predecessors in combinatorics:* Edwards, *Pascal's Arithmetical Triangle,* e.g., 20–22; Bennett, "Instruments, Mathematics, Natural Knowledge," 81–83; Rabinovitch, "Rabbi Levi Ben Gershon and the Origins of Mathematical Induction." *Harriot's clarity via symbolism:* See esp. the opening pages of his *Magisteria magna,* BL Add MS 6782 ff108–111 (f110 is shown in plate 5 in Chapter 15 of this book).

2. *Developing Taylor's series from exact interpolation formula:* The *differences* in the Harriot-Newton-Gregory interpolation formula are the changes in a function evaluated at two nearby points; they are related to the later concept of *derivatives,* which are such differences divided by the distance between the points (which in the limit is infinitesimally small); see Edwards, *Pascal's Arithmetical Triangle,* 13. See also Boyer, *The History of the Calculus and Its Conceptual Development,* 234. By replacing differences with derivatives in the interpolation formula, first Newton, and then, independently, Brook Taylor, developed what is now known as Taylor's series, which enables complicated mathematical functions to be written as simple sums of increasing powers of x (Gregory had earlier found a special case of it; cf. Whiteside, "The Mathematical Principles underlying Newton's 'Principia Mathematica,'" 120.)

3. *Lower to Harriot:* In Harriot, *Thomas Harriot's Doctrine of Triangular Numbers,* 13.

4. *Sanderson versus Ralegh, Harriot, et al.:* The extant evidence is discussed in Shirley, "Sir Walter Raleigh's Guiana Finances." See also McIntyre, "William Sanderson," who also points out that, initially because of debts to the moneylender George Pitt, Sanderson spent seven years in a debtors' prison, from 1613. She also discusses the sons' feud (drawing on a note in Quinn, *Roanoke Voyages,* 2, 577n3).

5. *Northumberland's 1611 interrogation, battle over fine:* Clucas, "'Noble virtue in extremes'"; councilor's quote, 281.

6. *Lower to Harriot:* February 6, 1610, transcribed in Rigaud, *Supplement to Dr Bradley's Miscellaneous Works,* 45.

7. *Usury: Aristotle, Henry VIII:* Jane Gleeson-White, *Double Entry* (Sydney: Allen and Unwin, 2012), 96, 25. *Ancients' compound interest (e.g., Mesopotamians, Indians):* Boyer, *A History of Mathematics,* 30, 211. *Tacit acceptance:* Gleeson-White, 54; Biggs, "Thomas Harriot on Continuous Compounding," 66.

8. *Harriot on compounding interest/exponential series/finite limit:* BL MS 6782 ff67–70; context and analysis of Harriot's work are in Biggs, "Thomas Harriot on Continuous Compounding," and Stedall, "Symbolism, Combinations, and Visual Imagery," 393–394 (including Taylor, on whom see also Persson, Rafeiro, and Wall, 4, and Edwards, *Pascal's Arithmetical Triangle,* 14). *Indian series:* Roy, "The Discovery of the Series Formula for π," 300–305; Krishnachandran, "On Finite Differences, Interpolation Methods and Power Series Expansions in Indian Mathematics"; Stillwell, *Mathematics and Its History,* 120.

9. *Botero:* Sokol mentions him ("Thomas Harriot...on Population"), but for Botero's commentary, see "Giovanni Botero on the Forces Governing Population Growth," *Population and Development Review* 11, no. 2 (1985): 335–340.

10. *Fibonacci:* In 1202, Leonardo of Pisa published the rabbit problem in his *Liber abaci,* which was heavily indebted to the work of the ninth- and tenth-century Arab mathematician Abu Kamil and his predecessor al-Khwarizmi (whose algebraic work gave us the term "algebra," from the Arabic *al-jabr.* Fibonacci likely relied on Gerard of Cremona's twelfth-century Latin translation of al-Khwarizmi, who is often called the "father of algebra" because of his systematic treatment of linear and quadratic equations, although his work was entirely rhetorical). See, e.g., Scott and Marketos, "On the Origin of the Fibonacci Sequence."

11. *Harriot's calculation of increasing unchecked population growth*: See, e.g., BL Add MS 6788 f536. He assumed a reproductive age of 20, and that each male and female pair produces a child a year, one year male and the next year female. After 20 years the original pair— Adam and Eve?—has produced 20 new offspring, 10 male and 10 female. Because we need to consider pairs of parents, consider just the males. The oldest child, assumed to be male, is 20 at t = 20 years. After 30 years, he will be 30, and he will have produced 10 children (5 male, 5 female) in the 10 years from t = 20 to t = 30. At t = 30, the next oldest male is 28 and has produced 8 children, 4 of them males, and so on down to a male aged 22, who has produced two children, i.e., 1 male child. The younger siblings cannot yet reproduce. After 30 years, then, there are the 20 original new children plus $5 + 4 + 3 + 2 + 1 = 15$ male grandchildren, and the same number of female grandchildren. Harriot uses the formula $1 + 2 + 3 + \ldots + n = n(n+1)/2$ to add such sums, recognizing that this is the nth triangular number. The total number of new people at t = 30 is thus the original 20 children, plus 10 new children from the original couple, plus 30 grandchildren = 60 people. The same method gives 150 people after 40 years. After 50 years, great-grandchildren are born, and the number of this new generation turns out to be the sum of tetrahedral numbers, whose general formula is $n(n+1)(n+2)/6$, as noted in chapter 17. And so on. He applied this method every decade for 100 years (and more on other folio pages).
12. *Population in James I's reign*: Sokol, "Thomas Harriot…on Population," 208n6.
13. *Harriot on population growth*: See, e.g., BL Add MS 6782 f31. Harriot's scattered manuscripts on the topic are deciphered and analyzed in Sokol, "Thomas Harriot…on Population." *2012/10 billion*: US Census Bureau figure cited in Jeff Wise, "About That Overpopulation Problem," *Slate* online magazine, January 9, 2013.
14. *Population in* History of the World : Pt. 1, ch. 8, par. 4.
15. *Harriot dating Easter, comparing translations, etc.*: See, e.g., Petworth House/Leconfield HMC 241 I ff59–73, BL Add MS 6789 ff487–91, etc. *Doing this for Ralegh*: Mandelbrote, "The Religion of Thomas Harriot," 256.
16. *Ralegh's* History : Connection with the "Golden Speech": Becker, *A Modern Theory of Language Evolution*, 48; see also May, *Sir Walter Ralegh*, 132. *"Too saucy"*: Contemporary observer Chamberlain, quoted in Rowse, *Ralegh and the Throckmortons*, 271. *Good government*: May, 131–133. *Bestseller*: Rowse, 271–272; Strathmann, *Sir Walter Ralegh*, 255. *The King James Bible* was intended to be a political coup for James—a vernacular Bible translated with due respect for Anglicanism, the English nation, and, of course, kings, unlike earlier Calvinist renderings; Christopher Hitchens, "When the King Saved God," *Vanity Fair*, April 1, 2011.
17. *Critique of* History: See, e.g., May, *Sir Walter Ralegh*, 88ff.; but see 87–88, 129–135 for a generally positive assessment of Ralegh's contribution to literature. Strathmann outlines generally positive seventeenth-century responses to the *History*, followed by later less positive ones (*Sir Walter Ralegh*, 255–262).
18. *Ralegh's* History: Quoted in Rowse, *Ralegh and the Throckmortons*, 267–269. For a detailed analysis see Strathmann, *Sir Walter Ralegh*.
19. *Northumberland's fine, John Ford*: Clucas, "'Noble virtue in extremes,'" 28off. Clucas also discusses a similar earlier dedication by the poet John Davies, which was censored before publication.
20. *Bond's conjecture*: Rickey and Tuchinsky, "An Application of Geography to Mathematics," 164; Pepper, "Harriot's Calculation of the Meridional Parts," 360–361. John Collins and N. Mercator, who had known of Harriot's work on interpolation, were instrumental in raising awareness of Bond's conjecture: apparently they did not know of Harriot's work, but their role highlights the kind of influence Harriot might have had, if he had had more, and more long-lived, disciples, even if he had not published.
21. *Harriot's logarithms*: He did not use the notation *ln* or the modern concept of a natural logarithm as a well-defined mathematical function. But to carry out his addition of secants he was using the basic idea of logarithms as powers to a base, and his base number can now be shown to be *e*. For a detailed account of Harriot's method, see Pepper, "Harriot's Calculation of the Meridional Parts," 377ff; independent of Napier, 390. On 377, Pepper

claimed that Harriot did not have a base for his logarithms, a judgment he later corrected ("Thomas Harriot and the Great Mathematical Tradition," 24). Of course, Harriot later did know of Napier's logarithms, but this part of his loxodrome work had been completed before 1614 when Napier published.

22. *Harriot's proof:* In "Harriot's Calculation of the Meridional Parts," Pepper has given a detailed analysis of Harriot's process, including (367ff.) his proof of the trigonometric identity $\tan\left(\dfrac{\pi}{4} - \dfrac{\varphi n}{2}\right) = \left(\tan\left(\dfrac{\pi}{4} - \dfrac{\varphi 1}{2}\right)\right)^{n}$, where φn and $\varphi 1$ are the nth and first latitudes, respectively. Pepper notes a slight error that made Harriot's process more cumbersome than it should have been (and not quite exact, according to Sadler), but concludes by expressing "admiration" for such an astonishing piece of work. Sadler is similarly impressed—as is anyone who studies Harriot's intricate process.

23. *Accuracy of Harriot's final tables:* Pepper shows that although Harriot used a slightly incorrect value for a key quantity, his tables are accurate to four decimal places, and three decimal places for very high latitudes, the latter degree of accuracy still "unique" among tables ("Harriot's Calculation of the Meridional Parts," 382–383).

24. *Harriot's polynomials:* Seltman gives a good overview of his method ("Harriot's Algebra"). See also Tanner, "Nathaniel Torporley's 'Congestor analyticus' and Thomas Harriot's 'De triangularis rationalium.'"

25. *Mayerne:* Trevor-Roper, "Harriot's Physician," 48 (and 57, "noble" plant, next paragraph); see also next note. Shirley *(Biography,* 436–441) reproduces Harriot's detailed observations of his symptoms, which show the "modern", scientific nature of his approach.

26. *Harriot and Mayerne:* My account of their "outsider" relationship, Harriot's letter, and Mayerne's diagnosis, tobacco, etc., is based on Trevor-Roper, "Harriot's Physician," 48–49, 57–58, and Tanner, "Henry Stevens and the Associates of Thomas Harriot," 92–94. Shirley (*Biography,* 442) had a different view of their relationship, suggesting Harriot's letter was a required declaration of faith demanded by the Huguenot Mayerne, but Trevor-Roper has studied Mayerne's papers and gives good (albeit speculative) reasons for supporting Tanner's view, that the famous postscript was elicited by sympathy, not antipathy, between the two men" (49). The declaration that Harriot appends in his letter speaks of his "three-fold" faith but goes on to refer not to the standard Christian trinity but to God, medicine, and the doctor as God's minister; Shirley (441–442) says this is a paraphrase of a then-common addition to Ecclesiastes. Harriot's letters to Mayerne (Shirley, 434–441) show him as respectful and affectionate, collaboratively engaged with his doctor in his healing.

27. *Ralegh's later essays:* E.g., *On the Seat of Government* (fragment), *Dialogue between a Counsellor of State and a Justice of Peace* (later published posthumously as *The Prerogative of Parliaments), Discourse on... War.* Strathmann says Ralegh's political writings were unsystematic and lacked a comprehensive philosophy of the state (*Sir Walter Ralegh,* 161). See also May, *Sir Walter Ralegh,* on customs duties (106, from Ralegh's *Inventions*) and natural and unnatural war (107).

28. *James in debt, Guiana proposals:* May, *Sir Walter Ralegh,* 22. *Ralegh bribing courtiers; seeing London's changes* (from a contemporary eyewitness): Coote, *A Play of Passion,* 352; Rowse, *Ralegh and the Throckmortons,* 308. *Raising money:* Thompson, *Sir Walter Ralegh,* 263, Rowse, 308. Both authors give an analysis of the politics and practice re Guiana. So does Beer, *Bess,* 192ff., from an interesting perspective more sympathetic to Bess than to Ralegh.

29. *Pocahontas visiting Northumberland:* Salmon, "Thomas Harriot and the English Origins of Algonkian Linguistics," 15.

30. *George Percy:* Quinn, "Thomas Harriot and the New World," 49.

31. *Smith: ritual initiation?* See, e.g., Quinn, *Set Fair for Roanoke,* 375. *Pocahontas, Rolfe:* Sarah Stebbins (drawing also on indigenous oral accounts), "Pocahontas: Her Life and Legend," on the National Park Service's *Historic Jamestowne* website.

32. *Kidnapped Powhatans, Mace:* Quinn, *Set Fair for Roanoke,* 356; Harriot teaching Mace, 375. *Powhatans on Thames:* Vaughan, "Sir Walter Ralegh's Indian Interpreters," 358. Vaughan says there is no evidence Ralegh or Harriot was involved in their imprisonment.

33. *Ralegh's journal*: Quoted in Vaughan, "Sir Walter Ralegh's Indian Interpreters," 368; that the chief was Harry is almost certain.
34. *Ralegh to Bess*: Letter 217, *The Letters of Sir Walter Ralegh*, 346.
35. *Scum*: Ralegh's *Apology*, quoted in Thompson, *Sir Walter Ralegh*, 274, and Rowse, *Ralegh and the Throckmortons*, 310.
36. *Wat's personality, Jonson*: Beer, *Bess*, 194–195.
37. *Details of the Guianan disaster* here and in the following paragraphs are from Ralegh's letters to Winwood (March 1618, letter 218), to Bess (March 1618, letter 219), and to Lord Carew (June 1618, letter 220), in *The Letters of Sir Walter Ralegh*; Rowse, *Ralegh and the Throckmortons*, 308–315; Coote, *A Play of Passion*, chapter 16 (including Ralegh's instructions, 360–361); Beer, *Bess*, 205–212.
38. *Copy of Ralegh's plans*: May, *Sir Walter Ralegh*, 22.
39. *Spanish dowry negotiations; Gondomar*: Thompson, *Sir Walter Ralegh*, 319, 267, 264–266.
40. *The Spanish knew Ralegh's plans*: Ralegh's letter to Bess (*The Letters of Sir Walter Ralegh*, letter 219, esp. 355) says that James had "commanded" him to set down in writing his plans, including his destination and the number of his ships and weaponry, which were sent by the ambassador to Philip. Steven May (*Sir Walter Ralegh*, 22) agrees that this means James supplied the itinerary to Gondomar; see also Nicholls, "Last Act?" 165, and Thompson, *Sir Walter Ralegh*, 277.
41. *Ralegh to Bess*: *The Letters of Sir Walter Ralegh*, letter 219.

22: THE END OF AN ERA

1. *Harriot's interests in 1617–18* are deduced from a list of books he purchased, as discussed first by Gatti ("The Natural Philosophy of Thomas Harriot,") and then by Mandelbrote ("The Religion of Thomas Harriot," 253–256), who makes the case for Harriot's theological reading being on behalf of Ralegh (for the *History* and other essays) and Northumberland. *Doctrinal debates*: Mandelbrote, 253ff.
2. *Harriot's* On Triangular Numbers: BL Add MS 6782 ff107ff., now published, with helpful annotations, in Harriot, *Thomas Harriot's Doctrine of Triangular Numbers*. These annotations are necessary—Harriot's masterpiece of economy needs a few more words of explaining! The editors (Beery and Stedall) translate *magisteria* simply as "doctrine." Pepper translates *magisteria* as "masterpiece" in Harriot's tract on collisions.
3. *Influence of Harriot's interpolation method*: For a detailed overview of the use Torporley and Warner made, and through them Pell, Collins, N. Mercator, and possibly Briggs, see Harriot, *Thomas Harriot's Doctrine of Triangular Numbers*, 20–47, 52 (Newton independent, 50; Cavendish quote, 3; Collins and Gregory, 43; Wallis method, 47–49; note [49] that Newton's final method was "even simpler than Harriot's and very much superior to Wallis's"). *Wallis, analytic math from Harriot*: Boyer, *The History of the Calculus and Its Conceptual Development*, 168–169; via Harriot's mss and *Praxis*: Stedall, "Rob'd of Glories", 483.
4. *Ralegh's Apology (and other documents)*: Coote, *A Play of Passion*, 367; Thompson, *Sir Walter Ralegh*, 316–324.
5. *Bess planning the escape*: Beer, *Bess*, 213. *Ralegh plotting with Huguenots?* Some biographers— e.g., Coote, *A Play of Passion*, 355 (quoting V. T. Harlow, *Ralegh's Last Voyage*)—say that his intention had always been that the French, not the English, could attack the Spaniards if necessary, and Ralegh would therefore be seen to have kept his word. Rowse says, of a "possible Anglo-French attack on the Spanish empire," that "there were no limits to Ralegh's ambition" (*Ralegh and the Throckmortons*, 314). But Thompson disagrees with Harlow (and presumably Rowse): he believes Ralegh was, almost without exception, "the most honest man of his time" (*Sir Walter Ralegh*, 304), and in the preceding and following pages gives a very detailed account of the saga—including the fact that James had been part of the initial planning of a French-English venture in Guiana (325).
6. *Betrayal*: Thompson, *Sir Walter Ralegh*, 324–327. The betrayer was Lewis Stukeley, "the Judas of Devonshire."

7. *Advice of commissioners*: Nicholls discusses the process ("Last Act?" 172–174), including the commissioners' reactions to Coke's recommendation (cf. next paragraph). *King of Spain; signing death warrant*: Coote, *A Play of Passion*, 369, 370.

8. *Due process in Ralegh's condemnation?* Nicholls says there was, and that James "understandably" saw disloyalty in the fact that Ralegh did have a French commission and offer of asylum ("Last Act?" 171). But Thompson says that proof against his "alleged disloyalty on negotiations with Frenchmen and others is *that he came back*" (*Sir Walter Ralegh*, 277). Thompson puts more blame (if there *was* an offence) on James' wife, who supported Ralegh (334). *Bacon*: Quoted in Thompson, 325. Thompson and other Ralegh biographers give detailed accounts of the evidence.

9. *Chief Justice Montagu*: Quoted in Coote, *A Play of Passion*, 372, and Nicholls, "Last Act?" 174.

10. *Harriot's notes*: BL Add MS 6789 f533r; also transcribed in Shirley, "Sir Walter Ralegh and Thomas Harriot," 16–17 and plate 1. See also Fleck, "'At the time of his death,'" 17. *Last of the Elizabethans* is Thompson's subtitle for *Sir Walter Ralegh*. Ralegh's complete speech is transcribed in Oldys and Birch, *Lives*, 691–696.

11. *Eyewitness*: Courtier Roger Twysden, quoted in Beer, *Bess*, 225.

12. *Lady Ralegh* to Nicholas Carew, on the day of the death of "my noble husband," quoted in Beer, *Bess*, 226.

13. *Bess Ralegh: promoting the legend*: Beer, *Bess*, 231–234. The 1603 farewell letter was printed or copied many times in the seventeenth century (Strathmann, *Sir Walter Ralegh*, 138). *Lawsuits*: Beer, *Bess*, 235ff.; *Carew restored*: 253–254.

14. *No dated telescopic or experimental observations after Ralegh's death*, aside from the comet: Batho, "Thomas Harriot's Manuscripts," 43.

15. *Kepler witch trial*: Rublack, "The Astronomer and the Witch."

16. *Newton's acknowledgment of Wren, Wallis, and Huygens, and Galileo*: *Principia*, first scholium in book 1, after the laws of motion and their corollaries. What a pity Harriot wasn't named, too! Lohne mentions Newton's citation of Wren et al. but not Kepler; he also suggests that Newton might have known, through Collins, that Harriot had worked on collisions, but that no one knew of Harriot's complete tract until relatively recently ("Essays on Thomas Harriot," 189).

17. *Harriot also treated oblique collisions*, whereas Wren, Wallis, and Huygens focused on direct impacts (Smith, "Optical Reflection and Mechanical Rebound," 18). Wallis did also consider inelastic collisions, and Huygens treated oblique collisions briefly in an unpublished manuscript (Lohne, "Essays on Thomas Harriot," 213, fig. 5). Wallis and Huygens had moved further toward the idea of conservation of momentum. Nevertheless, Harriot's tract was decades ahead of its time and gave the "purest conceptual distillation" of reflection as rebound before Newton (Smith, 16–17).

18. *Harriot's tract*: Petworth House/Leconfield HMC VIa ff23–31. Pepper ("Harriot's Manuscript on the Theory of Impacts") and Lohne ("Essays on Thomas Harriot") give extensive analyses, and Lohne produces an English translation of the first half of the Latin original. *Translations of preface*: Lohne, 201; for a slightly different version, Pepper, 133.

19. *Authoritative*: Harriot used the word *magister*, which Lohne ("Essays on Thomas Harriot," 201) has translated as "masteries" and Pepper ("Harriot's Manuscript on the Theory of Impacts," 133) translates as "masterpieces" or (133n8) as "magisterial". Beery and Stedall translate *magisteria* as "doctrine" in another context (in Harriot, *Thomas Harriot's Doctrine of Triangular Numbers*).

20. *The modern scholar* was Lohne ("Essays on Thomas Harriot," 190). Pepper also points out that Harriot did not *generalize* the relationship between two unequal balls, which is surprising as he was "master of algebra and its notation" ("Harriot's Manuscript on the Theory of Impacts", 135). Pepper analyzes the weakness in Harriot's argument here (135–136), as does Lohne (e.g., 195).

21. *Harriot originating study of mechanical collisions*: Pepper ("Harriot's Manuscript on the Theory of Impacts"), 138; Smith, "Optical Reflection and Mechanical Rebound," 8; Lohne, "Essays on Thomas Harriot," 206. For instance, *Galileo did not study collisions*: Cf. *Two New Sciences*; see also Schemmel, *The English Galileo* 1:233. But Lohne notes (208) the posthumously published Added Day in Galileo's *Two New Sciences*, which includes a brief verbal discussion of impacts (in contrast to Harriot's detailed mathematical treatment). *Bowling alley in Tower*: Shirley, "The Scientific Experiments of Sir Walter Ralegh, the Wizard Earl, and the Three Magi in the Tower," 54, drawing on Northumberland's account books for 1613–14. *Experimental evidence?* Lohne, 193. *Billiards* was popular among the English aristocracy in the late 1580s (Smith, n29).

22. *Reflection of balls cf. light and atoms?* Pepper, "Harriot's Manuscript on the Theory of Impacts," 141–142. *Optical reflection: Kepler, al-Haytham, et al.*: Smith, "Optical Reflection and Mechanical Rebound," 6–10. *Mechanical rebound more complex*: For instance, Harriot proved that the simple law of light reflection, where the angle of incidence equals the angle of reflection, did not always hold in mechanical collisions. Moreover, optical treatments assumed an incoming ray hitting a fixed surface such as a mirror, whereas mechanical collisions required an analysis of the rebound of two moving (or movable) balls. *Note*: Pepper (135n11) attributes the unusual (for Harriot) lack of algebra in this tract was to the fact that he was writing for Northumberland, who was not a mathematician. Pepper shows some manuscript examples revealing Harriot's ability to use algebra in such an analysis.

23. *Harriot and the laws of motion*: Lohne gives a pertinent comparison of statements by Harriot and Newton ("Essays on Thomas Harriot," 205–207). *Implicit conservation of momentum*: It was intuitive rather than a law, but Pepper shows how, apart from Harriot's aforementioned mistake, his equations are virtually identical to those derived from the law of conservation of momentum and Newton's "experimental law of impacts" ("Harriot's Manuscript on the Theory of Impacts," 141n14). Smith discusses conservation of components of motion in the optical analyses of al-Haytham and Kepler, on whom Harriot drew for his more complex analysis of the rebound of solid balls ("Optical Reflection and Mechanical Rebound," 13).

24. *Descartes*: Just as Harriot was likely motivated by atomism, however, so Descartes began not from a Newtonian universalizing imperative but from an attempt to analyze the interactions between light and matter (Smith, "Optical Reflection and Mechanical Rebound," 188–190); 190 gives a flaw found in Descartes's work but not in Harriot's, which illustrates the slow process of science, a process requiring many different ways of inching toward a correct analysis of natural laws.

25. *Harriot needed to clarify his assumptions: the case of free fall*: Schemmel has done a painstaking analysis of Harriot's results in *The English Galileo*: for instance, on 1:136 he shows that although Harriot did not explicitly state that in a vacuum all bodies fall at the same rate, this result nevertheless underpins his analysis. See also Lohne, "Essays on Thomas Harriot," part 2.

26. *Could Harriot have had an impact on future science without a grand Galilean synthesis?* As I've noted earlier, in actually working out his theory of gravity, Newton's most important influences were individual results, notably Kepler's three laws of planetary orbits, Wren et al.'s laws of collision, and Galileo's law of free fall and projectiles; cf. Newton's own acknowledgments in *Principia*, first scholium after the laws of motion in book 1. Indeed, Harriot and his circle "counter[ed] the Aristotelianism that remained supreme at all universities" (Bucciantini, Camerota, and Guidice, *Galileo's Telescope*, 131). Also, while Galileo discussed some questions that Harriot did not consider, so Harriot made advances in some topics on which Galileo did little or no work: e.g., collisions (Schemmel, *The English Galileo* 1:233), the rainbow (Boyer, *The Rainbow*, 194–195), and algebra.

23: ALL THINGS MUST PASS

1. From *John Donne's Devotions upon Emergent Occasions*, Meditation 17; the most famous lines are better known today as a poem rather than the original prose. *Bells tolling, etc.*: Ashley, *Elizabethan Popular Culture*, 9. *Bess and Donne*: Beer, *Bess*, 180, 193, 245.

2. *Harriot to Mayerne.* Reprinted in Mandelbrote, "The Religion of Thomas Harriot," 247. *Harriot was likely at Buckner's* to be closer to his doctors but his final decline was quick.

3. *Harriot's will* is published in Stevens (who discovered it in the 1880s), *Thomas Harriot.* See also Tanner, "The Study of Thomas Harriot's Manuscripts," pt. 1, "Harriot's Will."

4. *Exchange rate.* According to the National Archives website (www.nationalarchives.gov.uk/ currency-converter/), £1 in 1580 was worth about £204 in 2017; by 1620, £1 was worth about £131.52 in 2017.

5. *Harriot's epitaph* survives only from a record of it; translated from the original Latin in Shirley, *Biography*, 474. Shirley says (473) Buckner probably organized the plaque.

6. *Theology?* Maclean ("Harriot on Combinations," 70) suggests that this meant metaphysics: he points out that "the standard edition of Aristotle's works of 1619 refers to his metaphysics as theology."

EPILOGUE: RESURRECTING HARRIOT

1. *Aylesbury as a student of Harriot.* E.g., Aylesbury's letter to Harriot, BL Add MS 6789 f443; Tanner, "Nathaniel Torporley's 'Congestor analyticus' and Thomas Harriot's 'De triangularis rationalium,'" 395; Shirley, Biography, 415–6; and "Henry Stevens and the Associates of Thomas Harriot," 97–98.

2. *"Erred signally".* Torporley, *Analytical Corrector*, quoted in, e.g., Rosen, "Harriot's Science," 5.

3. *Torporley, Aylesbury's list.* Stedall, "Rob'd of Glories," 460.

4. *Briggs to Kepler.* Quoted in Harriot, *Thomas Harriot's Doctrine of Triangular Numbers*, 28.

5. *On the process of producing* Praxis: Lohne, "Essays on Thomas Harriot," 215; Stedall, "Rob'd of Glories," 461ff., and "Reconstructing Thomas Harriot's Treatise on Equations."

6. *Torporley's* Corrector *is analyzed*—and its preamble and introduction translated from the Latin—in Stedall, "Rob'd of Glories"; the translation is on 471–473. The preamble under Torporley's title suggests that his original intention had been to refute Harriot's atomism, but the manuscript instead turns to a generally perceptive defense of Harriot's mathematical achievements and fury at how they were diminished by the editors of the *Praxis. Published* Praxis *flawed*: Seltman, "Harriot's Algebra," 153, 155, 185; Stedall, "Rob'd of Glories" and "Reconstructing Thomas Harriot's Treatise on Equations," 60 (and 59–60 for Warner, Torporley's *Corrector*). Seltman and Stedall are the scholars most familiar with Harriot's mathematics.

7. *Warner on Harriot on refraction*: Having learned the law from Harriot, Warner later made his own table of refractions from his own experiments (Jacquot, "Harriot, Hill, Warner and the New Philosophy," 117): Pell describes an experimental setup used by Warner *after* Harriot's death (Goulding, "*Chymicorum in morem,*" 30n9). Warner's treatise on refraction was posthumously published in 1644 (Jacquot, 117).

8. *Papers returned*: Harriot, *Thomas Harriot's Doctrine of Triangular Numbers*, 38.

9. *Zach's discovery.* Lohne, "The Fair Fame of Thomas Harriott," 69–70.

10. *Robertson*: His report is in Rigaud, *Supplement to Dr Bradley's Miscellaneous Works*, 17–18, 57–61, and excerpted in Lohne, "The Fair Fame of Thomas Harriott," 72. Also discussed in Roche, "Harriot, Oxford, and Twentieth-Century Historiography," 239–240; Stedall, "Rob'd of Glories," 480.

11. *Division of Harriot's papers (BL, Petworth, etc.)*: Batho, "Thomas Harriot's Manuscripts," 291ff. *Rigaud's flawed analysis* of Harriot's sunspot results and techniques is analyzed in North, "Thomas Harriot and the First Telescopic Observations of Sunspots," 131–133, 146–150.

12. *Rigaud on Harriot.* In addition to Rigaud's fascinating original report (*Supplement to Dr Bradley's Miscellaneous Works*), see Lohne, "The Fair Fame of Thomas Harriott," Shirley, *Biography*, 14–29, and Roche, "Harriot, Oxford, and Twentieth-Century Historiography," who also give excerpts from Robertson's and Rigaud's correspondence.

13. *Greatest British mathematical scientist before Newton*: In terms of depth, method, and diversity. This is also the judgment of others—e.g., Whiteside's review of Shirley, 61; Seltman, "Harriot's Algebra," 184; Flood and Wilson, *The Great Mathematicians*, 80.

14. *Stevens*: His book is listed in my bibliography, but for analyses of its contents see also Tanner, "Henry Stevens and the Associates of Thomas Harriot" and "The Study of Thomas Harriot's Manuscripts." *Other early publications*: In 1928, Florian Cajori published "A Re-evaluation of Harriot's *Artis Analyticae Praxis*" in *Isis*, and in the nineteenth and early twentieth centuries a number of books and encyclopedia entries referred briefly to Harriot (cf. bibliography in Shirley, *Renaissance Scientist*).

15. *Shirley*: See my bibliography for most of his works, and also the bibliography in Shirley, *Renaissance Scientist*. See also Roche, "Harriot, Oxford, and Twentieth-Century Historiography," 241–243, for a brief discussion of Shirley's and others' pioneering efforts. *The other 1950s scholars* referred to are also listed in my bibliography.

16. *The Risner Witelo*: Lohne doesn't say whose copy it was, but Pepper assumes it was from Percy's library ("Thomas Harriot: A Biography," 213). Would Harriot annotate his patron's book? He had his own impressive library. Either way, it was the very book Harriot had studied. *Lohne's discovery*: Lohne, "Thomas Harriot (1560–1621)," 114–115.

17. *Refutations of Rigaud*: Lohne, "The Fair Fame of Thomas Harriot"; North, "Thomas Harriot (1560–1621)," 131–133, 140, 144–147, 154–155.

18. *Early seminar organizers* include John North, Alistair Crombie, Tanner, Shirley and his fellow enthusiasts (who called themselves "Harrioteers"), Gordon Batho, and Robert Fox (who initiated the Oriel Lectures); see Roche, "Harriot, Oxford, and Twentieth-Century Historiography," 241–243, and Shirley's preface (*Renaissance Scientist*, viii).

19. *Digitization*: For a personal account, see Goulding and Schemmel, "The Manuscripts of Thomas Harriot."

20. *The Oriel Harriot Lectures* were established in 1990 by Robert Fox, with generous assistance from Cecily Tanner; lectures and seminars had been established at Oxford and elsewhere since the late 1960s (cf. Roche, "Harriot, Oxford, and Twentieth-Century Historiography," 241–243.

21. *The plaque* in the Bank of England was unveiled on July 2, 1971, the 350th anniversary of Harriot's death; Shirley, *Biography*, 475. The inscription is in Latin; my English translation is from Shirley, 474. The epitaph's authorship by Northumberland is not certain, although the prominent reference to Syon is telling, and Batho (like Stevens, cf. Shirley 472–473) assumes it ("Thomas Harriot and the Northumberland Household," 46). Harriot's bones likely lie in a vault in Nunhead cemetery, southeast London. The remains in the Christopher-le-Stocks church and cemetery in Threadneedle Street were eventually reinterred at Nunhead.

 In 2009, a special plaque honoring Harriot was erected at Syon, near the site of his house, which unfortunately no longer exists. But the main house and grounds still give a wonderful sense of what it must have been like in Harriot's time.

BIBLIOGRAPHY

PRIMARY SOURCES

Acosta, José de. *The Natural and Moral History of the East and West Indies.* Translated by Edward Grimston. London, 1604. Originally published in Spanish in 1590.

Al-Haytham. *Book of Optics (Kitab-al-manazir).* In Risner, *Opticae thesaurus.*

Batho, G. R., ed. *The Household Papers of Henry Percy, Ninth Earl of Northumberland (1564–1632).* London: Royal Historical Society, 1962.

Bourne, William. *Regiment for the Sea.* London, 1574.

de Bry, Theodor, ed. *America, Part I.* An edition of Harriot's *Brief and True Report* with engravings of White's drawings and paintings. Frankfurt, 1590.

Dee, John. *John Dee: Essential Readings,* Selected and introduced by Gerald Suster. Berkeley, CA: North Atlantic Books, 2003.

Galilei, Galileo. *Dialogue Concerning the Two Chief World Systems—Ptolemaic and Copernican.* 1632. Translated by Stillman Drake. 2nd ed. Berkeley: University of California Press, 1967.

Galilei, Galileo. *Dialogues and Mathematical Demonstrations Relating to Two New Sciences.* 1638. Translated as *Two New Sciences* by Stillman Drake. 2nd ed. Toronto: Wall and Emerson, 1989 and 2000.

Gilbert, William. *De magnete.* London, 1600.

Hakluyt, Richard. *The Principal Navigations, Voyages, Traffiques and Discoveries of the English Nation.* Vol. 8. Glasgow: James MacLehose and Sons, Glasgow, publishers to the University (Oxford), 1904. Originally published 1589.

Harriot, Thomas. *A briefe and true report of the new found land of Virginia: Of the commodities there found and to be raysed, as well merchantable, as others for victual, building and other necessarie uses for those that are and shall be the planters there; and of the nature and manners of the naturall inhabitants.* Rpt. in Hakluyt, *Principal Navigations,* 349–386; and, with extensive footnotes by Quinn, in Quinn, *Roanoke Voyages* 1:317–387.

Harriot, Thomas. *The Greate Invention of Algebra: Thomas Harriot's Treatise on Equations.* Edited by Jacqueline Stedall. Oxford and New York: Oxford University Press, 2003.

Harriot, Thomas. Manuscripts. In the British Library and Petworth House; now available at ECHO (European Cultural Heritage Online), edited by Jacqueline Stedall, Matthias Schemmel, and Robert Goulding.

Harriot, Thomas. *Thomas Harriot's "Artis analyticae praxis": An English Translation with Commentary.* Edited and translated by Muriel Seltzman and Robert Goulding. New York: Springer, 2007.

Harriot, Thomas. *Thomas Harriot's Doctrine of Triangular Numbers: The "Magisteria magna."* Edited by Janet Beery and Jacqueline Stedall. Zurich: European Mathematical Society, 2009.

Harriot, Thomas, and Johannes Kepler. Correspondence. In J. Kepler, *Gesammelte Werke,* vol. 15, *Briefe: 1604–1607.* Munich: Beck, 1951.

Kepler, Johannes. *Conversation with the Starry Messenger.* Prague: Daniel Sedesanus, 1610. Translated by Edward Rosen and available online at http://digitalcollections.library.cmu.edu/awweb/awarchive?type=file&item=393654.

Kinney, Arthur F., ed. *Elizabethan and Jacobean England: Sources and Documents of the English Renaissance.* Chichester, West Sussex; Malden, MA: Wiley-Blackwell, 2011.

Lambert, B. *The History and Survey of London and Its Environs*. London: T. Hughes, 1806.

Oldys, William, and Thomas Birch. *The Lives of the Author*. Vol. 1 of *The Works of Sir Walter Ralegh, Kt., now first collected: To which are prefixed "The Lives of the Author."* 1829. Rpt. New York: Burt Franklin, [1964]. Includes a transcript of Ralegh's 1603 trial.

Patterson, Annabel, ed. *The Trial of Nicholas Throckmorton*. Toronto: Center for Reformation and Renaissance Studies, 1998.

Peckham, George. *The Western Planting*, in Hakluyt, *Principal Navigations*, 89–131.

Ptolemy, Claudius. *Ptolemy's Almagest*. Translated and annotated by G. J. Toomer. London: Duckworth, 1984.

Quinn D. B., ed. *The Roanoke Voyages: 1584–1590*. 2 vols. London: Hakluyt Society, 1955.

Quinn, D. B. *Set Fair for Roanoke: Voyages and Colonies, 1584–1606*. Chapel Hill: University of North Carolina Press, 1985.

Ralegh, Walter. *The Discovery of the Large, Rich and Beautiful Empire of Guiana*. London: Robert Robinson, 1596.

Ralegh, Walter. *History of the World*. London: W. Burre, 1614.

Ralegh, Walter. *The Letters of Sir Walter Ralegh*. Edited by Agnes Latham and Joyce Youings. Exeter: University of Exeter Press, 1999.

Ralegh, Walter. *The Poems of Sir Walter Ralegh: A Historical Edition*. Edited by Michael Rudick. Tempe: Arizona Center for Medieval and Renaissance Studies, in conjunction with Renaissance English Text Society, 1999.

Rigaud, S. P. *Supplement to Dr Bradley's Miscellaneous Works, with an account of Harriot's Astronomical Papers*. Oxford: at the University Press, 1833.

Risner, Friedrich. *Opticae thesaurus*. Basel, 1572.

Salter, Richard, ed. *Elizabeth I and Her Reign: Documents and Debates*. Basingstoke: Macmillan Education, 1988.

Witelo. *Perspectiva*. In Risner.

Wright, Edward. *Certaine Errors in Navigation*. London, 1610.

Note on sources from compilations of other original documents, edited by Kinney, Quinn, and Salter: Early authors whose work is reprinted in these compilations and cited in my narrative will be referenced in the endnotes rather than in this bibliography. These include proclamations, letters, and speeches from Elizabeth I, James I, and Philip II; reports and letters on America by William Barlowe and Ralph Lane; reports on the state of the nation by William Harrison and Robert Cecil; religious pamphlets, and many other contemporary accounts, letters, and reports. Similarly, contemporary letters reprinted in secondary sources are referenced in the endnotes.

SECONDARY SCHOLARLY SOURCES RELATED TO HARRIOT

Alexander, Amir. "The Imperialist Space of Elizabethan Mathematics." *Studies in the History and Philosophy of Science* 26 (1995): 559–591.

Alexander, Amir. "Lunar Maps and Coastal Outlines: Thomas Hariot's Mapping of the Moon." *Studies in the History and Philosophy of Science* 29 (1998): 345–366.

Batho, G. R. "The Possible Portraits of Thomas Harriot." In Fox, *Harriot* 1: appendix A.

Batho, G. R. "Thomas Harriot and the Northumberland Household." Durham Thomas Harriot Seminar, Occasional Paper 1, 1983. Also published in Fox, *Harriot* 1:28–47.

Batho, G. R. "Thomas Harriot's Manuscripts." In Fox, *Harriot* 1: appendix B.

Beer, Anna. "Thomas Harriot and Sir Walter Ralegh's Wife." Durham Thomas Harriot Seminar, Occasional Paper 34, 2004.

Bennett, J. A. "Instruments, Mathematics, Natural Knowledge: Thomas Harriot's Place on the Map of Learning." In Fox, *Harriot* 1:137–152.

Biggs, Norman. "Thomas Harriot on Continuous Compounding." *BSHM Bulletin: Journal of the British Society for the History of Mathematics* 28, no. 2 (2013): 66–74.

Booth, Michael. "Thomas Harriot's Translations." *Yale Journal of Criticism* 16, no. 2 (2003): 345–361.

Brioist, Pascal. "Thomas Harriot and the Mariner's Culture: On Board a Transatlantic Ship in 1585." In Fox, *Harriot* 2: 183–200.

Chapman, Allan. "The Astronomical Work of Thomas Harriot." *Quarterly Journal of the Royal Astronomical Society* 36 (1995): 97–107.

Chapman, Allan. "A New Perceived Reality: Thomas Harriot's Moon Maps." *Astronomy and Geophysics* 50 (February 2009): 1.27–31.

Clucas, Stephen. "Thomas Harriot and the Field of Knowledge in the English Renaissance." In Fox, *Harriot* 1:93–136.

Dawson, Scott. "The Vocabulary of Croatoan Algonquian." *Southern Quarterly* 51, no. 4 (Summer 2014): 48–53.

Fantazzi, Charles. "Harriot's Latin." In Fox, *Harriot* 2: appendix B.

Fishman, Ronald S. "Perish, Then Publish: Thomas Harriot and the Sine Law of Refraction." *Archives of Ophthalmology* 18 (March 2000): 405–409.

Fox, Robert, ed. *Thomas Harriot*, vol. 1, *An Elizabethan Man of Science*. Farnham, Surrey; Burlington, VT: Ashgate, 2000. Essays from the Thomas Harriot Lectures at Oriel College, Oxford, 1990–99.

Fox, Robert, ed. *Thomas Harriot*, vol. 2, *Mathematics, Exploration, and Natural Philosophy in Early Modern England*. Farnham, Surrey; Burlington, VT: Ashgate, 2012. Essays from the Thomas Harriot Lectures at Oriel College, Oxford, 2000–9.

Gatti, Hilary. "Giordano Bruno: The Texts in the Library of the Ninth Earl of Northumberland." *Journal of the Warburg and Courtauld Institutes* 46 (1983): 63–77.

Gatti, Hilary. "The Natural Philosophy of Thomas Harriot." In Fox, *Harriot* 1:64–92.

George, Frank. "Hariot's Meridional Parts." *Journal of Navigation* 9 (1956): 65–69.

Goulding, Robert. "*Chymicorum in morem:* Refraction, Matter Theory, and Secrecy in the Harriot–Kepler Correspondence." In Fox, *Harriot* 2:27–52.

Goulding, Robert. "Thomas Harriot's Optics, Between Experiment and Imagination: The Case of Mr Bulkeley's Glass." *Archive for History of Exact Sciences* 68, no. 2 (2014): 137–178.

Goulding, Robert, Matthias Schemmel, and dedicated to the memory of Jacqueline Stedall. "The Manuscripts of Thomas Harriot (1560–1621)." *Journal of the British Society for the History of Mathematics* 32, no. 1 (2017): 17–19.

Greenblatt, Stephen. "Invisible Bullets: Renaissance Authority and Its Subversion, *Henry IV* and *Henry V.*" *Glyph: Textual Studies* 8 (1981): 40–60.

Henry, John. "Thomas Harriot and Atomism: A Reappraisal." *History of Science* 20, no. 4 (1982): 267–296.

Henry, John. "Why Harriot Was *Not* the English Galileo." In Fox, *Harriot* 2:113–138.

Herr, Richard B. "Solar Rotation Determined from Thomas Harriot's Sunspot Observations of 1611 to 1613." *Science*, n.s., 202, no. 4372 (December 8, 1978): 1079–1081.

Hunneyball, Paul M. "Sir William Lower and the Harriot Circle." Durham Thomas Harriot Seminar, Occasional Paper 31, 2002.

Jacquot, Jean. "Thomas Harriot's Reputation for Impiety." *Notes and Records of the Royal Society of London* 9, no. 2 (May 1952): 164–187.

Jacquot, Jean. "Harriot, Hill, Warner and the New Philosophy." In Shirley, *Thomas Harriot: Renaissance Scientist*, 107–128.

Jarrett, Joseph. "Algebra and the Art of War: Marlowe's Military Mathematics in *Tamburlaine 1* and *2.*" *Cahier Élisabéthains: A Journal of English Renaissance Studies* 95, no. 1 (2018): 19–39.

Kargon, Robert. "Thomas Harriot, the Northumberland Circle, and Early Atomism in England." *Journal of the History of Ideas* 27, no. 1 (January–March 1966): 128–136.

Lohne, J. A. "Essays on Thomas Harriot." *Archive for History of Exact Sciences* 20, no. 3/4 (September 1979): 189–312.

Lohne, J. A. "The Fair Fame of Thomas Harriott: Rigaud versus Baron von Zach." *Centaurus* 8, no. 1 (1963): 69–84.

Lohne, J. A. "Thomas Harriot (1560–1621): The Tycho Brahe of Optics." *Centaurus* 6, no. 2 (1959): 113–121.

Maclean, Ian. "Harriot on Combinations." In Fox, *Harriot* 2:65–88.

Mandelbrote, Scott. "The Religion of Thomas Harriot." In Fox, *Harriot* 1:246–279.

McAlindon, Tom. "Testing the New Historicism: 'Invisible Bullets' Reconsidered." *Studies in Philology* 92, no. 4 (Autumn 1995): 411–438.

McIntyre, Ruth A. "William Sanderson: Elizabethan Financier of Discovery." *William and Mary Quarterly* 13, no. 2 (April 1956): 184–201.

Moran, Michael G. "A Fantasy-Theme Analysis of Arthur Barlowe's 1584 Discourse on Virginia." *Technical Communication Quarterly* 11, no. 1 (2002): 31–59.

Nicholls, Mark. "Last Act? 1618 and the Shaping of Sir Walter Ralegh's Reputation." In Fox, *Harriot* 2:165–182.

Nicholls, Mark. "The 'Wizard Earl' in Star Chamber: The Trial of the Earl of Northumberland, June 1606." *Historical Journal* 30, no. 1 (1987): 173–189. Includes complete transcript.

North, John D. "Stars and Atoms." In Fox, *Harriot* 1:186–228.

North, John D. "Thomas Harriot and the First Telescopic Observations of Sunspots." In Shirley, *Thomas Harriot: Renaissance Scientist,* 126–165.

Oberg, Michael Leroy. "Gods and Men: The Meeting of Indian and White Worlds on the Carolina Outer Banks, 1584–1586." *North Carolina Historical Review* 76, no. 4 (October 1999): 367–390.

Pepper, Jon V. "Harriot's Calculation of the Meridional Parts as Logarithmic Tangents." *Archive for History of Exact Sciences* 4, no. 5 (1968): 359–413.

Pepper, Jon V. "Harriot's Earlier Work on Mathematical Navigation: Theory and Practice." In Shirley, *Thomas Harriot: Renaissance Scientist,* 54–90.

Pepper, Jon V. "Harriot's Manuscript on the Theory of Impacts." *Annals of Science* 33, no. 2 (1976): 131–151.

Pepper, Jon V. "A Letter from Nathaniel Torporley to Thomas Harriot." *British Journal for the History of Science* 3, no. 3 (1967): 285–290.

Pepper, Jon V. "Some Clarifications of Harriot's Solutions of Mercator's Problem." *History of Science* 14 (1976): 235–244.

Pepper, Jon V. "Thomas Harriot: A Biography." *British Journal for the History of Science* 19, no. 2 (1986): 212–216.

Pepper, Jon V. "Thomas Harriot and the Great Mathematical Tradition." In Fox, *Harriot* 2: 11–26.

Pumfrey, Stephen. "Harriot's Maps of the Moon: New Interpretations." *Notes and Records of the Royal Society* 63 (2009): 163–168.

Pumfrey, Stephen. "Patronizing, Publishing and Perishing: Harriot's Lost Opportunities and His Lost Work 'Arcticon.'" In Fox, *Harriot* 2:139–164.

Quinn, D. B. "Thomas Harriot and the New World." In Shirley, *Thomas Harriot: Renaissance Scientist,* 36–53.

Quinn, D. B. "Thomas Harriot and the Problem of America." In Fox, *Harriot* 1: 9–27.

Quinn, D. B. "Thomas Hariot and the Virginia Voyages of 1602." *William and Mary Quarterly* 27, no. 2 (April 1970): 268–281.

Quinn, D. B., and John W. Shirley. "A Contemporary List of Harriot References." *Renaissance Quarterly* 22, no. 1 (Spring 1969): 9–26.

Roche, John J. "Harriot, Oxford, and Twentieth-Century Historiography." In Fox, *Harriot* 1: 229–245.

Roche, John J. "Harriot's 'Regiment of the Sun' and Its Background in Sixteenth-Century Navigation." *British Journal for the History of Science* 14, no. 3 (1981): 245–262.

Rosen, Edward. "Harriot's Science: The Intellectual Background." In Shirley, *Thomas Harriot: Renaissance Scientist,* 1–15.

Sadler, D. H. "Nautical Triangles Compendious," pt. 2, "Calculating the Meridional Parts." *Journal of the Institute of Navigation* 6, no. 2 (1953): 141–147.

Salmon, Vivian. "Thomas Harriot and the English Origins of Algonkian Linguistics." Durham Thomas Harriot Seminar, Occasional Paper 8, 1993.

Schemmel, Matthias. *The English Galileo: Thomas Harriot's Work on Motion as an Example of Preclassical Mechanics.* 2 vols. Dordrecht: Springer, 2008.

Schemmel, Matthias. "Thomas Harriot as an English Galileo: The Force of Shared Knowledge in Early Modern Mechanics." In Fox, *Harriot* 2:89–112.

Seaton, E. "Thomas Harriot's Secret Script." *Ambix* 5 (1956): 111–114.

Seltman, Muriel. "Harriot's Algebra: Reputation and Reality." In Fox, *Harriot* 1:153–185.

Shirley, John W. "The Scientific Experiments of Sir Walter Ralegh, the Wizard Earl, and the Three Magi in the Tower, 1603–1617." *Ambix* 4 (1949–51): 52–66.

Shirley, John W. "Sir Walter Ralegh and Thomas Harriot." In Shirley, *Thomas Harriot: Renaissance Scientist*, 16–35.

Shirley, John W. "Sir Walter Raleigh's Guiana Finances." *Huntingdon Library Quarterly* 13, no. 1 (November 1949): 55–69.

Shirley, John W., ed. *Thomas Harriot: Renaissance Scientist*. Oxford: Clarendon Press, 1974.

Shirley, John W. *Thomas Harriot: A Biography*. Oxford: Clarendon Press, 1983.

Silverberg, Joel. "Nathaniel Torporley and his *Diclides coelometricae* (1602): A Preliminary Investigation," *Proceedings of the Canadian Society for History and Philosophy of Mathematics* 25:154–175, http://www.academia.edu/5890336/Nathaniel_Torporley_and_his_Diclides_Coelometricae.

Siegmund-Schultze, Reinhard. "Pulling Harriot out of Newton's Shadow: How the Norwegian Outsider Johannes Lohne Came to Contribute to Mainstream History of Mathematics." In *Historiography of Mathematics in the 19th and 20th Centuries*, edited by Volker R. Remmert, Martina Schneider, and Henrik Fragh Sørensen, 219–241. Cham, Switzerland: Birkhäuser, 2016.

Smith, Russell. "Optical Reflection and Mechanical Rebound: The Shift from Analogy to Axiomatization in the Seventeenth Century," pts. 1 and 2. *British Journal for the History of Science* 41, no. 1 (2007): 1–18; 42, no. 2 (2008): 187–207.

Smith, Russell. "Shining a Light on Harriot and Galileo: On the Mechanics of Reflection and Projectile Motion." *History of Science* 53, no. 3 (2015): 296–319.

Sokol, B. J. "Invisible Evidence: The Unfounded Attack on Thomas Harriot's Reputation." Durham Thomas Harriot Seminar, Occasional Paper 17, 1995.

Sokol, B. J. "The Problem of Assessing Thomas Harriot's 'A Briefe and True Report' of His Discoveries in America." *Annals of Science* 51, no. 1 (1994): 1–16.

Sokol, B. J. "Thomas Harriot—Sir Walter Ralegh's Tutor—on Population." *Annals of Science* 31, no. 3 (1974): 205–212.

Stedall, Jacqueline. "John Wallis and the French: His Quarrels with Fermat, Pascal, Dulaurens and Descartes." *Historia Mathematica* 39 (2012): 265–279.

Stedall, Jacqueline. "Notes Made by Thomas Harriot on the Treatises of François Viète." *Archive for History of the Exact Sciences* 62, no. 2 (2008): 179–200.

Stedall, Jacqueline. "Notes Made by Thomas Harriot (1560–1621), on Ships and Ship Building." *Mariner's Mirror* 99, no. 3 (2013): 325–327.

Stedall, Jacqueline. "Reconstructing Thomas Harriot's Treatise on Equations." In Fox, *Harriot*, 2:53–64.

Stedall, Jacqueline. "Rob'd of Glories: The Posthumous Misfortunes of Thomas Harriot and His Algebra." *Archive for History of Exact Sciences* 54, no. 6 (June 2000): 455–497.

Stedall, Jacqueline. "Symbolism, Combinations, and Visual Imagery in the Mathematics of Thomas Harriot." *Historia Mathematica* 34 (2007): 380–401.

Stevens, Henry. *Thomas Hariot: The Mathematician, the Philosopher, and the Scholar* (sometimes known as *Thomas Hariot and His Associates*). 1900. Rpt. New York: Lenox Hill (Burt Franklin), 1972.

Swan, Diccon. "The Portrait of Thomas Harriot." In Fox, *Harriot* 2: appendix C.

Tanner, R. C. H. "Henry Stevens and the Associates of Thomas Harriot." In Shirley, *Thomas Harriot: Renaissance Scholar*, 91–106.

Tanner, R. C. H. "Nathaniel Torporley's 'Congestor analyticus' and Thomas Harriot's 'De triangularis rationalium.'" *Annals of Science* 34, no. 4 (1977): 393–428.

Tanner, R. C. H. "The Ordered Regiment of the Minus Sign: Off-beat Mathematics in Harriot's Manuscripts." *Annals of Science* 37, no. 2 (1980): 127–158.

Tanner, R. C. H. "The Study of Thomas Harriot's Manuscripts," pt. 1, "Harriot's Will." *History of Science* 6, no. 1 (1967): 1–16.

Taylor, E. G. R., and D. H. Sadler. "The Doctrine of Nauticall Triangles Compendious." *Journal of the Institute of Navigation* 6 (1953): 131–147.

Trevor-Roper, Hugh. "Harriot's Physician: Theodore de Mayerne." In Fox, *Harriot* 1:48–63.

Vaughan, Alden T. "Sir Walter Ralegh's Indian Interpreters, 1584–1618." *William and Mary Quarterly* 59, no. 2 (April 2002): 341–376.

Whitaker, Ewen A. "The Digges-Bourne Telescope Revisited." *Journal of the British Astronomical Association* 119, no. 2 (2009): 64–65.

Young, Sandra. "Narrating Colonial Violence and Representing New World Difference: The Possibilities of Form in Thomas Harriot's 'A Briefe and True Report.'" *Safundi* 11, no. 4 (2010): 343–360.

OTHER SECONDARY SCHOLARLY SOURCES

Aaboe, Asger. "Observation and Theory in Babylonian Astronomy." *Centaurus* 24, no. 1 (1980): 14–35.

Adhikari, Surendra, and Erik R. Ivins. "Climate-Driven Polar Motion: 2003–2015." *Science Advances* 2, no. 4 (1 April 1, 2016), htpps://doi.org:10.1126/sciadv.1501693.

Aldana, Gerardo. "Discovering Discovery: Chich'en Itza, the Dresden Codex Venus Table and 10th Century Mayan Astronomical Innovation." *Journal of Astronomy in Culture* 1 no. 1, (2016): 57–76.

Almeida, Bruno. "On the Origins of Dee's Mathematical Programme: The John Dee–Pedro Nunes Connection." *Studies in History and Philosophy of Science Part A* 43, no. 3 (2012): 460–469.

Atwell, William S. "International Bullion Flows and the Chinese Economy circa 1530–1650." *Past and Present* 95, no. 1 (May 1982): 68–90.

Batchelder, Ronald W., and Nicolas Sanchez. *The Encomienda and the Optimizing Imperialist: An Interpretation of Spanish Imperialism in the Americas.* UCLA Economics Working Paper 501, September 1988.

Batho, G. R. "A Difficult Father-in-Law: The Ninth Earl of Northumberland." *History Today* 6, no. 11 (November 1956): 744–751.

Batho, G. R. "The Execution of Mary, Queen of Scots." *Scottish Historical Review* 39, no. 127 (April 1960): 35–42.

Beer, Anna. "Ralegh's History of Her World." *Women's Writing* 12, no. 1 (2005): 29–42.

Bennett, J. "The Mechanics' Philosophy and the Mechanical Philosophy." *History of Science* 21, no. 1 (1986): 1–28.

Berggren, J. L. "Ptolemy's Maps of Earth and the Heavens: A New Interpretation." *Archive for History of Exact Sciences* 43, no. 2 (1991): 133–144.

Boner, Patrick J. "Life in the Liquid Fields: Kepler, Tycho and Gilbert on the Nature of the Heavens and Earth." *History of Science* 46, no. 3 (2008): 275–297.

Borelli, Arianna. "The Weatherglass and Its Observers in the Early Seventeenth Century." In *Philosophies of Technology: Francis Bacon and His Contemporaries,* edited by Claus Zittel, 67–130. Leiden and Boston: Brill, 2008.

Boyer, Allen D. "The Trial of Sir Walter Ralegh: The Law of Treason, the Trial of Treason, and the Origins of the Confrontation Clause." *Mississippi Law Journal* 74 (2004–5): 869–901.

Boyer, Carl B. "Kepler's Explanation of the Rainbow." Paper presented at the symposium Use of Historical Material in Elementary and Advanced Instruction, American Association of Physics Teachers, Barnard, NY, February 4, 1950.

Britton, John P. "Ptolemy's Determination of the Obliquity of the Ecliptic." *Centaurus* 14. no. 1 (1969): 29–41.

Britton, John P. "An Early Function for Eclipse Magnitudes in Mesopotamian Astronomy." *Centaurus* 32, no. 1 (1989): 1–52.

Bucciantini, Massimo, Michele Camerota, and Franco Giudice. *Galileo's Telescope: A European Story.* Translated by Catherine Bolton. Cambridge: Harvard University Press, 2015.

Burrow, Colin. "Higher Learning." In Kinney, *Elizabethan and Jacobean England,* 501–510.

Burstyn, Harold L. "Galileo's Attempt to Prove That the Earth Moves." *Isis* 53, no. 2 (June 1962): 161–185.

Bushnell, Rebecca. "Early Education." In Kinney, *Elizabethan and Jacobean England,* 493–500.

Cajori, Florian. "The History of Zeno's Arguments on Motion: Phases in the Development of the Theory of Limits." *American Mathematical Monthly* 22, no. 3 (March 1915): 77–82.

Caudano, Anne-Laurence. "Carolingian Astronomy." Review of *Ordering the Heavens*, by Bruce S. Eastwood. *Journal for the History of Astronomy* 39 (2008): 275–277.

Cerasano, S. P. "Philip Henslowe, Simon Forman, and the Theatrical Community of the 1590s." *Shakespeare Quarterly* 44, no. 2 (Summer 1993): 145–158.

Chamberlain, R. S. *The Pre-Conquest Tribute and Service System of the Maya as Preparation for the Spanish* Repartimiento-Encomienda *in Yucatán.* Miami: University of Miami Press, 1951.

Clucas, Stephen. "'Noble virtue in extremes': Henry Percy, Ninth earl of Northumberland, Patronage and the Politics of Stoic Consolation." *Renaissance Studies* 9, no. 3 (September 1995): 267–291.

Clulee, Nicholas H. "Astrology, Magic, and Optics: Facets of John Dee's Early Natural Philosophy." *Renaissance Quarterly* 30, no. 4 (Winter 1977): 632–680.

Cohen, Adam Max. "Tudor Technology in Transition." In *A Companion to Tudor Literature*, edited by Kent Cartwright. Chichester, West Sussex; Malden, MA: Wiley-Blackwell, 2010.

Collinson, Patrick. "Post-Reformation Religion: Uniformity, Nonconformity, Diversity, and Conflict." In Kinney, *Elizabethan and Jacobean England*, 191–198.

Cullen, C. "A Chinese Eratosthenes of the Flat Earth: A Study of a Fragment of Cosmology in Huai Nan Tzu." *Bulletin of the School of Oriental and African Studies, University of London* 39, no. 1 (1976): 106–127.

Damerow, Peter, et al. *Exploring the Limits of Preclassical Mechanics.* New York: Springer, 2004.

Eastwood, Bruce S. "Grosseteste's 'Quantitative' Law of Refraction: A Chapter in the History of Non-Experimental Science." *Journal for the History of Ideas* 28, no. 3 (July–September 1967): 403–414.

Eastwood, Bruce S. "Johannes Scottus Eriugena, Sun-Centered Planets, and Carolingian Astronomy." *Journal for the History of Astronomy* 32 (2001): 281–324.

Eastwood, Bruce S. *Ordering the Heavens: Roman Astronomy and Cosmology in the Carolingian Renaissance,* Leiden and Boston: Brill, 2007.

Eastwood, Bruce, and Gerd Graßhoff. "Planetary Diagrams for Roman Astronomy in Mediaeval Europe, ca. 800–1500." *Transactions of the American Philosophical Society*, n. s., 94, no. 3 (2004): i–xiv, 1–158.

Finney, Ben. "Rediscovering Polynesian Navigation through Experimental Voyaging." *Journal of Navigation* 46, no. 3 (1993): 383–394.

Fleck, Andrew. "'At the time of his death': Manuscript Instability and Walter Ralegh's Performance on the Scaffold." *Journal of British Studies* 48 (January 2009): 4–28.

Gaspar, Joachim, and Henrique Leitao. "Squaring the Circle: How Mercator Constructed His Projection in 1569." *Imago Mundi* 66, no. 1 (2014): 1–24.

Gingerich, Owen, and James R. Voelkel. "Tycho and Kepler: Solid Myth versus Subtle Truth." *Social Research* 72, no. 1 (Spring 2005): 77–106.

Goldstein, Bernard R., and Alan C. Bowen. "A New View of Early Greek Astronomy." *Isis* 74, no. 3 (September 1983): 330–340.

Goldstein, Bernard R., and Giora Hon. "Kepler's Move from *Orbs* to *Orbits*: Documenting a Revolutionary Scientific Concept." *Perspectives on Science* 12, no. 1 (2005): 74–111.

Grafton, Anthony. "Kepler as a Reader." *Journal of the History of Ideas* 53, no. 4 (October–December 1992): 561–572.

Graney, Christopher. "The Inquisition's Semicolon." arXiv:1402.6168 [physics.hist-ph], February 2014.

Grofe, Michael J. "Measuring Deep Time: The Sidereal Year and the Tropical Year in Maya Inscriptions." *"Oxford IX" International Symposium on Archaeoastronomy: Proceedings of the International Astronomical Union* 7, no. S278 (January 2011): 214–230.

Hadfield, Andrew. "Another Look at Serena and Irena." *Irish University Review* 26, no. 2, special issue, "Spenser in Ireland: 'The Faerie Queene,' 1596–1996" (Autumn–Winter 1996): 291–302.

Hales, Thomas C. "An Overview of the Kepler Conjecture." arXiv:math/9811071v2 [math. MG], May 2002.

Hammer, Paul E. J. "A Reckoning Reframed: The 'Murder' of Christopher Marlowe Revisited." *English Literary Renaissance* 26, no. 2 (1996): 225–242.

Hammer, Paul E. J. "Shakespeare's Richard II, the Play of 7 February 1601, and the Essex Rising." *Shakespeare Quarterly* 59, no. 1 (Spring 2008): 1–35.

Hayden, Brian, and Suzanne Villeneuve. "Astronomy in the Upper Palaeolithic?" *Cambridge Archaeological Journal* 21, no. 3 (2011): 331–355.

Heath, F. G. "Origins of the Binary Code." *Scientific American*, August 1972, 76–83.

Hellman, C. Doris. "The Role of Measurement in the Downfall of a System: Some Examples from Sixteenth-Century Comet and Nova Observations." *Vistas in Astronomy* 9, no. 1 (1967): 43–52.

Henry, John. "Animism and Empiricism: Copernican Physics and the Origins of William Gilbert's Experimental Method." *Journal of the History of Ideas* 62, no. 1 (January 2001): 99–119.

Hinkle, Roscoe C. "Medieval Islamic Spain (al-Andalus) as a Civilizational Bridge between Later Antiquity and Early Modernity." *Comparative Civilizations Review* 61 (Fall 2009): 88–104.

Hutchinson, Keith. "Galileo, Sunspots, and the Orbit of the Earth." *Isis* 81, no. 1 (March 1990): 68–74.

Jenkins, Raymond. "Spenser and Ireland." *ELH* 19, no. 2 (June 1952): 131–142.

Jenkins, Raymond. "Spenser with Lord Grey in Ireland." *PMLA* 52, no. 2 (June 1937): 338–353.

Johnson, Francis R. "The Influence of Thomas Digges on the Progress of Modern Astronomy in Sixteenth-Century England." *Osiris* 1 (1936): 390–410.

Johnston, Stephen. "Mathematical Practitioners and Instruments in Elizabethan England." *Annals of Science* 48, no. 4 (1991): 319–344.

Jones, Alexander, and John M. Steele. "A New Discovery of a Component of Greek Astrology in Babylonian Tablets: The 'Terms.'" Institute for the Study of the Ancient World, *ISAW Papers* 1 (2011), http://doi.org/2333.1/k98sf96r.

Kak, Subhash. "Babylonian and Indian Astronomy: Early Connections." arXiv:physics/0301078 [physics.hist-ph], February 17, 2003.

King, Peter. "Mediaeval Thought Experiments: The Metamethodology of Mediaeval Science." In *Thought Experiments in Science and Philosophy*, edited by Tamara Horowitz and Gerald J. Massey. Savage, MD: Rowman & Littlefield, 1991.

Kinney, Arthur F. "The Essex Rebellion: A New Account." *Papers of the Bibliographical Society of America* 66, no. 1 (1972): 296–302.

Kosso, Peter. "And Yet It Moves: The Observability of the Rotation of the Earth." *Foundations of Science* 15 (2010): 213–225.

Krishnachandran, V. N. "On Finite Differences, Interpolation Methods and Power Series Expansions in Indian Mathematics." Presentation at the International Conference on Mathematical Modelling and Applications to Industrial Problems, March 2011, https://www.slideshare.net/PlusOrMinusZero/on-finite-differences-interpolation-methods-and-power-series-expansions-in-indian-mathematics.

Leitao, Henrique, and Joaquim Alves Gaspar. "Globes, Rhumb Tables, and the Pre-History of the Mercator Projection." *Imago Mundi* 66, no. 2 (2014): 180–195.

Lindberg, David C. "Continuity and Discontinuity in the History of Optics: Kepler and the Medieval Tradition." *History and Technology* 4 (1987): 431–448.

Lozovsky, Natalia. Review of *Ordering the Heavens*, by Bruce S. Eastwood. *Speculum* 83, no. 3 (July 2008): 692694.

Marrioli, Cesare S. "A Fruitful Exchange/Conflict: Engineers and Mathematicians in Early Modern Italy." *Annals of Science* 70, no. 2 (2013): 197–228.

Mayer, Thomas F. "The Censoring of Galileo's *Sunspot Letters* and the First Phase of His Trial." *Studies in History and Philosophy of Science, Part A* 42, no. 1 (2011): 1–10.

Mayer, Thomas F. "The Roman Inquisition's Precept to Galileo (1616)." *British Journal for the History of Science* 43 (2010): 327–351.

McConica, James. "Humanism and Aristotle in Tudor Oxford." *English Historical Review* 94, no. 371 (April 1979): 291–317.

Miller, David Marshall. "*O Male Factum:* Rectilinearity and Kepler's Discovery of the Ellipse." *Journal for the History of Astronomy* 39 (2008): 43–63.

Mueller, Janel. "The Correspondence of Queen Elizabeth I and King James VI." Lecture, May 5, 2000, Fathom Archive, University of Chicago Library Digital Collections, http://fathom.lib.uchicago.edu/1/777777122584.

Neugebauer, O. "The Alleged Babylonian Discovery of the Precession of the Equinoxes." In Neugebauer, *Astronomy and History: Selected Essays.* New York: Springer, 1983.

Nicholls, Mark. "Sir Walter Ralegh's Treason: A Prosecution Document." *English Historical Review* 110, no. 438 (September 1995): 902–925.

Nicholls, Mark. "Strategy and Motivation in the Gunpowder Plot." *Historical Journal* 50, no. 4 (2007): 787–807.

Norris, Ray P., Cilla Norris, Duane Hamacher, and Reg Abrahams. "Wurdi Youang: An Australian Aboriginal Stone Arrangement with Possible Solar Indications." *Rock Art Research* 30, no. 1 (2013): 55–65.

Owens, Rebekah. "Thomas Kyd and the Letters to Puckering." *Notes and Queries* 53, no. 4 (December 2006): 458–461.

Padget, Cindy D. "The Lost Indians of the Lost Colony: A Critical Legal Study of the Lumbee Indians of North Carolina." *American Indian Law Review* 21 (1997): 365–390.

Parramore, Thomas C. "The 'Lost Colony' Found: A Documentary Perspective." *North Carolina Historical Review* 78, no. 1 (January 2001): 67–83.

Parsons, Christopher M. "Wildness without Wilderness: Biogeography and Empire in Seventeenth Century French North America." *Environmental History* 22 (2017): 643–667.

Pásztor, Emília. "Prehistoric Astronomers? Ancient Knowledge Created by Modern Myth." *Journal of Cosmology* 14 (2011), http://journalofcosmology.com/Consciousness159.html.

Péoux, Gérald. "Atmospheric Refraction and the Ramus Circle: Aspects of a Late Sixteenth-Century Dispute." *Annals of Science* 67, no. 4 (2010): 457–484.

Persson, Lars-Erik, Humberto Rafeiro, and Peter Wall. "Historical Synopsis of the Taylor Remainder." *Note di Matematica* 37, no. 1 (2017): 1–21.

Pingree, David. "Astronomy and Astrology in India and Iran." *Isis* 54, no. 2 (June 1963): 229–246.

Prins, Johannes Lambertus Maria, "Walter Warner (ca. 1557–1643) and His Notes on Animal Organisms." Doctoral thesis, University of Utrecht, 1992.

Puglisi, Michael J. "Capt. John Smith, Pocahontas and a Clash of Cultures: A Case for the Ethnohistorical Perspective." *History Teacher* 25, no. 1 (November 1991): 97–103.

Pumfrey, Stephen. "John Dee: The Patronage of a Natural Philosopher in Tudor England." *Studies in History and Philosophy of Science Part A* 43, no. 3 (2012): 449–459.

Pumfrey, Stephen. "'O tempora, O magnes!': A Sociological Analysis of the Discovery of Secular Magnetic Variation in 1634." *British Journal for the History of Science* 22, no. 2 (1989): 181–124.

Pumfrey, Stephen. "The *Selenographia* of William Gilbert: His Pre-Telescopic Map of the Moon and His Discovery of Lunar Libration." *Journal for the History of Astronomy* 42 (2011): 193–203.

Pumfrey, Stephen. "'Your astronomers and ours differ exceedingly': The Controversy over the 'New Star' of 1572 in the Light of a Newly Discovered Text by Thomas Digges." *British Journal for the History of Science* 44, no. 1 (2011): 29–60.

Quinn, D. B. "Renaissance Influences in English Colonization: The Prothero Lecture." *Transactions of the Royal Historical Society* 26 (1976): 73–93.

Rabinovitch, Nachum L. "Rabbi Levi Ben Gershon and the Origins of Mathematical Induction." *Archive for History of Exact Sciences* 6, no. 3 (1970): 237–248.

Ragep, F. Jamil. "Islamic Response to Ptolemy's Imprecisions." In *Ptolemy in Perspective*, edited by Alexander Jones, 121–134. Dordrecht and New York: Springer, 2010.

Randles, W. G. L. "Pedro Nunes' Discovery of the Loxodromic Curve (1537)" *Journal of Navigation* 50, no. 1 (1997): 85–96.

Rashed, Roshdi. "A Pioneer in Anaclastics: Ibn Sahl on Burning Mirrors and Lenses." *Isis* 81, no. 3 (September 1990): 464–491.

Richardson, Catherine. "Social Life." In Kinney, *Elizabethan and Jacobean England*, 294–300.

Rickey, V. Frederick, and Philip M. Tuchinsky. "An Application of Geography to Mathematics: History of the Integral of the Secant." *Mathematics Magazine* 53, no. 3 (May 1980): 162–166.

Robertson, Karen. "Negotiating Favour: The Letters of Lady Ralegh." In *Women and Politics in Early Modern England, 1450–1700*, edited by James Daybell, 99–113. London: Routledge, 2017.

Roy, Ranjan. "The Discovery of the Series Formula for π by Leibniz, Gregory, and Nilakantha." *Mathematics Magazine* 63, no. 5 (December 1990): 291–306.

Rudes, Blair A. "Giving Voice to Powhatan's People: The Creation of Virginia Algonquian Dialogue for *The New World*." *Southern Quarterly* 51, no. 4 (Summer 2014): 29–37.

Ruggles, Clive. *Ancient Astronomy: An Encyclopedia of Cosmologies and Myth*. Santa Barbara, CA: ABC-CLIO, 2005.

Ruggles, Clive. "Astronomy and Stonehenge." *Proceedings of the British Academy* 92 (1997): 203–229.

Sabra, A. I. Review of Rashed's *Géometrie et dioptique au Xe siècle*. *Isis* 85, no. 4 (December 1994): 685–686.

Schaefer, Bradley E. "The Thousand Star Magnitudes in the Catalogues of Ptolemy, Al Sufi, and Tycho Are All Corrected for Atmospheric Extinction." *Journal for the History of Astronomy*) 44 (2013): 47-A97.

Scott, T. C., and P. Marketos. "On the Origin of the Fibonacci Sequence." *MacTutor History of Mathematics* (website maintained by J. O'Connor and E. Robertson, University of St. Andrews, Scotland), March 23, 2014.

Scott-Warren, Jason. "What Can We Learn from Early Modern Drama?" *Historical Journal* 55, no. 2 (June 2012): 553–562.

Silva, Jorge Nuno. "On Mathematical Games." *BSHM Bulletin* 26 (2011): 80–104.

Simpson, L. B. *The Encomienda in New Spain*. Berkeley: University of California, 1929.

Sims, Lionel. "The 'Solarization' of the Moon: Manipulated Knowledge at Stonehenge." *Cambridge Archaeological Journal* 16, no. 2 (2006): 191–207.

Smith, A. Mark. "Le 'De aspectibus' d'Alhazen: Révolutionnaire ou réformiste?" *Revue d'Histoire des Sciences* 60, no. 1 (January–June 2007) : 65–81.

Smith, A. Mark. "Ptolemy's Search for a Law of Refraction: A Case Study in the Classical Methodology of 'Saving the Appearances' and Its Limitations." *Archive for History of Exact Sciences* 26, no. 3 (1982): 221–240.

Steele, J. M. "A Comparison of Astronomical Terminology, Methods and Concepts in China and Mesopotamia, with Some Comments on Claims for the Transmission of Mesopotamian Astronomy to China." *Journal of Astronomical History and Heritage* 16, no. 3 (2013): 250–260.

Steele, John M. "Eclipse Prediction in Mesopotamia." *Archive for History of Exact Sciences* 54, no. 5 (2000): 421–454.

Thirsk, Joan. "Seeking National Prosperity and Personal Survival." In Kinney, *Elizabethan and Jacobean England*, 403–410.

Topper, David. "Galileo, Sunspots, and the Motions of the Earth." *Isis* 90, no. 4 (December 1999): 757–767.

Voelkel, James R. *The Composition of Kepler's "Astronomia Nova."* Princeton, NJ: Princeton University Press, 2001.

Waldvogel, Jörg. "Jost Bürgi's Artificium of 1586 in Modern View, an Ingenious Algorithm for Calculating Tables of the Sine Function." *Elemente der Mathematik* 71 (2016): 88–99.

Walker, Jearl. "Multiple Rainbows from Single Drops of Water and Other Liquids." *American Journal of Physics* 44, no. 5 (May 1976): 421–433.

Wallis, Helen M. "A Newly Discovered Molyneux Globe." *Imago Mundi* 9 (1952): 78.

Wasserstein, A. "Thales' Determination of the Diameters of the Sun and Moon." *Journal of Hellenic Studies* 75 (1955): 114–116.

Wright, J. Leitch, Jr. "Sixteenth Century English-Spanish Rivalry in La Florida." *Florida Historical Quarterly* 38, no. 4 (April 1960): 265–279.

Whiteside, D. T. "Essay Review: In Search of Thomas Harriot." *History of Science* 13 (1975): 61–70.

Whiteside, D. T. "The Mathematical Principles underlying Newton's 'Principia Mathematica.'" *Journal for the History of Astronomy* 1 (1970): 116–138.

Whiteside, D. T. "Newton's Early Thoughts on Planetary Motion: A Fresh Look." *British Journal for the History of Science* 2, no. 6 (1964): 117–137.

Wittmann, A. "The Obliquity of the Ecliptic." *Astronomy and Physics* 73 (1979): 129–131.

Woodhead, Christine. "England, the Ottomans and the Barbary Coast in the Late Sixteenth Century." *State Papers Online, 1509–1714*. Reading: Gale Cengage Learning, 2009.

Zuccato, Marco. Review of Eastwood's *Ordering the Heavens*, by Bruce S. Eastwood. *Isis* 9, no. 4 (December 2008): 823–824.

GENERAL REFERENCES

Alexander, Amir. *Infinitesimal: How a Dangerous Mathematical Theory Shaped the Modern World*. London: Oneworld, 2014.

Ashley, Leonard R. N. *Elizabethan Popular Culture*. Bowling Green, OH: Bowling Green State University Popular Press, 1988.

Becker, Carl J. *A Modern Theory of Language Evolution*. Lincoln, NE: iUniverse, 2004.

Beer, Anna. *Bess: The Life of Lady Ralegh, Wife to Sir Walter*. London: Constable, 2004.

Boyer, Carl B. *A History of Mathematics*. Revised by Uta Merzbach. New York: John Wiley & Sons, 1991.

Boyer, Carl B. *The History of the Calculus and Its Conceptual Development*. New York: Dover, 1959.

Boyer, Carl B. *The Rainbow: From Myth to Mathematics*. New York and London: Sagamore Press, 1959; rpt. Princeton, NJ: Princeton University Press, 1987; rpt. Hampshire and London: Macmillan Education, 1987.

Coote, Stephen. *A Play of Passion: The Life of Sir Walter Ralegh*. London: Macmillan, 1993.

Cotter, Charles. *A History of Nautical Astronomy*. London: Hollis & Carter, 1968.

Custalow, Linwood, and Angela L. Daniel. *The True Story of Pocahontas: The Other Side of History*. Golden, CO: Fulcrum, 2007.

Duncan, David Ewing. *The Calendar*. London: Fourth Estate, 1998.

Edwards, A. W. F. *Pascal's Arithmetical Triangle*. London: Charles Griffin; New York: Oxford University Press, 1987.

Einstein, Albert. "Johannes Kepler." In Einstein, *Ideas and Opinions*, 262–266. New York: Three Rivers Press, 1982.

Flood, Raymond, and Robin Wilson. *The Great Mathematicians: Unravelling the Mysteries of the Universe*. London: Arcturus, 2011.

Gauquelin, Michel. *The Cosmic Clocks*. Chicago: Regnery, 1967.

Greenblatt, Stephen. *The Swerve: How the World Became Modern*. New York: Norton, 2011.

Hart, Roger. *Battle of the Spanish Armada*. London: Wayland, 1973.

Heggie, Douglas. *Megalithic Science: Ancient Mathematics and Astronomy in Northwest Europe*. London: Thames & Hudson, 1981.

Honan, Park. *Christopher Marlowe: Poet and Spy*. Oxford and New York: Oxford University Press, 2005.

Jackson, Kevin. *Columbus: The Accidental Hero*. Kindle Single. Amazon.com, 2014.

Jackson, Kevin. *Mayflower: The Voyage from Hell*. Kindle Single. Amazon.com, 2013.

Kassell, Lauren. *Medicine and Magic in Elizabethan London: Simon Forman: Astrologer, Alchemist, and Physician*. London: Clarendon, 2005.

Kupperman, Karen Ordahl. *Settling with the Indians: The Meeting of English and Indian Cultures in America, 1580–1640*. Totowa, NJ: Rowman & Littlefield, 1980.

Martin, Julian. *Francis Bacon, the State, and the Reform of Natural Philosophy*. Cambridge and New York: Cambridge University Press, 1992.

May, Steven. *Sir Walter Ralegh.* Boston: Twayne, 1989.

Mazur, Joseph. *Enlightening Symbols: A Short history of Mathematical Notation and Its Hidden Powers.* Princeton, NJ: Princeton University Press, 2014.

McCrum, Robert, William Cran, and Robert MacNeil. *The Story of English.* New York: Penguin, 1987.

Montgomery, Scott L. *Science in Translation: Movements of Knowledge through Cultures and Time.* Chicago and London: University of Chicago Press, 2000.

Nasr, Seyyed Hossein, with photographs by Roland Michaud. *Islamic Science: An Illustrated Study.* London: World of Islam Festival Publishing, 1976.

Nicholls, Mark, and Penry Williams. *Sir Walter Raleigh: In Life and Legend.* Bloomsbury Academic, 2011.

Ousby, Ian. *The Cambridge Guide to Literature in English.* Cambridge and New York: Cambridge University Press, 1995.

Penrose, Roger. *The Emperor's New Mind.* Oxford: Oxford University Press, 1989.

Perry, Maria. *Elizabeth I: The Word of a Prince: A Life from Contemporary Documents.* London: Folio Society, 1990.

Pierce, Donna. "The Mission: Evangelical Utopianism in the New World (1523–1600)." In *Mexico: Splendors of Thirty Centuries.* New York: Metropolitan Museum of Art, 1990.

Resnikoff, H., and R. Wells Jr. *Mathematics in Civilization.* New York: Dover, 1984.

Riskin, Jessica. *The Restless Clock: A History of the Centuries-Long Argument over What Makes Living Things Tick.* Chicago: University of Chicago Press, 2016.

Rowse, A. L. *The Elizabethan Renaissance: The Life of the Society.* London: Macmillan, 1971.

Rowse, A. L. *Ralegh and the Throckmortons.* London: Macmillan, 1962.

Rublack, Ulinka. "The Astronomer and the Witch: How Kepler Saved His Mother from the Stake." *The Conversation,* October 21, 2015, https://theconversation.com/the-astronomer-and-the-witch-how-kepler-saved-his-mother-from-the-stake-49332. Based on Rublack's book *The Astronomer and the Witch* (Oxford and New York: Oxford University Press, 2015).

Shah, Idries. *The Way of the Sufi.* Harmondsworth: Penguin, 1975.

Singman, Jeffrey L. *Daily Life in Elizabethan England.* Westport, CT, and London: Greenwood Press, 1995.

Smart, W. M. *Textbook on Spherical Astronomy.* Cambridge and New York: Cambridge University Press, 1936.

Sobel, Dava. *Galileo's Daughter.* London: Fourth Estate, 1999.

Stillwell, John. *Mathematics and Its History.* New York: Springer-Verlag, 1989.

Strathmann, Ernest. *Sir Walter Ralegh: A Study in Elizabethan Skepticism.* New York: Octagon Books, 1973.

Struik, Dirk. *A Concise History of Mathematics.* New York: Dover, 1967.

Sweet, Timothy. *American Georgics: Economy and Environment in American Literature.* Philadelphia: University of Pennsylvania Press, 2002.

Tharoor, Shashi. *Inglorious Empire: What the British Did to India.* Melbourne: Scribe, 2017.

Thompson, Edward. *Sir Walter Ralegh: The Last of the Elizabethans.* London: Macmillan, 1935.

Wallace, Willard Mosher. *Sir Walter Raleigh.* 1959. Princeton, NJ: Princeton University Press, 2015.

Woolley, Benjamin. *The Queen's Conjuror.* London: HarperCollins, 2001.

Wright, Frances W. *Celestial Navigation.* Cambridge, MD: Cornell Maritime Press, 1969.

INDEX